DRAWING REQUIREMENTS MANUAL

FIFTH EDITION

EDITED BY
JEROME H. LIEBLICH

PUBLISHED BY

A Division of Information
Handling Services

2625 Hickory St., P.O. Box 2504
Santa Ana, California 92707 U.S.A.
Phones: (714) 540-9870 (213) 624-1216
Telex: 692-373 (800) 854-7179

© GLOBAL ENGINEERING DOCUMENTS, 1983

ISBN 0-912702-18-4

FOREWORD

This DRM contains engineering drawing requirements, based on DOD-D-1000, **DOD-STD-100** and supporting documents, as issued by the Department of Defense, National Bureau of Standards, and by various trade societies. It also recognizes a number of industry practices, not covered by any such documents. If **DOD-STD-100**, or subsidiaries are revised, substantive changes will be reflected in supplements to the DRM.

While these practices are generally acceptable to DOD and to Industry, any given contract may invoke peculiar customer requirements. In that case, the contract requirements must be given precedence.

Drawing examples are complete only to the extent necessary to illustrate the requirements of the accompanying text.

Comments and suggestions for improving this DRM are solicited by the publisher.

JEROME H. LIEBLICH
President, Global Engineering Documents
A Division of Information Handling Services

Section	#
SPECIFICATIONS & STANDARDS	1
GENERAL REQUIREMENTS	2
DRAWINGS & PART NUMBERING	3
DRAWING TYPES	4
FORMAT	5
TITLES	6
NOTES	7
DRAFTING PRACTICES	8
PARTS LIST	9
DIMENSIONS & TOLERANCES 1973	10
DIMENSIONS & TOLERANCES 1982	10M
SURFACE TEXTURE	11
CASTINGS	12
FORGINGS	13
IDENTIFICATION MARKINGS	14
WELDING	15
THREADS	16
ELECTRICAL & ELECTRONICS	17
CHANGES	18
PROTECTIVE COATINGS	19
ABBREVIATIONS	20
TOOLING	21
SHEET METAL	22
PACKAGING	23
FACILITIES	24
DEFINITIONS	25

A SEPERATE AND COMPLETE INDEX FOR SECTION 10M IS PROVIDED IN FRONT OF THAT SECTION. THE PURPOSE IS TO MAINTAIN SEPERATE INTERPRETATIONS FOR ANSI Y14.5 FOR OLDER DRAWINGS PREPARED TO THE ANSI Y14.5-1973 VERSION AS OPPOSED TO NEW DRAWINGS BEING PREPARED TO ANSI Y14.5-1982 VERSION.

FOR METRIC INFORMATION SEE SECTION 16 AND APPENDIX A AND B

CONTENTS

INTRODUCTION

1. SPECIFICATIONS AND STANDARDS DATA
1.1 Scope
1.2 Applicable Documents
1.3 Company Standards, Specifications and Procedures
1.4 Definitions
1.5 Reference on Drawings

2. DRAWING LEVELS
2.1 Scope
2.2 Applicable Documents
2.3 Definitions
2.4 Drawing Requirements
2.5 Drawing Levels
2.6 Responsibilities of Engrg. Drawing Management & Related Functions

3. DRAWING AND PART NUMBERING SYSTEM
3.1 Scope
3.2 Applicable Documents
3.3 Definitions
3.4 Drawing Number Assignment
3.5 Part Number Assignment
3.6 Dash Number Callout On The field Of Drawing
3.7 Item Number
3.8 Numbering of Charts, Graphs and Other Data

4. TYPES OF ENGINEERING DRAWINGS
4.1 Scope
4.2 Applicable Documents
4.3 Definitions
4.4 General Drawing Control Requirements

5. ENGINEERING DRAWING FORMAT
5.1 Scope
5.2 Applicable Documents
5.3 Definitions
5.4 Drawing Arrangement
5.5 Title Block
5.6 Supplementary Data Blocks
5.7 Parts List
5.8 Revision Block
5.9 Multisheet Drawings
5.10 Book-Form Drawings
5.11 Print Folds
5.12 Proprietary Notices
5.13 Security Classification Marking

6. DRAWING TITLES
6.1 Scope
6.2 Applicable Documents
6.3 Definitions
6.4 Title Assignment
6.5 General Rules

7. DRAWING NOTES
7.1 Scope
7.2 Applicable Documents
7.3 Definitions
7.4 General Application
7.5 Local Notes
7.6 General Notes

8. DRAFTING PRACTICES
8.1 Scope
8.2 Applicable Documents
8.3 Scale of Drawing
8.4 Lettering
8.5 Types of Lines
8.6 Drafting Symbols
8.7 Section or Sectional Views
8.8 Projection Procedures
8.9 Descriptive Geometry

CONTENTS (continued)

9.	**PARTS LIST**		11.	**SURFACE TEXTURE**
9.1	Scope		11.1	Scope
9.2	Applicable Documents		11.2	Applicable Documents
9.3	Definitions		11.3	Definitions
9.4	General Requirements for Integral Parts List		11.4	Surface Texture Control Applied to Drawings
9.5	Parts List Location		11.5	Recommended Value Ratings
9.6	Drawing Application		11.6	Recommended Ratings for Standard Processes
9.7	Commercial Products			
9.8	Customer Furnished and Controlled Items		12.	**CASTING DRAWINGS**
			12.1	Scope
9.9	Parts Lists Separate from the Drawing		12.2	Applicable Documents
			12.3	Defintions
9.10	Parts List Security Classification		12.4	Desing Practice
9.11	Drawing Callout - General Notes		12.5	Markings
			12.6	Material Requirements
10.	**DIMENSIONS AND TOLERANCES (ANSI Y14.5 - 1973)**		12.7	Drawing Title
			12.8	Drawing Notes
			12.9	Reference Data
10.1	Scope.			
10.2	Applicable Documents		13.	**FORGING DRAWINGS**
10.3	Definitions		13.1	Scope
10.4	General Requirements		13.2	Applicable Documents
10.5	Method of Dimensioning		13.3	Definition
10.6	General Rules and Interpretations of Dimensions and Tolerances		13.4	Design Practices
			13.5	Markings
10.7	Geometric Tolerances		13.6	Material Requirements
10.8	Datum Targets		13.7	Drawing Title
10.9	Drawing Application and Interpretation		13.8	Drawing Notes
			13.9	Stocked Forging
10.10	Positional Tolerance		13.10	Reference Data
10.11	**Metric Tables**			
			14.	**IDENTIFICATION MARKING**
10M	**DIMENSIONS AND TOLERANCES (ANSI Y14.5M - 1982)**		14.1	Scope
			14.2	Applicable Documents
10.1	Scope		14.3	Definitions
10.2	Applicable Documents		14.4	General Requirements
10.3	Definitions		14.5	Markings
10.4	General Requirements		14.6	Drawing Application
10.5	Method of Dimensioning		14.7	Identification Plates, Decalcomanias, Labels and Other Marking Devices
10.6	General Rules and Interpretations of Dimensions and Tolerances			
10.7	Geometric Tolerances		14.8	Application of MIL-STD-130
10.8	Datum Targets			
10.9	Drawing Application and Interpretation		15.	**WELDING SYMBOLOGY**
			15.1	Scope
10.10	Positional Tolerance		15.2	Applicable Documents
10.11	Metric Tables			

CONTENTS (continued)

15.3	Definitions		19.2	Applicable Documents
15.4	Welding Symbols		19.3	Definitions
15.5	Application for Fusion Welds		19.4	Design Practice
15.6	Suggested Weld Rod Callouts		19.5	General Requirements
15.7	Typical Drawing Notes		19.6	Finish Block Entries
			19.7	Drawing Note Callouts
16.	THREAD REPRESENTATION		19.8	Parts List Entries
16.1	Scope		19.9	Reference Data
16.2	Applicable Documents			
16.3	Definitions		20.	ABBREVIATIONS
16.4	Drawing Application		20.1	Scope
16.5	Metric Threads		20.2	Applicable Documents
			20.3	Definitions
17.	ELECTRICAL AND ELECTRONIC DRAFTING		20.4	Drawing Application
			20.5	Rules for Use
17.1	Scope		20.6	Detail Requirements
17.2	Applicable Documents			
17.3	Definitions		**21.**	**TOOL DRAFTING PROCEDURE**
17.4	Electrical and Electronic Schematic Diagrams		**21.1**	**Scope**
			21.2	**Applicable Documents**
17.5	Wiring Diagram		**21.3**	**Definitions**
17.6	Electrical or Electronic Assembly Drawing		**21.4**	**Responsibilities of the Tool Designer**
17.7	Electrical or Electronic Detail Drawing		**21.5**	**Responsibilities of the Tool Design Checker**
17.8	Cable Assembly Drawing		21.6	Drawing Numbering System
17.9	Wiring Harness Drawing		21.7	Assignment of Detail Numbers
17.10	Interconnection Diagram		21.8	Drawing Format
17.11	Printed Circuits and Printed Wiring		21.9	Drawing Titles
			21.10	Drawing Notes
17.12	Logic Diagrams		21.11	Parts List
17.13	Drawing Callout - General Notes		21.12	Welding Procedures
			21.13	Threads
18.	DRAWING CHANGE PROCEDURE		21.14	Drawing Change Procedure
18.1	Scope			
18.2	Applicable Documents		22.	SHEET METAL DRAWINGS
18.3	Definition		22.1	Scope
18.4	General Change Procedures		22.2	Applicable Documents
18.5	Revision to Multi-Sheet Drawings		22.3	Definitions
			22.4	General
18.6	Advance Document Change Notice (ADCN)		22.5	Characteristics of Sheet Metal Bends
18.7	Document Change Notice (DCN)		22.6	Bend Relief Cutouts
18.8	Document Change Request		22.7	Joggles
			22.8	Beads
19.	PROTECTIVE COATINGS		22.9	Lightening Holes
19.1	Scope		22.10	Dimpling and Countersinking

CONTENTS (continued)

22.11	Dimensioning and Tolerancing Sheet Metal Parts		A2.2	Applicable Documents
22.12	Limitations in Forming Sheet Metal		A2.3	Definitions
22.13	Undimensioned Drawings		A2.4	Conversion of Linear Units
			A2.5	Conversion of Explicit Tolerances
23.	PACKAGING DRAWINGS		A2.6	Conversion of Temperature Units
23.1	Scope		A2.7	Frequently Used Conversions
23.2	Applicable Documents			
23.3	Definitions			
23.4	General Information		A3.	WRITING SI
23.5	Drawing Application		A3.1	Scope
23.6	Parts List		A3.2	Applicable Documents
23.7	General Dimensioning Practices for Packaging		A3.3	Definitions
			A3.4	Spelling
23.8	Packaging Identification and Marking		A3.5	Obsolete Terms, Units, Prefixes
			A3.6	Writing Unit Names and Unit Symbols
24.	FACILITY DRAWINGS		A3.7	Writing Prefix Names and Symbols
24.1	Scope			
24.2	Applicable Documents		A3.8	Writing Plurals
24.3	Definitions		A3.9	Proper Spacing
24.4	Revisions (Changes)		A3.10	Indicating Division
24.5	Identification Marking		A3.11	Mixing Words, Symbols and Units
24.6	Drawing Scales			
24.7	Drawing Format and Sizes		A3.12	Use of Linear Dimensions on Drawings
25.	DEFINITIONS		A3.13	Use of the Period
APPENDIX 1	METRIC APPLICATION		A3.14	Use of the Raised Dot
			A3.15	Use of the Comma
A1.	INTERNATIONAL SYSTEM OF UNITS (SI)		A3.16	Writing Decimals
A1.1	Scope			
A1.2	Applicable Documents		A4.	CONVERSION OF AN INCH DRAWING TO A METRIC DRAWING
A1.3	Definitions			
A1.4	SI—The System of International Units		A4.1	Modify Existing Drawing Form
A1.5	SI Units		A4.2	Identify as a Metric Drawing
A1.6	Non-SI Units			
A1.7	Obsolete Units and Prefixes		A4.3	Equivalent Dimensions
A1.8	SI Prefixes		A5.	MISC METRIC INFORMATION
A1.9	SI Symbols		A5.1	Metric Paper Sizes
A2.	CONVERSION OF UNITS			
A2.1	Scope		APPENDIX 2	METRIC GLOSSARY

INDEX

- A -

ABBREVIATIONS	20.0
ABSOLUTE ZERO	A2-5
ACCOUNTABILITY CONTROL NUMBER	5.13.5
ACME THREADS	16.3.8
	16.4.2.2.2
ADVANCE DOCUMENT CHANGE NOTICE (ADCN)	18.3.1
	18.4
	18.6
approval signatures	18.4.5
	18.6.6.1
changes to	18.6.4
description of revision	18.6.3
incorporation of	18.3.7
maximum allowed	18.6.2
pre-released	18.3.9
	18.6.6
processing completed	18.6.5
AERONAUTICAL PIPE THREADS	16.3.7
ALLOWANCE	10.3.1
ALTERED ITEM DRAWING	4.1
ALTERED AND/OR SELECTED PARTS IDENTIFICATION	4.1
AMERICAN NATIONAL	
form threads	16.3.4
taper pipe threads	16.3.6
AMPERE	A1.3.4
ANGLE	
draft	12.3.3
	13.3.5
plane, unit of(metric)	A1.3.8
solid, unit of(metric)	A1.3.9
true, intersecting planes	8.9.10
	8.9.11
true, line and plane	8.9.9
ANGULAR DIMENSIONS	10.4.1.2
ANGULARITY	10.9.8
ANODIZE	19.3.2
	19.7.1
APPLICATION BLOCK	5.6.7
APPROVAL	
blocks	5.5.4
ADCN	18.6.6.1
DCN	18.4.3
checker's signature	2.6.6.1
check prints	2.6.6.2
requirements	2.6.6
	2.6.2.4
ARC WELD SYMBOLS	15.4
ARROWHEADS	8.5.4.1
microfilm	5.4.3
ASSEMBLY	6.3.4
circuit card	17.3.4

ASSEMBLY (continued)	
component board	17.3.6
cross section	8.7.5.10
	8.7.7.2
drawing	4.1
detail	4.1
inseparable	4.1
tabulated	4.1
harness	17.3.18
marking of	14.6.4.2
	14.8.3
multiple, parts list	9.6.3.1.2
opposite, parts list	9.6.3.1.1
parts list	9.6.3
printed wiring board	17.3.42
serialization	14.3.8
	14.6.6
shown, parts list	9.6.3.1.1
terminal board	17.3.49
thread	16.4.1.3
thread insert	16.4.1.4
AUTHORIZED CHECKER	2.6.6.1
AUXILIARY VIEWS	8.8.3
AXIS	10.3.2

- B -

BACKING WELD	15.4.4.1.2
BASIC	
dimensions	10.3.5.1
	10.7.4
name	6.4.5
size	10.3.10
BEADS	22.8
BEND, CENTER LINE	22.3.9
BEND ALLOWANCE	22.3.1
formulas	22.5.1.1
BEND ANGLE	22.3.2
BEND LINE	22.3.5
BEND RELIEF	22.3.4
cutouts	22.6
dimensioning	22.6.1
BEVEL	
closed	22.3.6
open	22.3.7
BILATERAL	
tolerance	10.3.13.1
BLANK	22.3.8
BLIND HOLES	16.4.6.5
BLOCK DIAGRAM	17.3.2
BLOCK	
drawing number	5.5.9
number of sheets	5.5.12
scale	5.5.10
	5.9.2.2
supplementary	5.6.1
BOARD PRINTED WIRING	17.3.42

1

INDEX

BOARD PRINTED WIRING (continued)		CHECKING	
tables applicable to	17.11.4.7	drawing	2.6.5
BOARD, TERMINAL	17.3.48	minimum requirements	2.6.4.1
assembly	17.3.49		2.6.4.2
BOOK-FORM DRAWING	5.3.3	policy	2.6.5
	5.10	responsibilities	2.6.4
BOOK STANDARD	1.3.1.3	CHEMICAL COATING	19.3.2
BREAK LINES	8.5.5	CIRCUIT, PRINTED	17.3.37
BROKEN-OUT SECTION	8.7.5.9	CIRCUIT, THIN FILM	17.3.51
BULK ITEM	9.4.4	CLASS, THREAD	16.4.2.1
	9.4.4.1		16.4.2.1.3
quantity	9.6.3.1.4	2A	16.4.2.1.3
	9.6.3.1.5	3A	16.4.2.1.3
BUTTRESS THREADS	16.3.9	CLASSIFICATION, SECURITY	5.13
		CLEANING	23.3.1
- C -			23.3.2
CABLE ASSEMBLY	17.3.3	CLEARANCE FIT	10.3.8.1
CALLOUT, THREAD	16.4.5	COATED	
CANCELLED DRAWING	18.3.6.3	surfaces	11.4.5
CANDELA	A1.3.7	threads	16.4.3
CAPACITANCE	17.4.1.3	COATING	
CAST AND MACHINE DRAWING	Page 12-10	chemical	19.3.2
CASTING	12.3.1	conformal	19.3.9
design practice	12.4	lubricant	19.3.6
drawings	12.0	metallic	19.3.4
markings	12.5	organic	19.3.3
material	12.6	protective	19.0
notes	12.8		19.3.1
part number	12.5.2	permanent	19.3.1.1
reference data	12.9	temporary	19.3.1.2
CATEGORIES DRAWING	2.1	vitreous	19.3.5
CELSIUS	B1-1	CODE	
CENTER LINE	11.3.3	color	17.5.3.1.1
CENTER LINE OF BEND	22.3.9		17.3.5
CENTER LINE, SYMMETRY	10.5.5	changing	18.4.13
CENTER LINES	8.5.2	handbook, federal supply	1.2
CENTER PLANE	10.3.3	parts list	9.6.1.2
CERAMIC COATING	19.3.5		.6.3.2
CHAMFERS	10.5.14	wire	17.3.54
	16.4.6.6		17.3.5
CHANGE		COMMERCIAL ITEM	9.8
direct drawing	18.3.2	COMMERCIAL PRODUCTS	9.7
design layout	18.4.3	COMPONENT	
general notes	18.4.10	active	17.3.1
letters	18.4.4	board assembly	17.3.36
new identification	18.9	discrete	17.3.8
request	18.8	passive	17.3.35
dash number	18.4.14.1	printed	17.3.39
identification	18.4.13	representation	17.5.1.1
to ADCN	18.6.4	terminal identification	17.5.1.2
undocumented	18.4.8	values	17.4.1.3
CHARTS, NUMBERING OF	3.8	CONCENTRICITY	10.9.10
CHECK PRINT	2.6.4.6	CONDUCTIVE PATTERN	17.3.7
	2.6.5.1	CONDUCTOR	
	2.6.5.2	identification	17.5.3
	2.6.6.2	numerical code	17.5.3.1.2

INDEX

CONDUCTOR (continued)	17.5.3.1.3
CONFIGURATION ELEMENT IDENTIFIER	14.3.1
CONFORMAL COATING	19.3.9
CONTACT AREA	11.3.14
CONTACT, PRINTED	17.3.40
CONTRACT NUMBER	5.5.2
CONVENTIONS, SECTION-LINING	8.7.7
COORDINATE DIMENSIONING	10.5.9
CORE	12.3.11
CORRECTIONS, DRAWING	2.6.3.5
COUNTERBORE	10.5.11
COUNTERSINK	10.5.13
	16.4.6.8
COUNTERSINKING SHEET METAL	22.10
CROSS-SECTION, THREADS	16.4.1.3
CROSS-SECTIONING	8.7.7.2
CROSS REFERENCE	
general note	7.6.8
zone	8.6
CUSTOMER	
furnished items	9.8
CUTTING PLANE LINES	8.5.11

- D -

DASH NUMBER	
assignment	3.4
changes	18.4.14.1
suffix	3.3.3
DATA, IDENTIFICATION	3.8
DATE	
drawing release	5.5.11
revision block	18.4.15
DATUM	10.3.4
casting	Page 12-6
forging	Page 13-6
implied	10.6.2
lines	8.5.9
multiple	10.7.6
reference	10.7.3
selection	10.6.2.3
surface	10.8.1
symbol	10.7.3
targets	10.8
	10.3.4.1
	12.3.6
	13.3.10
method	10.8.2
true position	10.10.4
DECIMALS	A3.16
writing metric	A3.6
	A3.7
	A3.16
comma, metric	A3.15

DECIMALS (continued)	
period, metric use of	A3.13
dot, raised, metric, use of	A3.14
DECIMAL SYSTEM	10.4.1
DESCRIPTION	
parts list	9.6.1.4
	9.6.3.4
revision, ADCN	18.6.3
revision, DCN	18.7.2
revision block	18.4.16.3
DESRIPTIVE GEOMETRY	8.9
DESIGN	
activity	5.5.4
layout drawing	4.1
responsibility	2.6.2
size	10.3.6
DESIGNATIONS, REFERENCE	17.3.9
	17.4.1.2
	17.5.2
DETINATION, WIRE	17.3.53
	17.5.3.2
DETAIL	
assembly drawing	4.1
cross section	8.7.71
drawing	4.1
electrical or electronic	17.7
printed circuit	17.11.1
tabulated	4.1
views	8.8.4
DETAILED PACKAGING INSTRUCTIONS	23.3.1
DEVELOPED LENGTH	22.3.10
DIAGRAM	
block	17.3.2
interconnection	17.3.24
	17.10
logic	17.3.26
	17.12
schematic	
electrical and electronic	17.3.12
	17.4
mechanical	4.1
single line	17.3.45
wiring	17.3.55
	17.5
DIAMETRAL DIMENSIONS	10.5.4
DIE, FORGING	13.3.2
DIMENSION	10.3.5
and tolerance	10.0
angular	10.4.1.2
basic	10.3.5.1
	10.7.4
diametral	10.5.4
dual	A4.3.4
limit	10.3.9
linear, metric	A3.12

INDEX

DIMENSION (continued)		DOD-D-1000		1.2
lines	8.5.3	DOD-STD-100		1.2
locating	10.3.5.2	DOUBLE JOGGLES		22.7.3
maximum	10.3.5.3	DRAFT		12.3.3
minimum	10.3.5.4			13.3.5
out-of-scale	8.3.5	match		13.3.6
profile	10.9.5.2	DRAFTING		
reference	10.3.5.5	practices		8.0
size	10.3.5.6	documents for		1.1
staggered	10.5.3	reference designations		17.4.1.2
true	10.5.19	symbols		8.6
typical	10.4.3	symbols, graphic		17.4.1.1
DIMENSIONAL TOLERANCES,		tooling drawings		21.0
DRAWING	5.6.2	responsibilities of		2.6.3
	10.4.2	symbols		8.6
DIMENSIONING		symbols(SI)		A1.10
casting	Page 12-8	DRAWING		
center line	10.4.3	altered or selected item		4.1
chain	10.4.3	applicable documents		1.2
	22.11.3.1	application block		5.6.7
chamfers	10.5.14	approval block		5.5.4
coordinate	10.5.9			5.6.8
counterbore	10.5.11	arrangement		5.0
countersink	10.5.13	assembly		4.1
drawn parts	22.11.2	detail		4.1
from datums	10.6.2.2	inseparable		4.1
hidden lines	10.4.3	parts list		5.7
holes	10.5.15	printed circuit master		17.11
keyways	10.5.16			17.11.1.2
knurls	10.5.17	separable		4.1
method of	10.5	tabulated		4.1
practices	10.4	assignment of number		3.4.1
rounded ends	10.5.8	book-form		5.3.3
spherical radii	10.5.10			5.10
spotface	10.5.12	cable		17.3.3
spun parts	22.11.3			17.8
tabular	22.11.3.2	cast and machined		12.10
tapers	10.5.18	casting		12.0
threads	16.4.6	categories		2.1
threads, combined	16.4.6.10	change procedure		18.0
true position	10.10	checker		2.6.4.1
tubes	10.5.7	checking		2.6.4
DIMPLING	22.3.11	code identification		5.5.8
	22.10	contract number		5.5.2
DISTANCE, POINT TO PLANE,		corrections		2.6.3.5
TRUE	8.9.8	design activity approval		5.5.4
DOCUMENT CHANGE		design activity		5.5.4
NOTICE(DCN)	18.3.4	name and address block		5.5.5
	18.7	design layout		4.1
DOCUMENT CHANGE REQUEST	18.3.5	detail		4.1
	18.8	electrical or electronic		17.7
DOCUMENT, RELEASE	2.7	printed circuit		17.11.1
DOCUMENTS, APPLICABLE TO		tabulated		4.1
DRAWINGS	1.2	drawn block		5.5.3
DOCUMENTS, REFERENCE NOT		duplicate		2.8
PERMITTED	1.5.5	electrical and electronic		17.0

INDEX

DRAWING (continued)

	17.3.11
	17.3.13
assembly	17.6
schematic diagram	17.3.12
electrical harness	17.9
engineering	4.4
envelope	4.1
equipment	21.3.1
facility	24.0
finish block	5.6.4
forge and machine	Page 13-10
forging	13.0
form	2.1
	2.5.4.1
format	5.0
tool drawing	21.8
formulation	4.1
identify as metric	A4.2
inactive	18.3.6.1
installation	4.1
interconnection	17.3.24
	17.10
interface control	4.1
isometric	8.8.8
kit	4.1
layout, printed circuit	17.11
	17.11.1.1
level	2.3.1
machining	
casting	Page 12-8
forging	Page 13-9
master	17.3.28
	17.11.3
master artwork	17.3.27
master lines	4.1
master pattern	17.3.29
	17.11.2
matched parts	4.1
modify to metric	A4.1
mono-detail	4.1
multisheet	5.3.2
	5.9
notes	7.0
number	3.3.1
block	5.5.9
sheet	5.5.12
	5.9.2.2
tool	21.6
numerical control	4.1
obsolete	18.3.6.2
one view	8.8.5
optics	4.1
out of scale	8.3.5
packaging	23.0
parts list	5.7

DRAWING (continued)

	5.9.3
	9.3.1
	9.5
pictorial	8.8.8
printed circuit master	17.11
	17.11.1.2
proposal	4.1
changes	18.4.7
reinstating	18.4.3
release date	5.5.11
replacement	18.3.6.2
requirements	2.1
	4.4.3
revisions	
authorization of	18.4.3
column	5.8
incorporation of	18.4.7
processing	18.6.5
	18.7.3
redrawn	18.4.3
workmanship	18.4.6.1
scale	8.3
block	5.5.10
schematic, mechanical	4.1
security classification	5.13
sheet metal	22.0
similar to block	5.6.5
size	5.3.1
block	5.5.7
sketch, freehand	2.4.1.1
source control	4.1
specification control	4.1
superseded	18.4.3
tabulated	
assembly	4.1
detail	4.1
parts list	9.6.2
textile	4.1
title	6.0
block	5.5
	5.5.1
	5.5.6
tool drawing	21.9
tolerance	5.6.2
tooling	21.3.1
treatment block	5.6.3
tube bend	4.1
types	4.1
weight block	5.6.6
wire table	17.5.4
wiring, electrical and electronic	17.5
without picture	8.8.7
zoning	5.4.5
	5.71
DRILL TAP	16.4.6.9

INDEX

DRYSEAL THREADS	16.3.10
DUAL DIMENSIONS	A4.3.4
DUPLICATE DRAWING	2.8

- E -

ELECTRIC CURRENT, UNIT OF (METRIC)	A1.3.4
ELECTRICAL AND ELECTRONIC	
drafting	17.0
wiring diagram	17.3.55
	17.5
ELECTRICAL AND ELECTRONIC drawing	
assembly	17.6
detail	17.7
harness	17.9
reference designations	17.5.2
schematic diagram	17.4
symbols	8.6.1
	17.3.10
	17.4.1.1
ELECTRONICS drawing	
assembly	17.6
detail	17.7
reference designations	17.5.2
schematic diagram	17.4
symbols	8.6.1
	17.3.10
	17.4.1.1
ELECTROPLATE	19.3.4
ENAMEL	19.3.3
porcelain	19.3.5
END ITEM	14.3.2
ENDS, ROUNDED, DIMENSIONING	10.5.8
ENGINEERING DRAWING	4.1
ENGINEERING RESPONSIBILITIES	2.6.2
ENTRIES, PARTS LIST, SEQUENCE	9.4.5
ENTRY SPACES, PARTS LIST	9.4.3
ENVELOPE DRAWING	4.1
EQUIPMENT DRAWING	21.3.1
EQUIPMENT LIST	17.4.1.5
ETCHING	17.3.15
EXTENSION LINES	8.5.8

- F -

FACILITY DRAWINGS	24.0
	24.3.7
assembly identification marking	24.5
drawings	
architectural	24.3.7.2.3
civil	24.3.7.2.1

FACILITY DRAWINGS (continued)	
connection diagram	24.3.1.1.2
construction	24.3.5
diagrammatic	24.3.1.1
electrical	24.3.7.2.5
erection	24.3.5.1
interconnection diagram	24.3.1.1.3
kit	24.3.3
legend blocks	24.7.3.7
logic diagram	24.3.1.1.5
mechanical	24.3.7.2.4
mechanical schematic diagram	24.3.1.1.6
notes	24.7.3.4
piping diagram	24.3.1.1.7
plan	24.3.5.2
plant equipment	24.3.4
plot plant	24.3.5.3
revision blocks	24.7.3.6
running(wire) list	24.3.6.2
scale	24.6
schematic diagram	24.3.1.1.1
single line diagram	24.3.1.1.4
structural	24.3.7.2.2
tabulation blocks	24.7.3.5
types	24.3.7.2
vicinity plan	24.3.5.4
format	24.7
line identification	24.5.2
package/set	24.3.8
part identification marking	24.5
parts list	24.7.3.2
reference drawing block	24.7.3.3
revision	24.4
symbol	24.4.3
sizes	24.7
specifications	24.3.9
FAHRENHEIT	A2.6.7
	A2.6.8
FASTENER	
fixed	10.10.3
floating	10.10.3
FEATURE	10.3.7
size, regardless of	10.3.16
FEDERAL SUPPLY CODE OF MANUFACTURERS (FSCM)	5.5.8
	9.6.3.2
	Page 25-6
FIELD WELD	15.4.4.3
FILLET WELD	15.4.4.1.2
FILM LUBRICANT	19.3.6
FIN	12.3.8
FIND NUMBER(ITEM NUMBER)	3.7
	9.3.2
FINISH	19.0

INDEX

FINISH (continued)	
block	5.6.4
	19.6
symbol, weld	15.4.3
FIT	10.3.8
clearance	10.3.8.1
interference	10.3.8.2
line	10.3.8.4
transition	10.3.8.3
FLANGE WELDS	15.4.4.1.2
	15.4.5.3
FLARE WELDS	15.4.5.2
FLASH	13.3.7
FLAT PATTERN	22.3.13
FLATNESS	10.7.9
	10.7.9.1
FLAWS	11.3.5
FLOW, GRAIN	13.3.9
FLUSH WELD	15.4.4.3
FORGING	13.3.1
design practice	13.4
die	13.3.2
dimensioning	Page 13-6
drawing	13.0
machining	Page 13.8
markings	13.5
methods	13.6.3
plane	13.3.4
stocked	13.3.11
	13.9
FORM	
drawing	2.1
	2.5.4
thread	16.4.2.1
tolerances	10.7.10
FORM, BLOCK LINE	22.3.12
FORMAT, DRAWING	5.0
FROMULAE TRUE POSITION	
TOLERANCE	10.10.3
FREE-STATE VARIATION	10.9.3.2
FSCM	5.5.8
FULL FORM THREADS	16.4.6.2
FULL SECTION	8.7.5.4
FUNCTIONAL	
INTERCHANGEABILITY	18.3.8.1

- G -

GATE	12.3.7
GENERAL REQUIREMENTS	2.0
sheet metal	22.4
GENERAL NOTES	7.3.2
	7.6
changes	18.4.10
cross reference	7.6.8
location	7.6.1
	21.10

GENERAL NOTES (continued)	
relisiting	18.4.10.2
removal	18.4.10.1
revision	18.4.10
GEOMETRIC SYMBOLS	10.7.2
GEOMETRIC TOLERANCES	10.7
GEOMETRY, DESCRIPTIVE	8.9
GOVERNMENT DOCUMENTS	1.2
GRAIN FLOW	13.3.9
GRAPHS NUMBERING OF	3.8
GRID	17.3.17
GROOVE WELD	15.4.4.1.1

- H -

HALF SECTION	8.7.5.5
HARDWARE, CUSTOMER	
FURNISHED	9.8
HARNESS ASSEMBLY	17.3.18
HEAT TREATMENT	7.6.10
block	5.6.3
HIDDEN LINES	8.5.10
HIGHWAY SYSTEM	17.3.19
HOLE	
blind, threaded	16.4.6.5
depth	10.5.15.1
dimensioning	10.5.15
indexing	17.3.22
	17.11.1.2
lightening	22.9
locating	10.5.15.2
plated through	17.3.36
	17.11.2.5
thru	10.5.15.1

- I -

IDENTIFICATION	
ADCN, pre-released	18.6.6.4
code, federal supply	1.2
	5.5.8
	9.6.1.2
	9.6.3.2
marking	14.0
new	18.4.17
packaging	23.8
part	14.3.5
revision	18.0
section	8.7.3
view	8.7.3
IDENTIFIER, CONFIGURATION	
ELEMENT	14.3.1
INACTIVE DRAWING	18.3.6.1
	18.4.3
INCOMPLETE THREADS	16.3.11
	16.4.6.3
INCORPORATE AN ADCN	18.3.7
INDEXING HOLES	17.3.21

INDEX

INDEXING HOLES (continued)	17.11.1.2
INDEXING NOTCHES	17.3.22
	17.11.1.2
INDUCTANCE	17.4.1.3
INSEPARABLE ASSEMBLY	4.1
INSERTS, THREAD	16.4.1.4
INSPECTION	
notes, radiographic	7.6.10
weld notes	15.7
INSTALLATION DRAWING	4.1
INTEGRATED CIRCUITS	17.3.23
INTERCHANGEABLE PARTS	18.3.8
INTERCHANGEABILITY	18.3.8
functional	18.3.8.1
physical	18.3.8.2
INTERCONNECTION DIAGRAM	17.3.24
INTERFACE CONTROL DRAWING	4.1
INTERFERENCE FIT	10.3.8.2
INTERSECTION OF PLANES	8.9.10
	8.9.11
INVISIBLE LINES	8.5.10
IRREGULARITIES, SURFACE	11.3.4
ISOMETRIC, DRAWINGS	8.8.8
ITEM	9.3.2
customer furnished	9.8
end	14.3.2
name	6.3.2
number	3.3.7
	3.5
parts list entry	9.6.1.9
	9.6.3.9
multiple assemblies	3.7.6

- J -

JOGGLE	22.3.14
dimensioning	22.7
double	22.7.3

- K -

KELVIN	A1.3.5
	Page B1-2
KEYWAYS	10.5.16
KILOGRAM	A1.3.2
KIT DRAWING	4.1
KNURLS	10.5.18

- L -

LACQUER	19.3.3
LAND	17.3.25
LAY	11.3.15
LAYOUT	
drawing	4.1
changes	18.4.3
printed circuit	17.11
	17.11.1.1
LEADER LINES	8.5.4
LEAST MATERIAL CONDITION	10.3.15
LEFT HAND THREAD DESIGNATION	16.4.2.1.4
LENGTH, UNIT OF (METRIC)	A1.3.1
LETTER, REVISION	18.4.4
LETTERING	8.4
LEVEL, DRAWING	2.5
LEVELS, PRESERVATION AND PACKAGING	23.3.4
LIGHTENING HOLES	22.9
LIMIT DIMENSIONS	10.3.9
LIMITS	10.3.9.1
	10.3.9.2
LINE	
mean	11.3.3
parting	12.3.2
	13.3.3
perpendicular distance to point	8.9.5
true angle to plane	8.9.9
true length of	8.9.4
LINES	8.5
bread	8.5.5
center	8.5.2
cutting plane	8.5.11
datum	8.5.9
dimension	8.5.3
radii	8.5.3
termination	8.5.3
extension	8.5.8
hidden	8.5.10
leader	8.5.4
crossing	8.5.4.4
termination	8.5.4.2
non-intersecting	8.9.6
object	8.5.12
outline	8.5.12
phantom	8.5.6
sectioning	8.5.7
types of	8.5
viewing plane	8.5.11
visible	8.5.12
LIST, EQUIPMENT	17.4.1.5
LMC	10.3.15
LOCAL NOTES	7.3.1
	7.5
LOCATING	
datum targets	10.8.3
dimensions	10.3.5.2
holes	10.5.15.2
general notes	7.6.1
parts list	5.7
	9.5.1
supplementary blocks	5.6.1
LOGIC DIAGRAM	17.3.26
	17.12

INDEX

PAPER SIZES (METRIC)	A5.1	PARTS LIST (continued)	
PARALLELISM	10.9.6	seperate, revision of	18.4.14
PART	14.3.4	subassembly	9.6.3.1.3
defined by words	8.8.7	tabulated drawing	9.6.2
identification	14.3.5	PARTS LIST, ASSEMBLY	
altered and/or selected parts	14.6.9	drawing	9.6.3
cast	12.5.3	multiple	9.6.3.1.2
forged	13.5.2	opposite	9.6.3.1.1
marking	14.6	shown	9.6.3.1.2
matched	4.1	PASSIVATE	19.3.8
number	3.3.2	PATTERN	12.3.9
callout	3.3.5	conductive	17.3.7
reference	3.3.6		17.11.2.4
parts list entry	9.6.1.3	PERMANENT MARKING	14.3.6
	9.6.3.3		14.6.5
opposite	3.3.4	PERPENDICULAR DISTANCE,	
on field of drawing	3.6	POINT TO LINE	8.9.5
serialization	14.3.8	PERPENDICULARITY	10.9.7
	14.6.6	PHANTOM LINES	8.5.6
shown	3.3.4	PHYSICALLY	
	3.6.1	INTERCHANGEABLE PARTS	18.3.8.2
PARTING LINE	12.3.2	PICTORIAL, DRAWING	8.8.8
	13.3.3	PIN HOLE	22.3.17
PARTS LIST		PIPE THREADS	16.3.6
assembly drawing	9.6.3		16.3.7
code identification number	9.6.1.2		16.4.1.2
	9.6.3.2		16.3.10
detail drawing	9.6.1	dryseal	16.3.2.1
entry code identification	9.6.1.2	PITCH THREAD	16.4.2.1
	9.6.3.2	PLACEMENT, OF SECTION	8.7.2
commercial products	9.7	PLACEMENT, OF VIEW	8.7.2
customer furnished item	9.6.3.4.5	PLANE	
	9.6.3.5.6	center	10.3.3
	9.8	forging	13.3.4
item number	9.6.1.9	true angle between two planes	8.9.10
	9.6.3.9		8.9.11
material	9.6.1.5	true angle to line	8.9.9
	9.6.3.5	true distance to point	8.9.8
nomenclature	9.6.1.4	true view of	8.9.7
	9.6.3.4	PLATED SURFACES	11.4.5
part number	9.6.1.3		19.3.4
	9.6.3.3	PLATING	19.3.4
quantity required	9.6.1.1	PLUG WELD	15.4.4.1.2
	9.6.3.1	POINT, PERPENDICULAR	
sequence of	9.4.5	DISTANCE TO LINE	8.9.5
specification	9.6.1.6	POINT, DATUM TAGET	10.8
	9.6.3.6		12.3.6
unit weight	9.6.1.7		13.3.10
	9.6.3.7	POLCIY, CHECKING	2.6.4
zone	9.6.1.8	PORCELAIN, ENAMEL	19.3.5
	9.6.3.8	PRACTICES, DIMENSIONING	10.4
location	9.5		10.5
quantity required	9.6.1.1	drafting	8.0
	9.6.3.1	PRE-RELEASE ADCN	18.3.9
seperate	9.3.3		18.6.6
	9.9		

INDEX

PRESERVATION AND PACKAGING	23.3.2	REFERENCE datum	10.7.3
level of	23.3.4	designations	17.3.9
PRESERVATION	19.3.7		17.4.1.2
PRIMER	19.3.3		17.5.2.1
PRINTED CIRCUIT OR WIRING	17.3.37	dimension	10.3.5.5
	17.11	part number callout	3.7.2
assembly	17.3.41	REGARDLESS OF FEATURE SIZE	10.3.16
	17.11	REGISTER	17.3.43
board	17.3.38	REINSTATING DRAWING	18.4
layout	17.11.1.1	RELEASE DATE	5.5.11
master drawing	17.11.3	RELEASE ENGINEERING	
	17.11.1.2	DOCUMENT	2.7
master pattern drawing	17.11.2	RELIEF, THREAD	16.4.6.7
PRINTED		RELISTING GENERAL NOTES	18.4.10.2
component	17.3.39	REMOVED SECTION	8.7.5.7
contact	17.3.40	REMOVING GENERAL NOTES	18.4.10.1
wiring	17.3.42	REPLACED DRAWING	18.3.6.2
PROCESS SPECIFICATION	1.3.2.2	REQUEST, DOCUMENT CHANGE	18.8
PROCESSES, WELDING	15.6.1	REQUIREMENTS, DRAWING	2.4
PROCUREMENT IDENTIFICATION			2.6.2
NUMBER	14.3.7		2.6.3
PRODUCT SPECIFICATION	1.3.2.1	RESISTANCE	17.4.1.3
PROFILE	11.3.2	RESISTANCE WELD	15.4.4.2
PROFILE, TOLERANCE		RESPONSIBILITIES	
DEFINITION OF	10.9.5	checking and checking	
PROFILE, TOLERANCE	10.9.5.1	management	2.6.4
application	10.9.5.2	drafting and drafting	
PROGRAM MANAGEMENT		management	2.6.3
RESPONSIBILITIES	2.6.2	engineering and engineering	
PROJECTION	8.8	management	2.6.2
orthographic	8.8.1	program management	2.6.1
procedures	8.8	tool design checker	21.5
third angle	8.8.1	tool design	21.4
weld	15.4.4.2	REVISION	18.0
PROPOSAL DRAWING	4.1	block	18.4.16
changes	18.4	description	18.6.3
PROPRIETARY NOTICE	5.12		18.7.2
PROTECTIVE COATINGS	19.0	drawing	18.3.5
	19.3.1	column	5.8
PTF-SAE SHORT THREADS	16.3.10.3	multisheet	18.5
PULLEY BRACKET LAYOUT	8.9.12	processing	18.6.5
			18.7.3
- Q -		general notes	18.4.10
QUANTITY, BULK MATERIAL	9.6.1.1	identification	18.4.4
	9.6.3.1.4	security classification	18.4.11
QUALITY REQUIRED, ASSEMBLY			18.4.12
DRAWING	9.6.3.1	seperate parts list	18.4.14
		REVOLVED SECTION	8.7.5.6
- R -		RFS	10.3.16
RADIAN	A1.3.8	RIB, SECTION THRU	8.7.5.3
RADIOACTIVE MATERIAL	5.13.8	RIGHT HAND THREAD	
RADIUS, SPHERICAL		DESIGNATION	16.4.2.1.4
DIMENSIONING	10.5.10	ROD, WELD	15.6
RANKLINE	A2.6.4.1	ROLLED THREAD	16.3.12
REDRAW, DRAWING REVISIONS	18.4	ROTATION	

12

INDEX

ROTATION (continued)	
section	8.7.2
view	8.7.2
ROUGHNESS	11.3.7
ROUGHNESS HEIGHT RATING	11.3.8
ROUNDED ENDS, DIMENSIONING	10.5.8
ROUNDNESS	10.9.3
RULES, ABBREVIATIONS, USE OF	20.5
RULES, TRUE POSITION DIMENSIONING	10.10.2
RUNOUT	10.9.9
RUNOUT, THREAD	16.3.11

- S -

SCALE, DRAWING	5.5.10
drawing	8.3
SCALE, SECTION	8.3.4
	8.7.6
SCHEMATIC DIAGRAM	
electrical and electronic	17.4
mechanical	4.1
SCREW THREADS	16.0
SCREW THREADS, METRIC	16.5
SEAM WELD	15.4.4.2
SECOND	A1.3.3
broken-out	8.7.5.9
from a section	8.7.1
full	8.7.5.4
half	8.7.5.5
identification	8.7.3
lining	8.7.7.2
offset	8.7.5.8
removed	8.7.5.7
revolved	8.7.5.6
rotation	8.7.2
scale of	8.7.6
thin	8.7.5.1
thru rib	8.7.5.3
title	8.7.4
unlined	8.7.5.2
views	8.7
welded assy	8.7.5.10
	15.5.5
SECTIONING	
conventions	8.7.7
lines	8.5.7
SECTIONS, TYPES OF	8.7.5
SECURITY CLASSIFICATION	5.13
group markings	5.13.4
markings on J size drawing	5.13.6
multisheet drawings markings	5.13.7
revision	18.4.11
	18.4.12
SECURITY CLASSIFICATION (continued)	
task code or accountability control number	5.13.5
SELECTED ITEM DRAWING	4.1
SELECTION OF VIEWS	8.8.2
SEPARABLE ASSEMBLY DRAWING	4.1
SEPARATE PARTS LIST	9.3.3
	9.9
revision	18.4.14
SEQUENCE, PARTS LIST	9.4.5
SERIAL NUMBER	14.3.8
casting	12.5.5
forging	13.5.4
SERIALIZATION	14.3.8
	14.6.6
SERIES, THREAD	16.4.2.1
SET	14.3.9
SET BACK	22.3.16
SHEET NUMBER	5.5.12
	5.9.2.2
SHEET METAL	22.0
bend formulas	22.5.1
bend radius, minimum	22.12.3
dimensioning	22.11
dimensioning chain	22.11.3.1
hole patterns	22.11.3.3
spun parts	22.11.3
drawing callouts	22.10.2
drawings	22.0
forming	22.12
general information	22.4
notes	22.10.2
	22.12.4
power brake forming	22.12
undimensional drawings	22.13
SHOWN	3.3.4
	3.5.6
	3.6.1
assembly, parts list	9.6.3
SI SYSTEM OF INTERNATIONAL UNITS	A1.4
non SI units	A1.6
obsolete SI units	A1.7
prefixes	A1.8
symbols	A1.9
SIGNATURE	
approval	2.6.6
ADCN and DCN	18.4.5
block	5.5.4
SIMILAR TO BLOCK	5.6.5
SINGLE LINE DIAGRAM	17.3.45
SINGLE VIEW DRAWING	8.8.5
SIZE	
basic	10.3.10
design	10.3.6

INDEX

SIZE (continued)
- dimension 10.3.5.6
- drawing 5.3.1
- 5.5.7
- flange weld 15.4.5.3
- lettering 8.4.4
- mean 10.3.11
- nominal 10.3.12
- 16.3.1
- stock 10.5.6
- symbol, weld 15.4.4
- thread 16.3.1
- 16.4.2.1
- tolerance 10.3.13
- 10.6.1

SLOT WELD 15.4.4.1.2
SOCKET, IDENTIFICATION 17.5.2.2
SOURCE CONTROL DRAWING 4.1
SPACING, VIEWS 8.8.6
SPECIAL THREADS 16.4.2.2
SPECIFICATION CONTROL DRAWING 4.1
SPECIFICATIONS
- company 1.3.2
- drafting practices 1.2
- drawing application 1.5.1
- government 1.2
- order of precedence 1.5.2
- parts list entry 9.6.1.6
- 9.6.3.6
- process 1.3.2.2
- product 1.3.2.1

SPHERICAL RADII, DIMENSIONING 10.5.10
SPOTFACE 10.5.12
SPUN PARTS, DIMENSIONING 22.11.3
STAGGERED DIMENSIONS 10.5.3
STANDARD
- book 1.31.3
- company 1.3.1
- design 1.3.1.1
- military 1.2
- order of precedence 1.5.2
- part 1.3.1.2
- series threads 16.3.5
- tolerances 10.4.2

STERADIAN A1.3.9
STOCK NUMBER 14.3.11
STOCK SIZE 10.5.6
STRAIGHTNESS 10.9.2
STUB ACME THREADS 16.3.8
16.4.2.2.2
SUBASSEMBLY, MARKING 14.8.3
- parts list 9.6.3.1.3
SUBSTANCE, UNIT OF (METRIC) A1.3.6
SUFFIX (DASH NUMBER) 3.3.3

SUPERSEDED DRAWING 18.4.3
SUPPLIER DRAWING 2.6.7.5
SUPPLIER DATA, REDRAWING 2.6.7.7
SUPPLIER-DEVELOPED ITEM 2.6.7.6
SURFACE 11.3.1
- coated 11.4.5
- control 11.1
- 11.4
- datum 10.6.2
- plated 11.4.5
- roughness 11.3.7
- symbol 11.4.7
- texture 11.0
SYMBOL 10.7.2
- angularity 10.7.2
- combined 10.7.7
- concentricity 10.7.2
- cross reference, general note 7.6.8
- cross reference, zone 8.6
- cylindricity 10.7.2
- datum, identifying 10.7.3
- datum target 10.8.2
- drafting 8.6
- drafting " SI" A1.10
- drawing application and interpretation 10.9
- electrical and electronic 8.6
- 17.3.10
- 17.4.1.1
- feature control 10.7.5
- finish, weld 15.4.3
- flatness 10.7.2
- forge 13.5.3
- former 10.7.11
- foundry 12.5.4
- graphic 17.4.1.1
- mechanical 8.6
- nonstandard 8.6.1
- parallelism 10.7.2
- perpendicularity 10.7.2
- roundness 10.7.2
- straightness 10.7.2
- surface roughness 11.4.7
- symmetry 10.7.2
- thread 16.4.1
- tolerancing, geometric 10.7.2
- true position 10.7.2
- weld 15.4.2
- welding 15.4
SYMBOLS, METRIC A3.6
- geometric, metric A4.4.6
- indicating division, metric A3.10
- metric prefixes A3.7
- A3.5
- metric spacing A3.9
- metric units A3.5.1

INDEX

SYMBOLS, METRIC (continued)		THREAD (continued)	
	A1.7	blind hole	16.4.6.5
mixing metric	A3.11	British Standard	16.4.2.2.2
obsolete, metric	A3.5	buttress	16.3.9
roughness, cutoff, metric	A4.4.11.1	callout	16.4.5
roughness, surface, metric	A4.4.9.1	class	16.4.2.1
waviness, metric	A4.4.10.1	class 2A	16.4.2.1.3
welding, metric	A4.4.7	class 3A	16.4.2.1.3
writing metric	A3.7	coated	16.4.3
	A3.6	designation	16.4.2
SYMMETRY	10.5.5	diameter	
	10.9.11	major	16.3.2.2
SYSTEM	14.3.12	minor	16.3.2.3
		pitch	16.3.2.4
- T -		dimensioning	16.4.6
TABLE, WIRE	17.5.4	dryseal	16.3.10
TABULATED DRAWING	4.1	form	16.3.3
parts list	9.6.2		16.3.4
TABULATION BLOCK, LOCATION	5.6.1		16.3.5
TAG	14.4.4		16.3.6
TAP DRILL	16.4.6.9		16.3.7
TAPPERED THREADS	16.4.1.2		16.3.8
TAPERS	10.5.18		16.3.9
TASK CODE NUMBER	5.13.5		16.3.10
TEMPERATURE	A2.6	full form	16.4.6.2
absolute zero	A2.6.4	incomplete	16.3.11
Celsius	A2.6.3		16.4.6.3
conversion	A2.6	inserts	16.4.1.4
fahrenheit	A2.6.7	left hand	16.4.2.1.4
	A2.6.8	length	16.4.6.4
kelvin	A1.3.5	metric	16.4.2.2.2
	A2.6.3		16.3
rankine	A2.6.4	multiple	16.4.2.1.2
relative	A2.5	nominal size	16.3.1
TEMPERATURE, UNIT OF			16.4.2.1.1
(METRIC)	A1.3.3	number of	16.4.2.1.2
	A1.3.5	pipe	16.3.6
TEMPORARY MARKING	14.6.4.2		16.3.7
	14.6.10		16.4.1.2
TEMPORARY PROTECTIVE		pitch	16.3.2.1
COATING	19.3.1.2		16.4.2.1
TERMINAL			16.4.2.1.2
area	17.3.47	pitch diameter	16.3.2.4
board	17.3.48	relief	16.4.6.7
assembly	17.3.49	representation	16.0
identification	17.5.1.2		16.4.1
pad	17.3.50	right hand	16.4.2.1.4
TEXTILE DRAWING	4.1	rolled	16.3.12
THIN FILM CIRCUITS	17.3.51	runout	16.3.11
THIN SECTIONS	8.7.5.1		16.4.6.2
THIRD ANGLE PROJECTION	8.8.1		16.4.6.3
THREAD			16.0
Acme	16.3.8	screw	
	16.4.2.2.2	series	16.4.2.1
American National	16.3.4	spark plug	16.4.2.2.2
assembly	16.4.1.3	special	16.4.2.2
		standard series	16.3.5

INDEX

THREAD (continued)	
	16.4.6.1.1
tap drill	16.4.6.9
unified	16.3.3
whitworth	16.4.2.2.2
TIE BAR	12.3.4
TITLE BLOCK	5.5
TITLE	
assignment	6.4
casting drawing	12.7
drawing	5.5.6
	6.0
forging drawing	13.7
section	8.7.4
tool drawing	21.9.1
view	8.7.4
TOLERANCE	10.0
	10.3.13
angularity	10.9.81
bilateral	10.3.13.1
block	5.6.2
concentricity	10.9.10.1
cylindricity	10.9.4.1
flatness	10.9.1.1
formula, true position	10.10.3
geometric	10.7
parallelism	10.9.6.1
perpendicularity	10.9.7.1
profile	10.9.5.1
roundness	10.9.3.1
free-state variation	10.9.3.2
runout	10.9.9.1
size	10.6.1
standard	10.4.2
straightness	10.9.2.1
symetry	10.9.11.1
true position	10.10.1
unilateral	10.3.13.2
zone	10.9
TOLERANCE, METRIC	
conversion to metric	A2.5
limit, metric	A4.3.2
metric	A2.5
plus or minus	A4.3.2
rounding	A2.5
TOOL DRAFTING	
checker	21.5
procedure	21.0
TOOL DESIGNER	21.4
TOOL DRAWING	
change	21.14
format	21.8
notes	21.10
parts list	21.11
title	21.9
TOOL NUMBER	21.3.2

TOOL NUMBER (continued)	
	21.6
TOOL PART NUMBER	21.3.2
	21.6
TOOL HOLE	22.3.17
	22.12.4
TOOLING	21.1
applicable documents	21.2
drawing	21.3.1
point (datum target)	10.8
	12.3.6
threads	21.13
welding procedures	21.12
TRANSITION FIT	10.3.8.3
TREATMENT BLOCK	5.6.3
TRUE ANGLE	
intersecting planes	8.9.10
	8.9.11
line and plane	8.9.9
TRUE	
dimensions	10.5.19
distance, point to plane	8.9.8
length of line	8.9.4
position	10.10
datums	10.10.4
rules	10.10.2
symbol	10.7.2
tolerance formula	10.10.3
view of plane	8.9.7
TUBE BEND DRAWING	4.1
TUBES, DIMENSIONING OF	10.5.7
TYPE DESIGNATOR	6.3.3
TYPES, DRAWINGS	4.0
	4.1
sections	8.7.5
TYPICAL	10.4.3

- U -

UNDERLINING	7.4.2
UNDIMENSIONED DRAWINGS,	
SHEET METAL	22.13
UNDOCUMENTED CHANGES	18.4.8
UNIFIED THREADS	16.3.3
UNILATERAL TOLERANCE	10.3.13.2
UNIT	14.3.14
UNITS, METRIC	
angular, metric	A4.4.1
conversion of	A2.4
international system (SI)	A1.4
metric	A1.3
US	14.3.15

- V -

VARIATION, FREE-STATE	10.9.3.2
VARNISH	19.3.3
VIEW	

16

INDEX

VIEW (continued)			ZONE (continued)	
auxiliary	8.8.3		drawing	5.7.1
detail	8.8.4		parts list entry	9.6.1.8
identification	8.7.3			9.6.3.8
not shown	8.8.7		tolerance	
rotation	8.7.2		cylindrical	10.10
section (sectional)	8.7		true position	10.10

VIEWS	
section (sectional)	
placement	8.7.2
from a section	8.7.1
selection of	8.8.2
spacing of	8.8.6
title	8.7.4
VIEWING PLANE	8.5.11
VISIBLE OUTLINE	8.5.12
VITREAOUS COATING	19.3.5

- W -

WAVINESS	11.3.11
WEB, SECTION THRU	8.7.5.3
WEIGHT	
drawing block	5.6.6
unit, parts list entry	9.6.1.7
	9.6.3.7
WELD	
all-around	15.4.4.3
GMAW	15.6.4
GTAW	15.6.4
rod	15.6
section thru	8.7.5.10
	15.5.5
symbols	15.4.4.1
WELDED MODULE	17.3.52
WELDING	
designations	15.0
notes	15.7
processes	15.6.1
symbols	15.4
WHITWORTH THREADS	16.4.2.2.2
WIRE	
desination	17.3.53
number	17.3.54
table	17.5.4
WIRING	
diagram	17.3.55
	17.5
printed	17.3.42
WORKMANSHIP, DRAWING	
REVISION	18.4.7.1

- X -

X-RAY, INSPECTION NOTES	7.6.5.1

- Z -

ZONE

Section 1 — SPECIFICATIONS AND STANDARDS DATA

1.1 SCOPE ---- This section lists various Government, and approved non-government documents used in the preparation of engineering drawings. Documents listed herein are applicable only as specified in the contract or DRM sections.

1.2 APPLICABLE DOCUMENTS

REFERENCED IN SECTIONS

DEPARTMENT OF DEFENSE (DOD)

MILITARY SPECIFICATIONS

DOD-D-1000	- Drawing, Engineering and Associated Lists	ALL
MIL-W-5088	- Wiring, Aircraft, Installation of	17
MIL-D-5480	- Data, Engineering and Technical: Reproduction Requirements for	ALL
MIL-P-7105	- Pipe Threads, Taper, Aeronautical National Form, Symbol ANPT, General Requirements	16
MIL-S-7742	- Screw Threads, Standard, Optimum Selected Series: General Specifications for	16
MIL-B-7838	- Bolt, Internal Wrenching, 160 KSI FTU	16
MIL-W-8160	- Wiring Guided Missile, Installation of, General Specification for	17
MIL-S-8879	- Screw Threads, Controlled Radius Root with Increased Minor Diameter: General Specification for	16
MIL-Q-9858	- Quality Program Requirements	ALL
MIL-M-9868	- Microfilming of Engineering Documents, 35mm, Requirements for	5
MIL-P-55110	- Printed Wiring Boards	17

MILITARY STANDARDS

(MIL-STD-8	- Supsd by: ANSI-Y14.5)	—
(MIL-STD-9	- Supsd by: ANSI-Y14.6)	16
(MIL-STD-10	- Supsd by: ANSI-B46.1)	—
MIL-STD-12	- Abbreviations for Use on Drawings, Specifications, Standards and in Technical Documents	20

1.2 APPLICABLE DOCUMENTS (CONTINUED)

Document	Title	REFERENCED IN SECTIONS
MIL-STD-14	- Architectural Symbols	24
(Mil-STD-15-1	- Supsd by: ANSI-Y32.2 / IEEE-315)	—
MIL-STD-15-2	- Electrical Wiring Equipment Symbols for Ships Plans.	24
(MIL-STD-15-3	- Supsd by: ANSI-Y32.9)	—
(MIL-STD-16	- Supsd by: ANSI-Y32.16)	—
MIL-STD-17-1	- Mechanical Symbols (Non-Aerospace)	4
MIL-STD-17-2	- Mechanical Symbols for Aeronautical, Aerospace & Spacecraft Use	4
(MIL-STD-20	- Supsd by: AWS-A3.0)	—
MIL-STD-27	- Supsd by: ANSI C37.20	—
MIL-STD-29	- Drawing Requirements for Mechanical Springs	GEN
MIL-STD-34	- Preparation of Drawings for Optical Elements and Optical Systems, General Requirements	4
DOD-STD-100	- Engineering Drawing Practices	ALL
(MIL-STD-106	- Supsd by: ANSI Y10.20)	GEN
MIL-STD-129	- Marking for Shipment & Storage	23
MIL-STD-130	- Identification Marking of US Military Property	4,14
MIL-STD-143	- Specifications & Standards, Order of Precedence for the Selection of	2
MIL-STD-171	- Finishing of Metal & Wood Surfaces	19
MIL-STD-196	- Joint Electronics Type Designation System	6,17
MIL-STD-275	- Printed Wiring for Electronics Equipment	17
MIL-STD-403	- Preparation for & Installation of Rivets & Screws, Rocket & Missile Structure	22
MIL-STD-429	- Supsd by: ANSI/IPC-T-50	—
MIL-STD-454	- Standard General Requirements for Electronic Equipment	17
MIL-STD-490	- Specification Practices	1
MIL-STD-681	- Identification Coding & Application of Hook Up and Lead Wire	17

1.2 APPLICABLE DOCUMENTS (CONTINUED)

REFERENCED IN SECTIONS

Document	Description	Ref.
(MIL-STD-806	- Supsd by: ANSI-Y32.14 / IEEE-91)	—
MIL-STD-1306	- Fluerics, Terminology & Symbols (Fluidics)	4
DOD-STD-1476	–Metric System, Application In New Design	2

CATALOGING HANDBOOKS:

H4	- Federal Supply Code for Manufacturers:	—
H4-1	- --Name to Code	5.5.8
H4-2	- --Code to Name	5.5.8
H6	- Federal Item Name Directory for Supply Cataloging	6, 14
H6A	- --Alphabetical Index of Names	6
H6B	- --Index of Item Name Codes to FIIN's & FIIG's	6
H6C	- --Abbreviations & Symbols	6

MANUALS:

DOD 4120.3-M	- Standardization Policies, Procedures and Instructions	GEN
DOD 5220.22-M	-Industrial Security Manual for Safeguarding Classified Information	GEN

GENERAL SERVICES ADMINISTRATION (GSA)

Federal Specs:

UU-P-561	- Paper, Tracing	5
CCC-C-531	- Cloth, Tracing	5

FEDERAL STANDARDS:

FED–STD–H28	- Screw Thread Standards for Federal Services:	16
FED-STD-75	- (See: ANSI-MH15.1)	23
FED-STD-595	- Color Chips	19

US ARMY

TM5-581	- Construction Drafting	4, 24

US NAVY

NAVFAC-DM-6	- Drawings & Specifications (Construction)	4, 24

1.2 APPLICABLE DOCUMENTS (CONTINUED)

REFERENCED IN SECTIONS

AMERICAN NATIONAL STANDARDS INSTITUTE (ANSI)
(Formerly: USA Stds Inst (USASI); Amer Stds Assn (ASA)
1430 Broadway, New York, N.Y. 10018

ANSI B1.13M	- Metric Screw Threads - M Profile	16
ANSI B1.21M	- Metric Screw Threads - MJ Profile	16
ANSI B4.1	- Preferred Limits & Fits for Cylindrical Parts	10
ANSI B46.1	- Surface Texture (Supsd: MIL-STD-10)	11
ANSI C37.20	- Switch Gear Assy, including Metal Enclosed Bus	24
ANSI MH15.1	- Glossary of Packaging Terms (FED-STD-75)	23
ANSI Y10.3	- Letter Symbols for Quantities Used in Mechanics of Solids (Partly supsd: MIL-STD-12)	20
ANSI Y10.5	- (See: IEEE-280)	—
ANSI Y10.19	- (See: IEEE-260)	—
ANSI Y10.20	- Mathematical Signs & Symbols for Use in Physical Science & Technology	GEN
ANSI Y14.1	- Drawing Sheet Size & Format (Partly Supsd: DOD-STD-100)	—
ANSI Y14.2M	- Line Conventions & Lettering (Partly supsd: DOD-STD-100)	8
ANSI Y14.3	- Multiview Drawings (Partly Supsd: (DOD-STD-100)	4
ANSI Y14.5	- Dimensioning & Tolerancing (Supsd: MIL-STD-8)	10, 17, 18, 19, 22
ANSI Y14.5M	- Dimension & Tolerancing	10M
ANSI Y14.6	- Screw Thread Representation (Supsd: MIL-STD-9)	16
ANSI Y14.6aM	- Metric Supplement to Y14.6	16
ANSI Y14.7.2	- Gears, Splines & Serrations	GEN
ANSI Y14.15	- Electrical & Electronics Diagrams	17
ANSI Y14.17	- Fluid Power Diagrams	GEN
ANSI Y14.36	- Surface Texture Symbols	11
ANSI Y32.2	- (See: IEEE-315)	17
ANSI Y32.3	- (See: AWS-A2.4)	—
ANSI Y32.4	- Graphic Symbols for Plumbing	4,24
ANSI Y32.9	- Graphic Symbols for Electrical Layout Diagrams Used in Architecture and Building Construction (Supsd: MIL-STD-15-3)	24
ANSI Y32.10	- Graphic Symbols for Fluid Power Diagrams	4
ANSI Y32.11	- Graphic Symbols for Process Flow Diagrams	
ANSI Y32.12	- (See: AWA-C2.6)	
ANSI Y32.14	- (See: IEEE-91) (Supsd: MIL-STD-806)	
ANSI Y32.16	- (See IEEE-200)	17
ANSI Y32.17	- (See: AWS-A2.4)	
ANSI Y32.18	- Symbols for Mechanical & Acoustical Elements as Used in Schematic Diagrams	

1.2 APPLICABLE DOCUMENTS (CONTINUED)

REFERENCED IN SECTIONS

AEROSPACE INDUSTRIES ASSOCIATION (AIA)
National Aerospace Standards Committee (NASC)
1321 Fourteenth St. N.W., Washington, D.C.

NAS 523	- Rivet Code	22
NAS 944	- Symbols, Hydraulic Test Equipment Drafting	4

AMERICAN SOCIETY OF MECHANICAL ENGINEERS (ASME)
United Engineering Center, 345 E. 47th St., New York, N.Y. 10017

NOTE: Applicable ASME Standards are cross referenced to ANSI Standard Numbers

AMERICAN SOCIETY FOR TESTING AND MATERIALS (ASTM)
1916 Race St., Philadelphia, PA 19103

ASTM E 380	- Metric Practice Guide	10

AMERICAN WELDING SOCIETY (AWS)
2501 N.W. 7th St., Miami, Florida 33125

AWS A2.0	- Welding Symbols (ANSI-Y32.3) (Supsd by: AWS-2.4)	15,21
AWS A2.2	- Non-Destructive Testing Symbols (ANSI-Y32.17) (Supsd by: AWS-2.4)	—
AWS A1.1	- Metric Practice Guide for the Welding Industry	15
AWS A2.4	- Symbols for Welding and Non-Destructive Testing	15,21
AWS A3.0	- Terms & Definitions (Welding) (Supsd by: MIL-STD-20)	—

INSTITUTE OF ELECTRICAL & ELECTRONICS ENGINEERS (IEEE)
United Engineering Center, 345 E. 47th St., New York, N.Y. 10017

IEEE 91	- Graphic Symbols for Logic Diagrams (ANSI-Y32.14)	14,24
IEEE 200	- Reference Designations for Electrical & Electronics Parts & Equipments	
IEEE 260	- Letter Symbols for Units Used in Science & Technology (ANSI-Y10.19) (Partly Supsd: MIL-STD-12)	—
IEEE 280	- Letter Symbols for Quantities Used in Electrical Science & Electrical Engineering (ANSI-Y10.5) (Partly Supsd: MIL-STD-12)	—
IEEE 315	- Graphic Symbols for Electrical & Electronics Diagrams (ANSI-Y32.2) (Supsd: MIL-STD-15-1)	17

INSTITUTE FOR INTERCONNECTING AND PACKAGING ELECTRONIC CIRCUITS (IPC)
3451 Church St., Evanston, IL 60203

ANSI/IPC-T-50 - Terms and Definitions for Interconnecting and Packaging Electronic Circuits	17

INTERNATIONAL STANDARDS ORGANIZATION
American National Standards Institute (ANSI)
1430 Broadway, New York, N.Y. 10018

ISO R724	- Metric Screw Threads, Basic Dimensions	—
ISO 1000	- SI Units and Recommendations For The Use Of Their Multiples And Of Certain Other Units	10

1.2 APPLICABLE DOCUMENTS (CONTINUED)

REFERENCED IN SECTIONS

NATIONAL AERONAUTICS & SPACE ADMINISTRATION (NASA)
U.S. Government Printing Office (GPO), Washington, D.C. 20402

SP 7012	- The International System of Units Physical Constants and Conversion Factors	10

NATIONAL BUREAU OF STANDARDS (NBS)
Washington, D.C. 20234

NBS Miscellaneous Publication 286	Units of Weights & Measure	10
NBS Special Publication SP 330	The International System of Units (SI)	10

SOCIETY OF AUTOMOTIVE ENGINEERS (SAE)
400 Commonwealth Dr., Warrendale, PA 15096

SAE HANDBOOK

AS 1290	- Graphic Symbols for Aircraft Hydraulic and Pneumatic Systems	GEN 4
AS 478	- Identification Markings	14
AS 1338	- Aerospace Metric 60° Screw Thread Profile and Tolerance Classes	16

1.3 COMPANY STANDARDS, SPECIFICATIONS AND PROCEDURES --
Company Standards shall meet the following requirements:

(a.) Is identified by name and address of issuing company, FSCM number, document nomenclature and number, and contract number.

(b.) Does not contain limited rights in technical data.

(c.) Provide the necessary design disclosure information for the level of drawing for which they are furnished.

(d.) Satisfy the same procurement requirements as for a specification control drawing when it defines a vendor item.

(e.) Drawing practices and symbols used are such that their intent and interpretation are clear and unambiguous whether of standard or non-standard origin.

(f.) Have clarity and legibility for the purpose of microfilming.

(g.) All documents referenced in a company standard shall also be supplied when not covered by a military, government, or non-government standard or specification.

1.3.1 STANDARDS —Standards are of three types: Design Standards, Part Standards, and Book Standards. The Company or Corporate Document Standardization Manual is used for their preparation when one exists.

1.3.1.1 STANDARD DESIGN — Standard Design is a standardized design feature of an item or process or a specially created shape (extruded, molded, drawn, etc.) having wide utility and use which is described and/or pictured on a "standard drawing" format. This is not a part number in that there is no part to identify. i.e. standard sizes for drilled holes.

1.3.1.2 PART STANDARD — A part or assembly of parts which has wide utility or recurring use.

1.3.1.3 BOOK STANDARD — Book Standards are comprehensive presentations of engineering test methods, procedures, criteria symbols and the like.

1.3.2 SPECIFICATION — Specifications are of Five types, System, Development, Product (Commodity) process, and material prepared in accordance with MIL-STD-490 and the Company or Corporate Document Standardization Manual when one exists.

1.3.2.1 SYSTEM SPECIFICATION — A system specification states the technical and mission requirements for a system as an entity, allocates requirements to functional areas and defines the interfaces between or among the functional areas. It establishes the performance, design, development and test requirements for the system.

1.3.2.2 DEVELOPMENT SPECIFICATION — A development specification state the requirements for the design or engineering development of a product during the development period.

1.3.2.3 PRODUCT (Commodity) SPECIFICATIONS — A product specification is a technical description of the design or performance characteristics of material, component, or performance characteristics. Performance characteristics should be placed on drawings through the medium of a specification.

1.3.2.4 PROCESS SPECIFICATION — A process specification is a technical description of the processing requirements necessary to produce a product. Such a specification is prepared only when it is not possible to state the requirements for the end product in other documents with sufficient completeness to assure that the product would be satisfactory. If the process applies to more than one product, a Company Standard will be prepared.

1.3.2.5 MATERIAL SPECIFICATION — A material specification is applicable to a raw material (chemical compound), mixtures (paints), or semi-fabricated material (electrical cable) which are used in the fabrication of a product.

1.3.3 COMPANY PROCEDURES — Company Procedures are of three types as follows:

1.3.3.1 CORPORATE — A statement of policy or general instruction establishing companywide uniformity and control.

1.3.3.2 PLANT — A policy or procedure governing activities within the responsibility of the Plant Manager.

1.3.3.3 SUPPLEMENT — A revision or amendment which does not change the basic content or the company procedure, but clarifies or updates it.

1.4 DEFINITIONS — Not applicable.

1.5 REFERENCES ON DRAWINGS —

1.5.1 When Government, non-government, or design activity specifications and standards are cited, show only the basci document number, do not list revision, amendment status, dates, etc. on drawings.

1.5.2 When specifications or standards are specified on drawings, follow the order of precedence for selection as established in MIL-STD-143.

1.5.3 Do not cite specifications or standards, with exceptions, deletions, additions or extraction of information by paragraph number. Refer to paragraphs by the paragraph title, since paragraph numbers may change.

1.5.4 A program contract should establish the requirements for applicable documents. Supersession or revision of a specified document does not necessarily mean that the document is no longer required on the contract.

1.5.5 Do not cite the following types of documents on engineering drawings:

1.5.5.1 DOCUMENTS:

    Procedure Manual
    Technical Manuals
    Catalogs
    Pamphlets
    Recordings
    Manuscripts
    Technical Reports
    Writings
    Policy Making Documents
    Maintenance Manuals
    Design Activity DRM

1.5.5.2 DRAWINGS OF:

Tools and Gages (except for tools and gages when contractually required for Interface Control between Companies).

Test Fixtures

1.5.5.3 Any document not subject to design activity change control, other than those released by military, industry societies, or associate contractors.

1.5.6 Do not create drawings for items covered by existing Government or nationally recognized industry (NAS, MS, AN, AMS, ASTM, etc.) specifications or standards, unless those standards fail to assign uniquely identifying part or identifying numbers.

NOTES

Section 2 — General Requirements

2.1 SCOPE--This section defines requirements for drawing levels and drafting requirements and the selection of applicable specifications and standards. Note: 'Categories' and 'Forms', formerly defined in DOD-D-1000, no longer apply.

2.2 APPLICABLE DOCUMENTS
DOD-D-1000   Drawings, Engineering & Associated Lists
DOD-STD-100   Engineering Drawing Practices
MIL-STD-143 Specifications & Standards, Order of Precedence for the Selection of
DOD-STD-1476 Metric System, Application in New Design

2.3 DEFINITIONS

2.3.1 DRAWING LEVEL — The provision of necessary design disclosure information to the degree required for the drawing level selected.

2.3.2 LEVEL 1, CONCEPTUAL & DEVELOPMENTAL DESIGN — Drawings that disclose engineering design information sufficient to enable the evaluation of an engineering concept as meeting stated requirements.

2.3.3 LEVEL 2, PRODUCTION PROTOTYPE & LIMITED PRODUCTION DESIGN — Drawings that disclose sufficient design information to support the manufacture of production prototype or a limited production model.

2.3.4 LEVEL 3, PRODUCTION DESIGN — Drawings that disclose sufficient engineering definition to enable any competent manufacturer to produce a quantity of items which are interchangeable with the original design without the need for:
—additional product design effort
—additional design data
—recourse to the design activity

2.4 DRAWING REQUIREMENTS

2.4.1 Engineering drawings required by contract are normally specified on DD Form 1423 and the level of drawings required to provide the necessary design disclosure. **Combination of Levels may be specified in the contract or order.**

2.4.1.1 On contracts where the drawing "level" is **not** specified, or on design activity IR&D and commercial work use the following guide:
- (a) Prepare on design activity standard engineering drawing format.
- (b) Utilize the design activity drawing and part numbering system.
- (c) Company, government and industry association standards or specifications may be omitted for parts, processes and materials, but their use is encouraged when economically feasible.
- (d) Use of standard parts is recommended.
- (e) Free-hand sketching is permissible and drawings need not be to an exact scale.
- (f) Drawing must be legible and reproducible.
- (g) Use of DRM drafting standards for symbols and conventions is recommended.
- (h) Utilization of simplified drafting techniques is encouraged.
- (j) Drawing must contain sufficient information necessary for fabrication, installation, assembly, procurement and inspection.

2.4.1.2 On contracts where the drawing "level" is specified for engineering drawings and associated lists, shall be prepared to the "level" so designated. With respect to engineering drawing preparation, the contract or order should list the following requirements:
- (a) whether a Government Design Activity or Contractor Design Activity designation and identity is to be placed in the Title Block of the engineering drawing or associated list.
- (b) whether Government Design Activity or Contractor Design Activity drawing numbers will be assigned.
- (c) when Government Design Activity drawing numbers are to be assigned, identify the assigning activity; and, if Government drawing formats are to be supplied will identify the source.
- (d) whether the mono-detail drawing system will be used.
- (e) kinds of associated list required.
- (f) assembly level at which associated lists will be prepared.
- (g) whether associated lists shall be integral with, or separate from, the engineering drawing.
- (h) drawing format, type, class and quality of stock.

2.4.1.3 EXISTING ENGINEERING DRAWINGS AND ASSOCIATED LISTS — Engineering drawings and associated lists prepared prior to the above requirement of paragraph 2.4.1.2 are acceptable if the engineering drawings and associated lists have been previously accepted by the Government; or meet the following requirements:
- (a) Are acceptable for entry into the Governments repository system, (i.e. are identified by name and address of design activity, code identification number, drawing nomenclature, drawing (part) number, in accordance with **DOD-STD-100,** and contract number).

(b) The format, legibility requirements, drawing practices and symbols used (including the use of legends/explanations for non-standard symbols) are such that their intent and interpretation of all requirements are clear and unambiguous.

(c) Provide the necessary design disclosure information for the level of drawing for which they are furnished.

2.5 DRAWING LEVELS

2.5.1 LEVEL 1, CONCEPTUAL AND DEVELOPMENT DESIGN — These drawings shall be legible and those most amenable to the mode of presentation. Sketches, layout drawings and combinations of types of engineering drawings may be used to convey the engineering concept in such a manner that the engineering information is understandable to cognizant engineers and scientists. Unless otherwise specified the requirements of DOD-STD-100 do not apply.

2.5.2 LEVEL 2, PRODUCTION PROTOTYPE AND LIMITED PRODUCTION — These selected engineering drawing types shall include, as applicable, parts lists, detail and assembly drawings, interface control data, logic diagrams, schematics, performance characteristics, critical manufacturing limits, and details of new materials and processes. Special inspection and test requirements necessary to determine compliance with requirements for the item shall be defined on the engineering drawings directly or by reference. Level 2 engineering drawings and associated lists shall be prepared in accordance with DOD-STD-100 unless otherwise specified in contract or order.

2.5.3 LEVEL 3, PRODUCTION — These selected engineering drawing types shall include details of unique processes, when essential to design and manufacture; detailed performance ratings; dimensional and tolerance data; critical manufacturing assembly sequences; toleranced input and output parameters; schematics; mechanical and electrical connections; physical characteristics, including form and finishes; details of material identification; inspection, test and evaluation criteria; necessary calibration information; and quality control data. Unless otherwise specified, a control drawing(s) as shown in Section 4 shall be prepared, in those instances where: (a) vendor or commercial items, are approved for use in the design, and are not covered by Government or nationally recognized industry association standards and specifications; (b) for items not developed at Government expense, (c) for items which the Government has not acquired unlimited rights. Level 3 engineering drawings and associated lists shall be prepared in accordance with DOD-STD-100.

2.5.4 The following conditions for the preparation of engineering drawings for Levels 2 and 3 shall be governed by the contract or order.

2.5.4.1 The application of the metric system in new design shall be in accordance with DOD-STD-1476.

2.5.4.2 Whether a Government Design Activity or Contractor Design Activity name, Federal Supply Code for Manufacturers (FSCM) number and drawing number will be placed in the title block of the engineering drawing(s) and associated list(s).

2.5.4.3 When Government Design Activity drawing numbers are to be assigned, identify the assigning activity; and, if Government drawing formats are to be supplied, identify the source.

2.5.4.4 Kinds of associated lists required.

2.5.4.5 Drawing assembly level at which associated lists will be prepared.

2.5.4.6 Whether the mono-detail drawing system will be used.

2.5.4.7 Whether parts lists shall be integral with, or separate from, the engineering drawing.

2.5.4.8 Drawing format material.

2.5.5 DRAWING LEVELS IN RELATION TO PREVIOUS DRAWING FORMS

2.5.5.1 Comparison of current **(1977) DOD-D-1000** and the former "NO" revision (1965). See Table 1.

MIL-D-1000 (1965) DRAWING FORM	DOD-D-1000B DRAWING LEVEL
FORM 1	LEVEL 3
FORM 2	LEVEL 2
FORM 3	LEVEL 1

TABLE 1.

2.6 RESPONSIBILITIES OF ENGINEERING DRAWING MANAGEMENT AND RELATED FUNCTIONS

2.6.1 PROGRAM MANAGEMENT

2.6.1.1 Program management is responsible for informing applicable documentation activities of the requirements that drawings must meet. These requirements are defined by Program Directive, Work Statement, Work Orders, etc. Table 2 lists the requirements for each of three drawing levels and includes the requirements or considerations for various end use.

DOD-D-1000 DRAWING LEVEL	DESIGN ACTIVITY STD DRWG FORMAT	DESIGN ACTIVITY DRAWING & PART NUMBER SYSTEM	USE OF FREE HAND SKETCHES	SIMPLIFIED DRAFTING PRACTICE	REPRODUCTION PER MIL-D-5480	FSCM NO. COMPANY OR GOVT
3	YES /1\	YES /1\	NO	NO	YES	YES
2	YES /1\	YES /1\	NO	YES	YES	NO /1\
1	YES	YES	YES	YES	NO	NO /1\

(CONTINUED)

DOD-D-1000 DRAWING LEVEL	SPECIFICATION USAGE FOR MAT'L & PROCESSES	SPECS SELECTION SEQUENCE PER MIL-STD-143	MICROFILM QUALITY	USE OF STD SYMBOLS, ABREV & CONVENTIONS	DIMENSIONING PER ANSI Y14.5
3	YES	YES	YES	YES	YES
2	YES	YES	YES	YES	YES
1	NO /2\	NO /2\	NO	NO /2\	NO /2\

/1\ UNLESS OTHERWISE SPECIFIED IN CONTRACT OR ORDER
/2\ RECOMMENDED IF FEASIBLY ECONOMICAL

TABLE 2

2.6.2 ENGINEERING AND ENGINEERING MANAGEMENT

2.6.2.1 The responsible engineer provides sufficient information such as sketches, layouts, etc., or other pertinent instructions necessary for drafting to complete the assigned task.

2.6.2.2 Engineers are responsible for all features of design.

2.6.2.3 Engineer's comments on check prints may be in any color other than those reserved for the checkers (red, yellow and blue).

2.6.2.4 When the responsible engineer affixes his signature to a drawing, he certifies that the drawing satisfies design requirements.

2.6.3 DRAFTING AND DRAFTING MANAGEMENT

2.6.3.1 The Drafting Group is responsible for the preparation of drawings and the method of presentation used to adequately describe the engineer's design requirements. Drafting Management is responsible for the interpretation of drawing standards, related specifications and specified drawing requirements, as well as the control of all original drawing vellums in process.

2.6.3.2 Drawings must be clear, concise, complete, and capable of only one interpretation. The drawing should portray the final product without specifying the method of manufacture, unless it is required for clarity or is a specific process necessary to meet the design requirements.

2.6.3.3 Each draftsman should thoroughly review his drawings to eliminate errors or omissions prior to submitting them for approval.

2.6.3.4 Applicable layouts, dimensional calculations, catalogs, design memoranda, reference prints, and other pertinent data should be provided to the checker together with the check prints.

2.6.3.5 Correction of the drawing after checking should be made by the original draftsman in order to avoid making the same or similar errors in the future. Final correction of the drawing should take precedence over other work. Design changes requested by the checker should be coordinated with the responsible engineer, prior to incorporation into the drawing.

2.6.3.6 Additional changes made while correcting an unreleased drawing may be marked on the check print in a color other than red, yellow or blue (reserved for checkers) and initialed by the responsible engineer or draftsman. The corrected drawing and marked-up check print are resubmitted to the checker to verify all changes.

2.6.4 CHECKING AND CHECKING MANAGEMENT

2.6.4.1 The checker shall assure that the documentation requirements specified on the Program Directive, Work Order, etc., have been fulfilled.

2.6.4.2 The checker shall verify the dimensional accuracy and completeness of drawings and assure conformance to the standard drafting procedures of applicable specifications and that all documents called out on a drawing are released and are currently in effect.

2.6.4.3 The checker's approval of correct dimensions, callouts, notes, etc., shall be indicated by a yellow (✓) or line drawn through the applicable data.

2.6.4.4 Incorrect data shall be either circled or marked in red and the correct information (if stated) lettered adjacent to the point in question. For extensive changes, a written note in red stating, "REVISE PER" (Provide suitable explanation and references to support reasons for change).

2.6.4.5 Checker's suggestions, general comments on features of design, simplification possibilities, or checker's notes for his own use, may be shown in blue or black.

2.6.4.6 The checking group should maintain a file of check prints for a minimum of six months to attest the information that has been checked and approved.

2.6.5 CHECKING POLICY

2.6.5.1 The engineering drawing checker shall be provided with a check print to convey his corrections and verifications. Using a check print avoids defacing the drawing and provides a record from which the checker can make a final check. The engineering drawing checker's print is the only one stamped "Check Print."

2.6.5.2 Check prints shall be returned to the checking department after the drawing is released.

2.6.5.3 Drawings or change documents shall not be changed after the checker's approval, prior to release, without his cognizance.

2.6.6 APPROVAL REQUIREMENTS

2.6.6.1 Only an authorized checker's signature shall appear in the Checker's Block.

2.6.6.2 The drawing shall be approved by the responsible engineer and draftsman before a drawing check is initiated.

2.6.6.3 Company standard part drawings and/or standard design drawings shall be checked the same as product drawings.

2.6.7 APPROVAL AND USE OF ENGINEERING DRAWINGS OTHER THAN THOSE DESIGNED BY THE DESIGN ACTIVITY

2.6.7.1 SPECIFICATION CONTROL DRAWING — A specification control drawing (see Section 4) shall only be prepared for vendor-developed item(s) when such are advertised or cataloged as available to the trade or the public on an unrestricted basis or procurable on order from a specialized segment of industry. Specification control drawings shall not be tailored to the characteristics of a single vendor's product to the exclusion of other equally suitable products. Conversely, specification control drawings shall not be so broad as to permit acceptance of products which will not perform in the equipment under all required conditions. Specification control drawings shall not be prepared for items developed at Government expense.

2.6.7.2 SOURCE CONTROL DRAWINGS — A source control drawing (see Section 4) shall be prepared only when the technological or other factors beyond the control of the contractor are evident which prevent the determination of performance, reliability and configuration control requirements needed to prepare a specification control drawing. Source control drawings shall not be prepared for items developed at Government expense.

2.6.7.3 ITEMS COVERED BY EXISTING SPECIFICATIONS AND STANDARDS — Engineering drawings shall not be prepared or submitted for items that are defined by Government specifications, standards or nationally recognized industry association specifications or standards.

2. 6. 7. 4 COMPANY STANDARD A Company Standard is a document that establishes the issuing company's engineering and technical information for describing items, materials, processes, methods, design and engineering practices without limited rights in technical data. All other requirements for their preparation shall be governed by drawing levels 2 and 3 and shall not be prepared or submitted for items that are identified by Government or Industry standards.

2.6.7.5 REFERENCE DOCUMENTS — Contractor documents referenced on engineering drawings are considered reference documents and shall be furnished as part of the level ordered as an integral part of the engineering drawing package. When first tier references do not provide the essential technical information the contractor's subordinate reference shall be provided to the extent necessary to meet the technical disclosure requirements of the level ordered. However, technical manuals, procedural manuals, maintenance manuals and company drafting manuals are not considered reference documents.

2.6.7.6 SUPPLIER DRAWING — A Supplier drawing is a cognizant design activity, manufacturer, wholesaler, or agent from whom are acquired drawings used in the performance of the contract.

2.6.7.6.1 This DRM will identify vendors supplying commercial items and vendor-developed items as "SUPPLIERS" and those supplying services and/or items fabricated to Company design drawings as "SUBCONTRACTORS."

2.6.7.7 SUPPLIER-DEVELOPED ITEM — Supplier-developed items are those products of industries which normally provide customer application engineering services for a commercial product line and their products are commercially available from a specialized segment of an industry. Typical examples of such items are: special motors, synchros, transformers, potentiometers, hydraulic valves, carburetors, potted servoamplifiers, key boards and tape readers.

2.6.7.8 REDRAWING SUPPLIER DATA ON COMPANY FORMAT — In the event it becomes necessary to document a supplier's item (s) on other than a control drawing, or altered or selected item drawing, the written permission of the supplier and approval by the Company Legal Department is recommended prior to preparing the drawing.

2.7 ENGINEERING DOCUMENT RELEASE — The process of transferring custody of an engineering document, or change thereto, from the preparing activity to a divisional control activity which is responsible for its production, distribution, storage, and maintenance of change history records.

2.8 Duplicate Drawing Restrictions—two or more identical drawings are not permitted

2.8.1 Duplicate drawings are permitted only to replace a worn or otherwise not maintainable original. Upon verification of the duplicate, the original shall be destroyed and the duplicate will take it's place.

2.8.2 A duplicate drawing may be used as a point of departure upon which changes are made to produce a new original drawing which will have a new identity and be separately maintained.

2.9 Department of Defense contracts, excluding construction or architect-service, generally control the selection of specifications and standards for the design of items materials and processes.

2.9.1 Whenever a contract does not explicitly specify specifications of standards to be used, then the choice of an applicable specification or standard is the responsibility of the design activity.

2.9.2 Order of precedence shall be executed by the design activity in the selection of applicable specifications and standards which fall within the group and sub-groups that follow:

(a.) Group I — Specifications and standards listed in the Department of Defense Index of Specifications and Standards (DODISS).

(1.) Sub-group 1: Coordinated Federal Specifications and Standards.

(2.) Sub-group 2: Coordinated Military Specifications and Standards.

(3.) Sub-group 3: Non-government specifications and standards of approved nationally recognized technical and trade societies for use in military contracts and listed in the DODISS.

(4.) Sub-group 4: Limited coordination Military and interim Federal specifications and standards listed in the DODISS.

(5.) Sub-group 5: Non-government specifications and standards having limited coordination status and listed in the DODISS.

(6.) Sub-group 6: Limited coordination Military and interim Federal specifications and standards issued by other government sources (e.g. NASA, FAA, etc.).

(b.) Group II — Non-government specifications and standards prepared by nationally recognized industry associations, trade or technical societies without military status and not listed in the DODISS. Copies are obtained from the associations concerned.

(c.) Group III — Government specifications and standards other than military or federal series and not listed in the DODISS (e.g. NASA, FAA, etc.).

(d.) Group IV — Individual company specifications and standards.

NOTES

Section 3 — DRAWING AND PART NUMBERING SYSTEM

3.1 SCOPE — This section establishes the numbering system for the identification of engineering drawings and parts.

3.2 APPLICABLE DOCUMENTS

DOD-STD-100   Engineering Drawing Practices

3.3 DEFINITIONS

3.3.1 DRAWING NUMBER — A nonsignificant number assigned each engineering drawing for identification purposes.

3.3.2 PART NUMBER — Part numbers are assigned to uniquely identify a specific item or part.

3.3.3 DASH NUMBER (SUFFIX) — The dash number is a numerical identification limited to three digits, suffixed to a drawing number to identify individual parts or assemblies.

3.3.4 SHOWN AND OPPOSITE (PARTS) — "Shown" applies to the part(s) as pictured. "Opposite" applies to the mirror-image of the "shown" part(s).

3.3.5 PART NUMBER CALLOUT — A part number callout is the citation of an item's P/N.

3.3.6 REFERENCE PART NUMBER CALLOUT — A reference part number callout is used to indicate an item that is shown and called out elsewhere.

3.3.7 ITEM NUMBER — An item number is a number assigned on an assembly drawing to a part, subassembly, etc., for the purpose of cross-referencing the callouts in the Parts List to their location on the field of the drawing and general or local notes.

3.4 DRAWING NUMBER ASSIGNMENT

3.4.1 A centralized administrative activity should control drawing number assignment to assume that numbers are not skipped or duplicated. Where a design activity is larger or dispersed, a block of numbers may be assigned to each local organization, where careful control must be exercised. Against each drawing number, record the following information:

    (a) Drawing Title (at least the noun name)
    (b) Name of draftsman
    (c) Name (or code number) of project
    (d) Date of drawing number assignment
    (e) Date of drawing release

3.4.2 If a drawing is not used in the end product, mark it "Not Used" and send it to file; make a corresponding entry in the assignment record logbook. Do not reassign the number to another drawing.

3.4.3 The drawing number shall comply with the following:
 (a) 15 characters maximum — correspondingly less if it must be prefixed and suffixed to form a part number (See paragraph 3.5.3).
 (b) Characters may include arabic numerals, upper-case letters and dashes, except:
  —Do not use the letters I, O, Q, S, X or Z
   The letters S and Z are permissable if they are part of an on going system.
  —Do not use punctuation such as: (), *, /, °, +, blank spaces
  —Do not include the code identification (FSCM) number
  —Do not include drawing format size letter and revision letter

  Example: DRAWING NUMBER   ABXXXXXXX
        Prefix (as applicable)
        Drawing Number

 (c) Significant drawing numbers are acceptable (but not recommended) so long as they provide complete, unique identification. Prefixes (other than those for associated lists) constitute a form of significant numbering.
 (d) For Associated List numbering, see Section 9.

3.5 PART NUMBER ASSIGNMENT

3.5.1 Each part number shall be or shall include the complete drawing number of the drawing which describes and controls the item.

3.5.2 Where more than one item is described on a drawing, differentiate among them by assigning each item a unique dash number, suffixed to the drawing number. (See Figure 1)

  Example: Part Number (2 or more per drawing)   ABXXXXXXX — XXX
        Prefix (as applicable)
        Drawing Number
        Dash
        Suffix (dash number)

3.5.3 Do not compose any part number (including the prefix and dash number) of more than 15 characters.

3.5.4 The characters used to compose a suffix or dash number are subject to the same restrictions as a drawing number (See paragraph 3.4.3.b)

Note: Requirements a through d apply to part numbers assigned by a design activity; government and industry standardizing activities unfortunately do not always follow these rules. The listing of their part numbers are covered under Parts Lists (see Section 9).

3.5.5 Assign dash (suffix) numbers in consecutive order, with -1 assigned to the detailed part or assembly.

Note: On a tabulated drawing, it may be convenient to assign part suffixes in a significant pattern. e.g., on a drawing of a family of instrument gears: "PART SUFFIX INDICATES THE NUMBER OF TEETH, -13 (13 TEETH) THROUGH -288 (288 TEETH). This can avoid extensive tabulation and confusion as to which gear has which suffix.

3.5.6 Indicate shown and opposite parts by suffixes in the field of the drawing. e.g.,

-1 SHOWN
-2 OPPOSITE

3.5.7 Shown and Opposite part dash numbers are entered above and to the left of the title block. (See **Figure 1**).

3.5.8 If the variation or difference between parts become confusing, prepare a separate drawing.

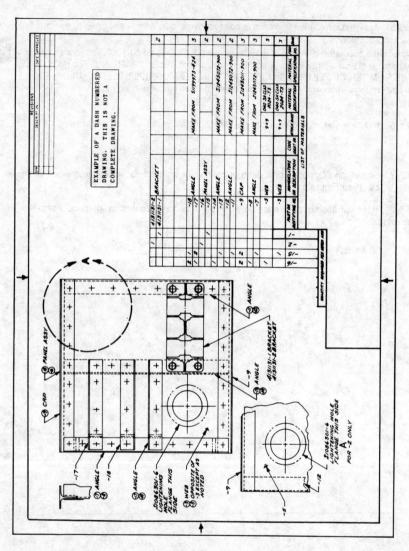

Figure 1

3.6 DASH NUMBER CALLOUT ON THE FIELD OF DRAWING

3.6.1 Enclose dash numbers on the field of a drawing within a .50 diameter circle. (See Figures 2 through 5)

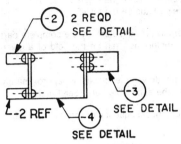

Detail Dash Number
Figure 2

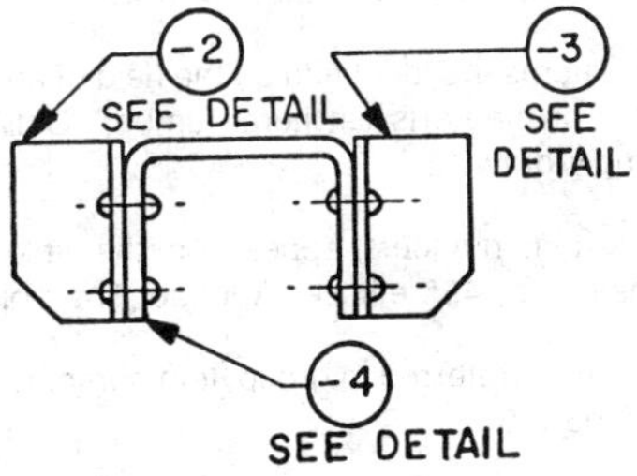

Shown and Opposite Detail Dash Number
Figure 3

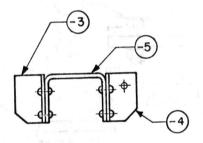

Shown and Opposite Assembly
Figure 4

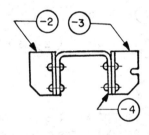

Shown and Opposite Assembly
With Minor Variation
Figure 5

3.7 ITEM NUMBER

3.7.1 Item Numbers are entered in the Item Number Column only on assembly drawings in numerically ascending order beginning with "1". No item number may be omitted or removed by a subsequent change to the drawing. (See Section 9)

3.7.2 For cross referencing items to other documents, the part number is used. For those parts called out on other drawings, using the part numbering syste

3.7.3 Items are identified on the field of an assembly drawing either by the complete part number or the parts list item number. Only one method is used on the field of a given assembly drawing.

3.7.4 Item numbers appear on the field of the drawing enclosed in approximately one-half inch, 45° ellipse. A leader line points to the applicable part. **(See Figure 6)**

3.7.5 It is preferred to group item numbers in numerical sequence or order of assembly. (See Figure 6)

3.7.6 ITEM NUMBERS ON MULTIPLE ASSEMBLY DRAWINGS

3.7.6.1 On multiple assembly drawings, any item number callout not applicable to all assemblies may specify the applicable assemblies. (See Figure 7)

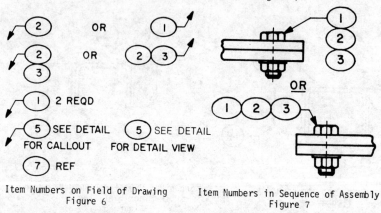

Item Numbers on Field of Drawing
Figure 6

Item Numbers in Sequence of Assembly
Figure 7

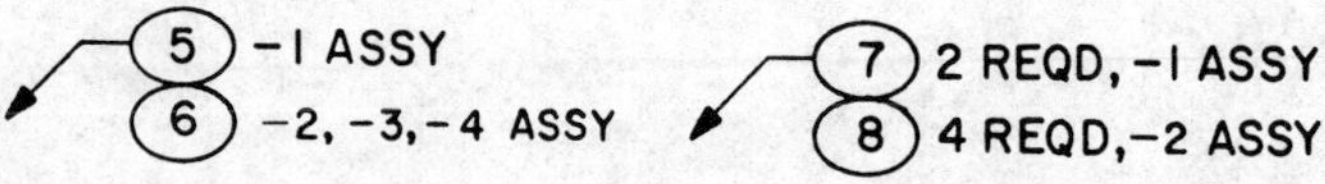

Item Numbers on Multiple Assembly Drawing
Figure 8

3-6

3.8 NUMBERING OF CHARTS, GRAPHS AND OTHER DATA

3.8.1 The numbering system for identifying and recording of departmental charts, graphs, study notes and data, other than Company product drawings and associated special equipment (tooling, etc.) is as follows:

```
                                    XXXX-XX-0000
         Department No.———————————————┘    │   │
         Year ——————————————————————————————┘   │
         Serial Control No.——————————————————————┘
```

3.8.2 CONTROL RESPONSIBILITY — A Division may designate that either a control center or each department be responsible for maintaining a log of serial control number assignments. Only one standardized method will apply at any one division. Each year the serial control number will start with the number 0001.

3.8.3 This numbering system is not to be used to identify hardware or company format drawings.

NOTES

Section 4 — TYPES OF ENGINEERING DRAWINGS

4.1 SCOPE — This section defines and illustrates the types of engineering drawings* normally prepared for end product* definition.

4.2 APPLICABLE DOCUMENTS

MIL-STD-17-1	Mechanical Symbols
MIL-STD-17-2	Mechanical Symbols for Aeronautical, Aerospace & Spacecraft Use.
MIL-STD-34	Preparation of Drawings for Optical Elements & Optical Systems
DOD-STD-100	Engineering Drawing Practices
MIL-STD-130	Identification Marking of US Military Property
MIL-STD-1306	Fluerics Terminology & Symbols
NAS 944	Symbols for Hydraulic Test Equipment Drafting
ANSI Y14.3	Projections
ANSI Y32.4	Graphic Symbols for Plumbing
ANSI Y32.10	Graphic Symbols for Fluid Power Diagrams
TM-5-581	Construction Drafting (US Army)
NAVFAC-DM-6	Drawings & Specifications (US Navy)

4.3 DEFINITIONS — Terms marked with an asterisk (*) are defined in Section 25.

4.4 GENERAL DRAWING CONTROL REQUIREMENTS

4.4.1 DRAWING CONTROL AND APPROVALS — Drawing control and approval signature requirements are specified either in corporate or plant procedures, by a contractual requirement, or as directed by the program manager's office.

4.4.2 DRAWING FORMAT — All drawings shall be prepared on reproducible material using company format with company code identification and drawing numbers, unless otherwise specified by contract.

4.4.3 DRAWING REQUIREMENTS — Contracts normaly specify when drawings are to be prepared to DOD-D-1000 including level. The minimum drawing requirements for the different levels are as specified in Section 2.

Figure No.	Title	Section 4 Page No.
1	Altered Item Drawing	5
2	Arrangement Drawing	7
3	Assembly Drawing	9
4	Book Form Drawing	11
5	Construction Drawing	13
6	Control Drawing	15
7	Correlation Drawing	17
8	Design Layout Drawing	19
9	Detail Assembly Drawing	21
10	Detail Drawing	23
11	Diagramatic Drawing	25
12	Elevation Drawing	27
13	Envelope Drawing	29
14	Erection Drawing	31
15	Expanded Assembly Drawing	33
16	Formulation Drawing	35
17	Inseparable Assembly Drawing	37
18	Installation Assembly Drawing	39
19	Installation Control Drawing	41
20	Installation Drawing	43
21	Interface Control Drawing	45
22	Kit Drawing	47
23	Matched Parts Drawing	49
24	Mechanical Schematic Diagram Drawing	51

Figure No.	Title	Section 4 Page No.
25	Modification Drawing	53
26	Monodetail Drawing	55
27	Multidetail Drawing	57
28	Numerical Control (N/C) Drawing	59
29	Optics Drawing	61
30	Photo Assembly or Installation Drawing	63
31	Piping Diagram	65
33	Plan Drawing	67
34	Plot (Plat) Plan Drawing	69
35	Proposal Drawing	71
36	Selected Item Drawing	73
37	Separable Assembly Drawing	75
38	Source Control Drawing	77
39	Specification Control Drawing	79
40	Tabulated Assembly Drawing	81
41	Tabulated Detail Drawing	83
42	Textile Drawing	85
43	Tube Bend Drawing	87
44	Undimensioned Drawing	89

ALTERED ITEM* DRAWING

DESCRIPTION & USE:
A drawing that shows the details of an alteration to an existing item.

REQUIREMENTS:
1. Show details of the alteration.
2. Specify in the altered item parts list the part number of the unlatered (original) item. Also show the code identification number* if the part number of the unaltered item was assigned by a vendor*.
3. Marking: specify obliteration of the item's unaltered part number. Specify marking with the altered part number, preceded by the design activity code identification number (See: MIL-STD-130)
4. Drawing submission must include, not only the altered item drawing, but also a document which completely defines the unaltered item. If the originating design activity drawing is available, this may be submitted; if not, prepare a Specification Control Drawing, If the unaltered item and its alteration are simple, they may be combined on a single drawing.
5. Add the notation "ALTERED ITEM DRAWING" above the title block in .25 high letters.

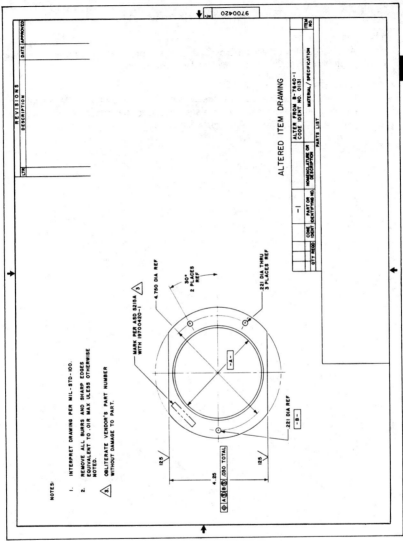

Altered Item Drawing
Figure 1

ARRANGEMENT DRAWING

DESCRIPTION & USE:
An arrangement drawing depicts in any projection or perspective, with or without controlling dimensions, the relationship of major units.

REQUIREMENTS:
Show sufficient views to convey the configuration and location of major units. Overall, locating and other general dimensions necessary to define the configuration may be shown. Major units shall be identified.

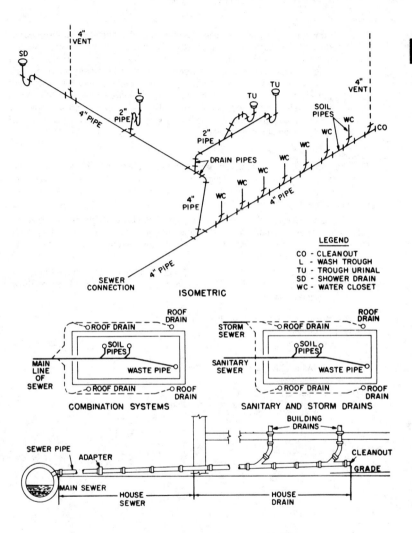

Arrangement Drawing
Figure 2

ASSEMBLY DRAWING

DESCRIPTION & USE:
    An assembly drawing depicts the assembled relationship of (a) two or more parts, (b), a combination of parts and subordinate assemblies, or (c) a group of assemblies required to form an assembly of higher order. The drawing shall contain sufficient views to show the relationship between each subordinate assembly and part. Cite subordinate assemblies and parts in the field of the drawing by their find (item) numbers cross-referenced to the identifying numbers in a parts list. When the assembled relationship and identification of parts are shown on an assembly drawing of a subordinate assembly, do not repeat on the assembly drawing of higher order or associated parts list.

REQUIREMENTS:
    Assembly drawings or associated parts list shall contain references to pertinent associated lists, installation drawings, wiring and schematic diagrams, etc. The division of an item into subordinate assemblies should be in accordance with practical assembly and disassembly procedures.

    NOTE 1: Show & identify electrical items on assembly drawings depicting where mounted; however, small electrical items mounted by means of wires affixed thereto may be shown and identified either on the assembly drawing or on the pertinent wiring diagram.

    NOTE 2: Itemize attaching parts (bolts, nuts, washers, etc.) required to mount and retain assemblies on foundations or on assemblies of higher order on the drawing showing the attachment.

TYPES OF DRAWINGS IN THE ASSEMBLY CATEGORY:
    Detail Assembly
    Tabulate Assembly
    Photo Assembly
    Inseparable Assembly
    Installation Assembly
    Expanded Assembly
    NOTE: Each of the above assembly drawings are depicted separately.

4

	SEE PAGE
DETAIL ASSEMBLY	4 - 21
TABULATED ASSEMBLY	4 - 81
PHOTO ASSEMBLY	4 - 63
INSEPARABLE ASSEMBLY	4 - 37
INSTALLATION ASSEMBLY	4 - 39
EXPANDED ASSEMBLY	4 - 33

NOTE: Each of the above assembly drawings are depicted separately.

Assembly Drawing
Figure 3

BOOK FORM DRAWING

DESCRIPTION & USE:
A BOOK-FORM drawing is an assemblage of related data disclosing by means of pictorial delineations, text or technical tabulations, or combinations thereof, the engineering requirements of an item, a family of items, or a system. A book-form drawing is used for a special purpose application in which it is expeditious to provide a document consisting of numerous small sheets, suitable for binding into book form.

REQUIREMENTS:
A book-form drawing shall preferably be prepared on A-size drawing formats. Other standard size formats may be used provided the final document size sheets are reduced to 11-inch height and can be folded to 8.5-inch width, with resultant legibility maintained. Book-form drawings shall not be prepared to circumvent the requirements for furnishing the types of drawings normally required for the delineation of an item or system. Book-form continuation sheets (sheet 2 and subsequent) need not include a revision block.

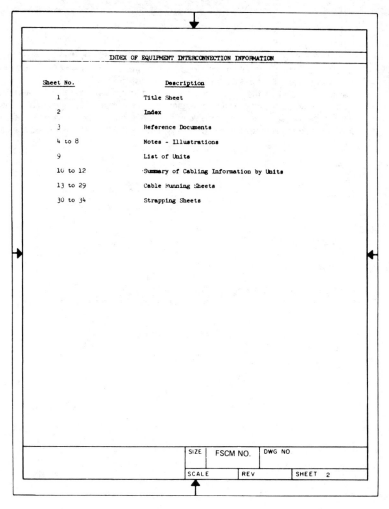

Book Form Drawing
Figure 4

NOTE: Illustrated is page two (2) of a multipage document in book form. See Section 5 for page 1 format.

CONSTRUCTION DRAWING

DESCRIPTION & USE:
    A construction drawing delineates the design of buildings, structures, or related construction, ashore or afloat, individually or in groups, and are normally associated with the architectural-construction-civil engineering operations. Construction drawings establish all the interrelated elements of an architectural-civil engineering design, including pertinent services, equipment, utilities and other engineering details. Maps, except those accompanying or used in conjunction with construction drawings, sketches, presentation drawings, perspectives, and renderings are not considered to be construction drawings.

REQUIREMENTS:
    See:
TM-5-581	Construction Drafting (US Army)
NAVFAC-DM-6	Drawings & Specifications (US Navy)

TYPES OF DRAWINGS IN CONSTRUCTION CATEGORY:
    Elevation drawing
    Erection drawing
    Plan drawing
    Plot drawing
    Vicinity plan or site drawing
    NOTE: Each of the above construction type drawings are depicted separately.

	SEE PAGE
ELEVATION DRAWING	4 - 27
ERECTION DRAWING	4 - 31
PLAN DRAWING	4 - 67
PLOT DRAWING	4 - 69
VICINITY PLAN OR SITE DRAWING	CITY MAPS

NOTE: Each of the above construction type drawings are depicted separately.

Construction Drawing
Figure 5

CONTROL DRAWING

DESCRIPTION & USE:
    A control drawing discloses configuration and configuration limitations; performance and test requirements; weight and space limitations; access clearance, pipe and cable attachments, etc. to the extent necessary that an item can be developed or procured on the commercial market to meet the stated requirements; or, for the installation and co-functioning of an item to be installed with related items. Control drawings are classified as:
   Envelope Drawing
   Specification Control Drawing
   Source Control Drawing
   Interface Control Drawing
   Installation Control Drawing
NOTE: Each of the above drawings are depicted separately.

	SEE PAGE
ENVELOPE DRAWING	4 - 29
SPECIFICATION CONTROL DRAWING	4 - 79
SOURCE CONTROL DRAWING	4 - 77
INTERFACE CONTROL DRAWING	4 - 45
INSTALLATION CONTROL DRAWING	4 - 41

NOTE: Each of the above control type drawings are depicted separately.

Control Drawing
Figure 6

CORRELATION DRAWING

DESCRIPTION & USE:
A correlation drawing depicts physical and functional engineering requirements within a subsystem. They are used to correlate interface engineering data between a subsystem* design activity* and others involved in the design of the subsystem.*

REQUIREMENTS:
Follow requirements for interface control drawings except that the delineation provided on correlation drawings shall be restricted to UNITS* within the subsystems. Correlation drawings shall contain no requirements intended to control interfaces of associated subsystems. Show the notation "CORRELATION DRAWING" in .25 high letters, adjacent to the title block. Data prescribed on correlation drawings which cannot be changed without affecting co-functioning subsystems shall be suitably identified.

4-16

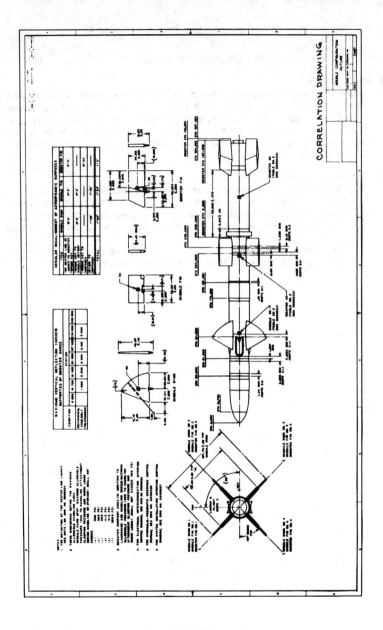

Correlation Drawing
Figure 7

DESIGN LAYOUT DRAWING

DESCRIPTION & USE:
A design layout drawing shows a design concept with design information requiring special consideration by the engineer and draftsman, i.e. special clearances, adjustments, equipment provisions, materials, processes, finishes, critical tolerances and other special features. It serves as an engineering work sheet from which engineering drawings are made. The drawing is not used for fabrication, procurement or inspection.

REQUIREMENTS:
1. Plastic film or other suitable stable material is recommended.
2. Specify all necessary design information, e.g., dimensions, tolerances, special clearances, adjustments, materials, finishes, processes, etc.
3. Identify the drawing as "DESIGN LAYOUT" above the title block in .25 high letters.
4. This drawing should be coordinated with responsible program management and supporting functions before starting working drawings.

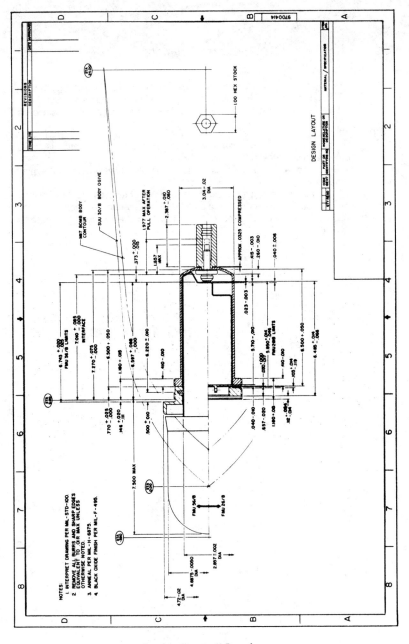

Design Layout Drawing
Figure 8

DETAIL ASSEMBLY DRAWING

DESCRIPTION & USE:
    A detail assembly drawing shows the assembled relationship of items, one or more of which is detailed on the assembly drawing. Separate engineering drawings are not required for items so delineated. Particularly useful for inseparable assemblies.

REQUIREMENTS:
1. Include all necessary information for procurement, fabrication and inspection of the details and assembly.
2. Show items which require extensive machining operations on a separate mono detail drawing.
3. Parts may be detailed in the assembled condition or as separate details on the drawing. Clarity for fabrication should be the prime consideration.
4. Meet all requirements of a mono detail drawing and an inseparable assembly drawing.

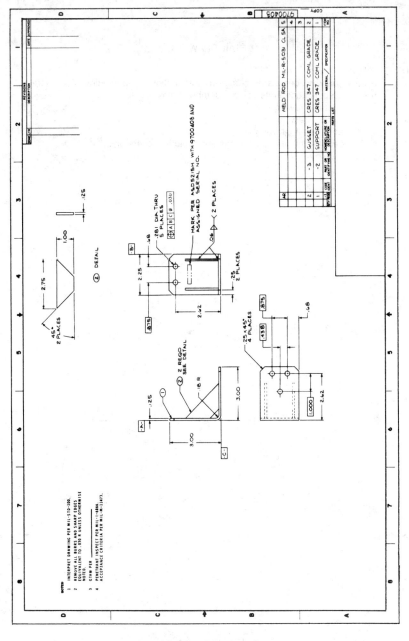

Detail Assembly Drawing
Figure 9

DETAIL DRAWING

DESCRIPTION & USE:
    A detail drawing depicts all necessary information for fabrication of an item(s), ready for application in its using assembly. Classified as:
- Mono detail Drawing
- Multi-detail Drawing
- Tabulated detail Drawing

NOTE: Each of the above drawings are depicted separately.

	SEE PAGE
MONO DETAIL DRAWING	4 - 55
MULTI-DETAIL DRAWING	4 - 57
TABULATED DETAIL DRAWING	4 - 83

NOTE: Each of the above detail type drawings are depicted separately.

Detail Drawing
Figure 10

DIAGRAMATIC DRAWING

DESCRIPTION & USE:
    A diagramatic drawing is a graphic presentation of an assembly (or higher-tier assembly), using standardized symbols, codes, and interconnecting lines to describe the function of an installation, assembly or system.

TYPES OF DRAWINGS CLASSIFIED IN THE DIAGRAMATIC CATEGORY
    Schematic Diagram
    Connection or Wiring Diagram
    Interconnection Diagram
    Single Line or One Line Diagram
    Logic Diagram
    Mechanical Schematic Diagram
    Piping Diagram
NOTE: Each of the above diagrams are depicted separately.

	SEE PAGE
SCHEMATIC DIAGRAM	17 - 21
CONNECTION OR WIRING DIAGRAM	17 - 22 & 23
INTERCONNECTING DIAGRAM	17 - 28
SINGLE LINE OR ONE LINE DIAGRAM	17 - 19
LOGIC DIAGRAM	17 - 36
MECHANICAL SCHEMATIC DIAGRAM	4 - 51
PIPING DIAGRAM	4 - 65

NOTE: Each of the above diagramatic type diagram drawings are depicted separately.

Diagramatic Drawing
Figure 11

ELEVATION DRAWING

DESCRIPTION & USE:
An elevation drawing depicts vertical projections of building or structures; or profiles of equipment such as aircraft, automotive and marine, or portions thereof.

REQUIREMENTS:
An elevation drawing shows configuration, shapes and sizes of features, walls, bulkheads, compartments, assignment of space, location and arrangement of machinery or fixed equipment. An elevation drawing may indicate materials of construction.

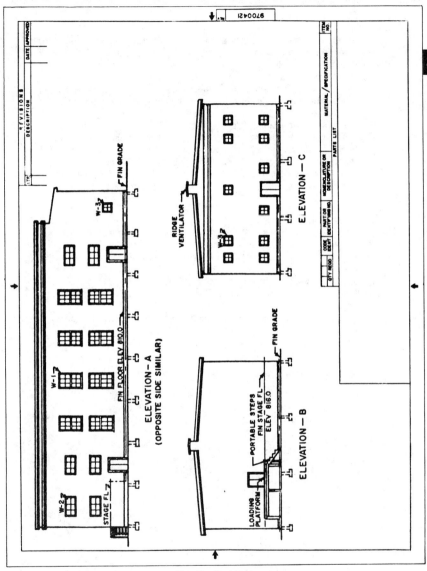

Elevation Drawing
Figure 12

ENVELOPE DRAWING

DESCRIPTION & USE:

An envelope drawing depicts an item which specifies a configuration & performance envelope, without details of internal construction. All features other than those shown on the drawing are left to the producer to meet the specified design requirements and performance data.

1. The term "design requirement" means the minimum performance requirements to be met to satisfy the design of the end item or a system designed for the end item.
2. The term "performance data" is a list of physical and functional characteristics under specified operating conditions (i.e. loads, speeds, etc.) and environmental conditions, as required to fully describe the essential operating characteristics under which the item must operate and perform. The characteristics shall be defined sufficiently to permit interchangeability of substitute items by other manufacturers if the specified performance data for the item is met.

REQUIREMENTS:

1. Disclosure requirements are the same as for a specification control drawing.
2. Define all interface* requirements in exact detail.
3. Add the notation "ENVELOPE DRAWING" above the title block in .25 high letters.
4. Delete any proprietary notice when applicable.
5. Identify the item on its using assembly by the envelope drawing - dash - suffix number.
6. Once a vendor part number is assigned to the item, the envelope drawing may be converted to a specification control drawing.

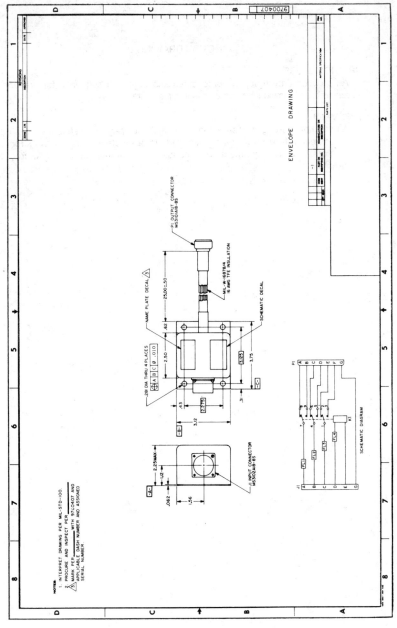

Envelope Drawing
Figure 13

ERECTION DRAWING

DESCRIPTION & USE:
An erection drawing shows procedures and operation sequence for erection or assembly of individual items or assemblies of items.

REQUIREMENTS:
An erection drawing shall show the location of each part in the structure, identification marking, types of fastenings required, approximate weight of heavy structural members, controlling dimensions, and any other information needed to erect the structure.

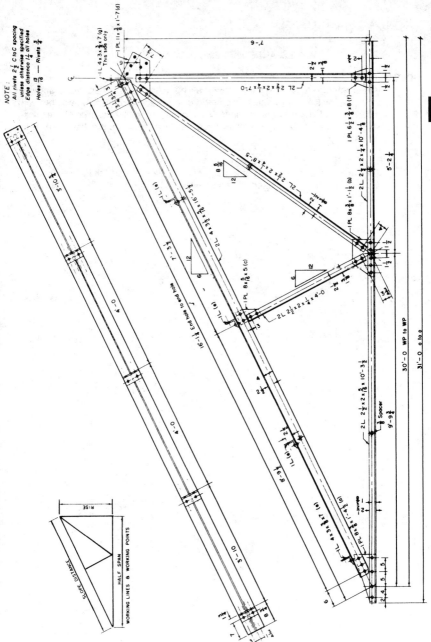

Erection Drawing
Figure 14

EXPANDED ASSEMBLY DRAWING

DESCRIPTION & USE:

(Often referred to as "exploded view drawing") An expanded assembly drawing shows assembly relationship in isometric or perspective drawing views, but with parts "exploded" along center lines which show assembly/disassembly relationship. This type of drawing is usually used for illustrated parts breakdown (IPB) or provisioning.

REQUIREMENTS:

The drawing shows each part or a group of inseparable parts with its axes in alignment to its actual assembled position.

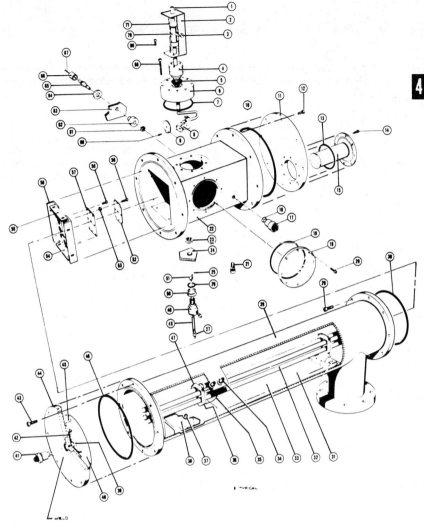

Expanded Assembly Drawing
Figure 15

FORMULATION DRAWING

DESCRIPTION & USE:
Lists the constituents of an explosive propellent, pyrotechnic, filler, etc. to identify the mixture, weight, volume, or particle size as used within the particular formulation.

REQUIREMENTS:
1. Show information either hand-lettered or typed with a standard typewriter, upper and lower case letters providing it meets the required legibility.
2. Disclose and identify the formulation, including weight, volume % composition, or particle size and type, with necessary batching and processing steps.

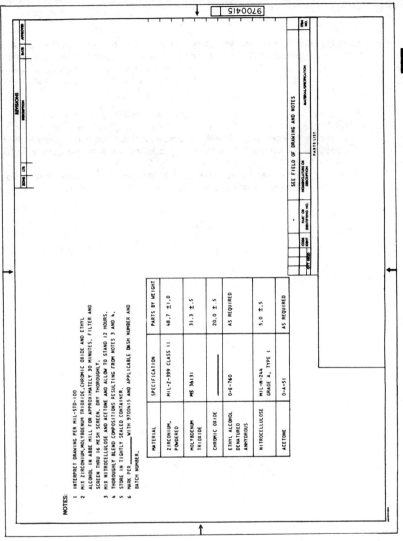

Formulation Drawing
Figure 16

INSEPARABLE ASSEMBLY DRAWING

DESCRIPTION & USE:
    An inseparable assembly drawing shows the assembled relationship of two or more parts, separately fabricated and permanently joined together by brazing, cementing, riveting, soldering, welding, etc., and not subject to disassembly. Their future use is considered as a single item if it be the end item or in subsequent assemblies.

REQUIREMENTS:
1. The drawing shall fully define the end product as assembled. Any dimensions that are not characteristic of the end product are informational and should be labeled "REF".
2. Part of an inseparable assembly may be defined on separate detail drawings or on the inseparable detail drawing (See: separate descriptions).

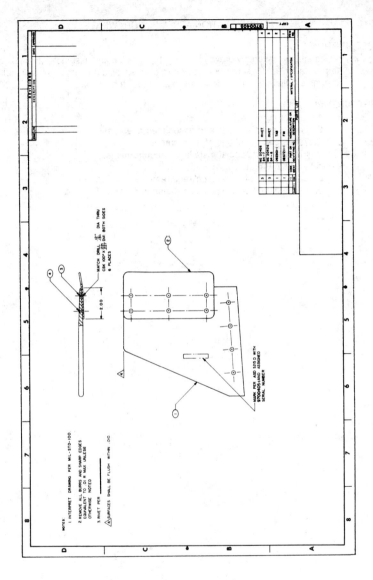

Inseparable Assembly Drawing
Figure 17

INSTALLATION ASSEMBLY DRAWING

DESCRIPTION & USE:
An installation assembly drawing shows the installed and assembled position of an item(s) relative to its supporting structure or to associated items. i.e. to install and assemble bellcranks, electrical wire harnesses, tubing, etc.

REQUIREMENTS:
1. List of items to be installed.
2. Locating dimensions and associated tolerances.
3. Types and quantities of attachment.
4. Process and special installation requirements.
5. Adjustment data.
6. Special test or inspection requirements.
7. Detail definition of special installation parts.

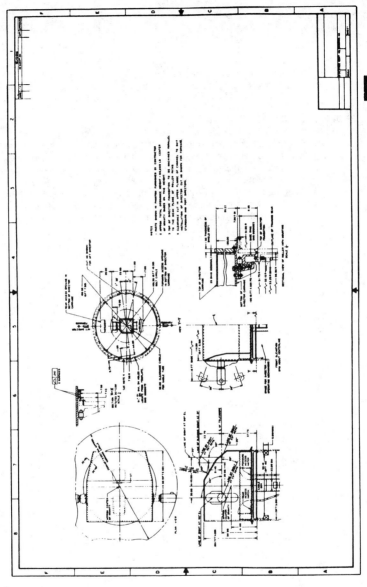

Installation Assembly Drawing
Figure 18

INSTALLATION CONTROL DRAWING

DESCRIPTION & USE:

An installation control drawing sets forth information for an item in terms of area, weight and space, access clearance, training clearances, pipe and cable attachments required for the installation and co-functioning of the item with related items.

REQUIREMENTS:

An installation control drawing shall include the following information as applicable:

1. Overall and principal dimensions in sufficient detail to establish the limits of space in all directions required for installation, operation and servicing.
2. The amount of clearance required to permit the opening of doors or the removal of plug-in units.
3. Clearance for travel or rotation of any moving parts, including the centers of rotation, angles of train in azimuth, elevation and depression, and radii from each pivot to the end of each rotating element involved in clearance determination.

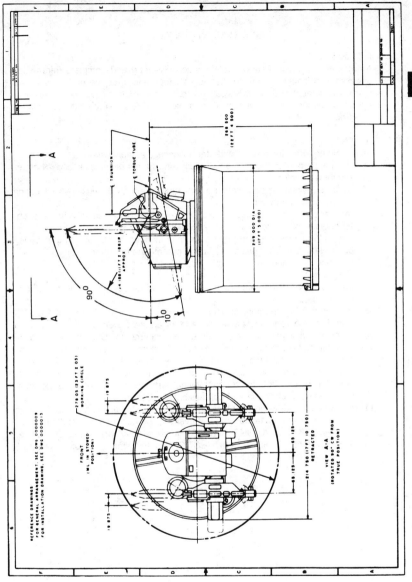

Installation Control Drawing
Figure 19

INSTALLATION DRAWING

DESCRIPTION & USE:
An installation drawing shows configuration envelope and complete information necessary to install an item relative to its supporting structure or to associated items. An installation drawing may show a specific completed installation. Installation drawings for one-of-a-kind installation may be revised to record the as-installed or as-built condition.

REQUIREMENTS:
An installation drawing shall include the following, as applicable:
1. Installed intem(s) shown in solid lines; other items (e.g., bulkheads & structures) in phantom.
2. Interface mounting and mating information, such as dimensions of location for attaching hardware.
3. Interface pipe and cable attachments required for the installation and co-functioning of the item to be installed with related items.
4. Information necessary for preparation of foundation plans, including mounting place details, drilling plans and shock mounting and buffer details.
5. Location, size and arrangement of ducts.
6. Weight of unit.
7. Location, type and dimensions of cable entrances, terminal tubes and electrical connectors.
8. Interconnecting and cabling data.
9. Reference notes to applicable lists and assembly drawings.
10. When not disclosed on other referenced documents: Overall and principal dimensions in sufficient detail to establish the limits of space in all directions required for installation, operation and servicing; the amount of clearance required to permit the opening of doors or the removal of plug-in units; clearance for travel or rotation of any moving parts, including the centers of rotation, angles of elevation and depression.
11. An installation drawing may include a parts list to establish the requirements for the installation hardware and, if desired, the items being installed.
12. The term "INSTALLATION DRAWING", in the second part of the drawing title.
13. A general note, "USE IN CONJUNCTION WITH (using assembly drawing.)"

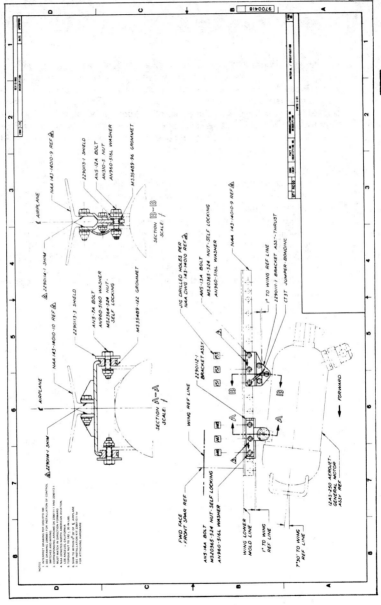

Installation Drawing
Figure 20

INTERFACE CONTROL DRAWING

DESCRIPTION & USE:
    An interface control drawing shows physical and functional interface engineering requirements of an item which affect the design or operation of co-functioning items. These drawings are used as design control documents, delineating interface engineering data coordinated for the purpose of: (a) establishing and maintaining compatibility between co-functioning items; (b) controlling interface designs thereby preventing changes to items requirements which would affect compatibility with co-functioning subsystems; (c) communicating design decisions and changes to participating activities.

REQUIREMENTS:
    An interface control drawing shall delineate, as necessary:
1. Configuration and all interface dimensional data applicable to the envelope, mounting and mating of the items.
2. Complete interface engineering requirements, such as mechanical, electrical, electronic, hydraulic, pneumatic, optical, etc., which affect the physical or functional characteristics of co-functioning items.
3. Any other characteristics which cannot be changed without affecting system design criteria.
4. The notation INTERFACE CONTROL DRAWING shall be shown, adjacent to the title block in .25 high letters.

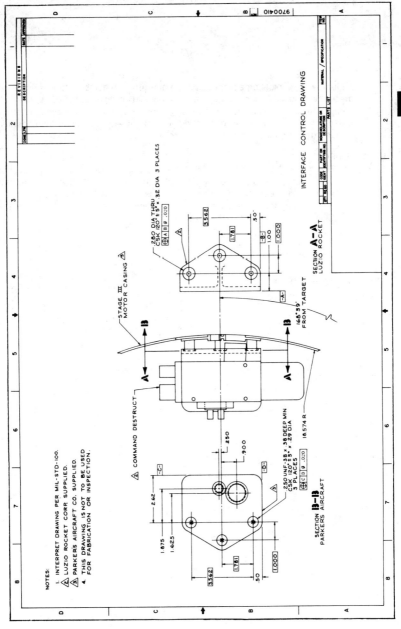

Interface Control Drawing
Figure 21

KIT DRAWING

DESCRIPTION & USE:
    A kit drawing depicts a packaged unit, item or group of items, instructions, photographs or drawings such as are used in modification, installation or survival. The items in a kit normally do not in themselves constitute a complete functional assembly. A kit drawing may be a listing of part numbers, a pictorial representation of parts or a combination of both.

REQUIREMENTS:
1. The drawing shall include a list of the entire kit content, including part numbers, titles, quantities required, and related installation drawings or other instructions necessary to complete the installation, modification, etc.
2. Include a general note on the drawing, indicating the kit's use. e.g., "USE IN CONJUNCTION WITH (NEXT ASSY. NO.)"

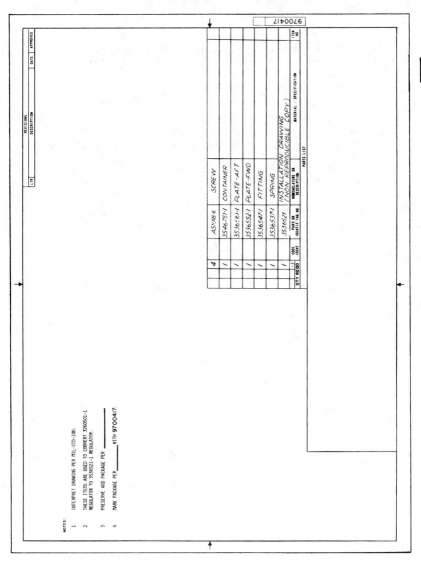

Kit Drawing
Figure 22

MATCHED PARTS DRAWING

DESCRIPTION & USE:
    A matched parts drawing that shows parts which are machine matched or otherwise mated, for which replacement as a matched set is essential.

REQUIREMENTS:
1. Include a parts list to reflect an assembly composed of the individual matched items.
2. Identify each part with its applicable part dash number and each part within the set with its assembly dash number.
3. Show the operating or mating characteristic of the matched set.
4. A general note: "FURNISH ONLY AS A MATCHED SET."
5. Identify each part within the set with the same serial number so that it can be distinguished from similar parts from other matched sets.

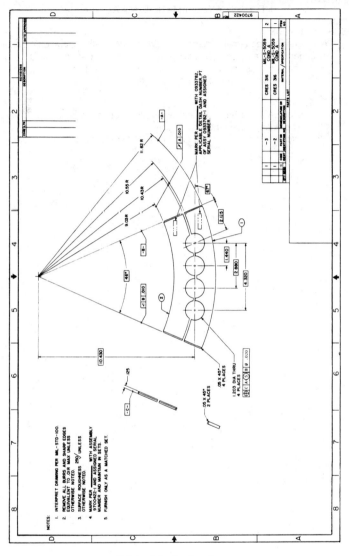

Matched Parts Drawing
Figure 23

MECHANICAL SCHEMATIC DIAGRAM DRAWING

DESCRIPTION & USE:
    A mechanical schematic diagram drawing shows the operational sequence or arrangement of a mechanical device. Its use is to illustrate the operation of an end item where conventional drawings are inadequate to fully explain, e.g. complex arrangement of gear trains, clutches, linkages, cams, etc.

REQUIREMENTS:
1. Describe any device element whose function is not clear.
2. Decode any non-standard symbols used.
3. Indicate directions of rotation, limits of travel, etc.

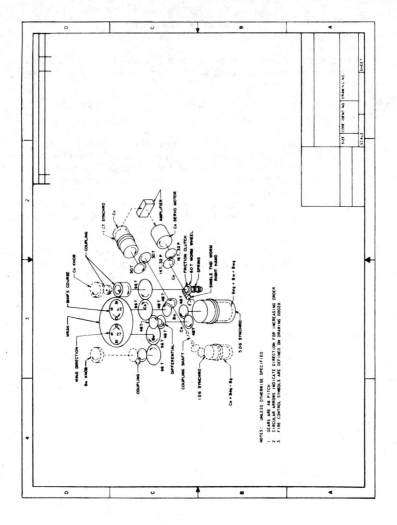

Mechanical Schematic Diagram
Figure 24

MODIFICATION DRAWING

DESCRIPTION & USE:
    A modification drawing delineates a change to a delivered item and used to add, remove or rework delivered equipment to improve safety, reliability, performance, etc.

REQUIREMENTS:
    Modification drawings contain complete information for accomplishing the change, including, as applicable:
1. Instructions for the removal or installation of affected parts.
2. Special notes, including identification by part drawing numbr of modified part.
3. Dimensions necessary to accomplish the modification. (These should be given from some specific point which is readily identified and accessible, rather than from some theoretical reference plant).
4. A parts list (integral or separate) containing all items required for the modification and items to be deleted, salvaged, and items to compile a kit, as applicable.
5. Listing of special tools or equipment required or supplied.
6. Effectivity. The serial number of the end products* which are to be modified.
7. The word "MODIFICATION" above the title block in .25 high letters.

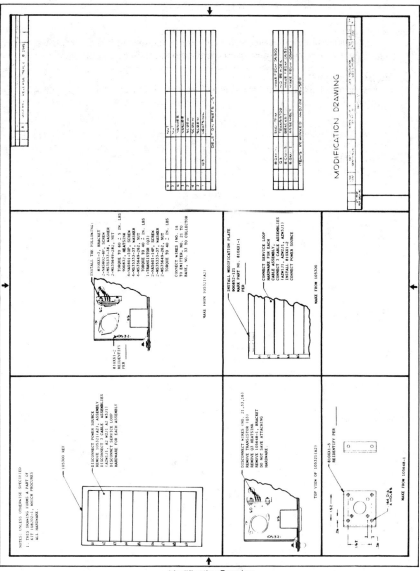

Modification Drawing
Figure 25

MONODETAIL

DESCRIPTION & USE:
A monodetail drawing delineates a single part to the extent it is a complete end item by itself or ready for use in the next assembly.

REQUIREMENTS:
Drawing contains all information necessary for fabrication, finish, marking and inspection.

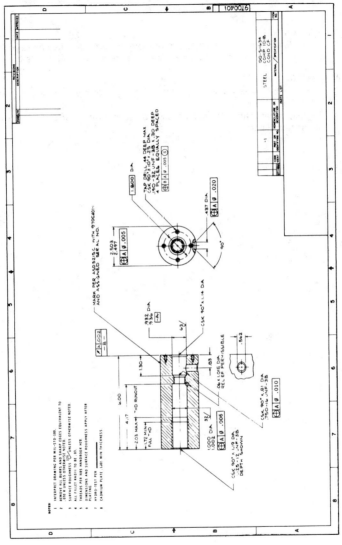

Mono-Detail Drawing
Figure 26

MULTIDETAIL DRAWING

DESCRIPTION & USE:
   A multidetail drawing delineates two or more uniquely identified parts on the same drawing. Items which can be simply defined are placed on a single drawing.

REQUIREMENTS:
   1. Each part depicted on a multidetail drawing shall meet all the requirements of a monodetail.
   2. Item identification is accomplished by dash (-) number assignment for each separate item.

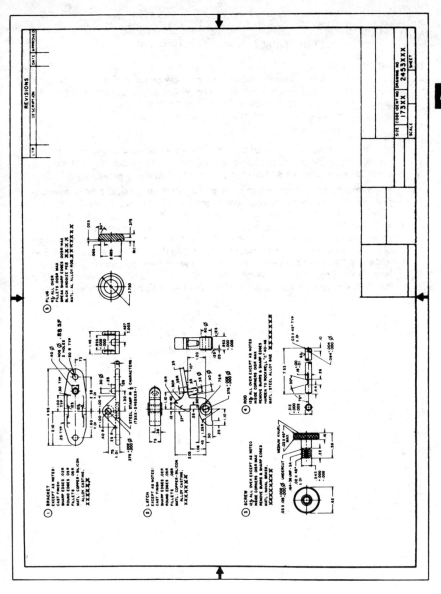

Multi Detail Drawing
Figure 27

NUMERICAL CONTROL (N/C) DRAWING

DESCRIPTION & USE:

A numerical control drawing shows the complete physical and functional engineering and product requirements of an item which will be partially or totally produced by numerical control means. This type of drawing is prepared for parts which are to be fabricated on numerically controlled machines.

REQUIREMENTS:

The drawing shall convey, directly or by reference, all engineering requirements for the part(s). For items capable of production by N/C or conventional methods the information shall be sufficient to permit manufacture by either method.

Integrate any special dimensioning requirements of numerical control machine planning and operation without compromising design requirements. Use the decimal inch system for location of features such as machined surfaces, holes, slots, etc. Features are normally defined in a rectangular coordinate system. Axes on the drawing shall intersect at an origin and should be common with the datum of the part view. The part view should preferably be drawn in the first quadrant (see ANSI Y14.3) so that positive values will result when the programmer prepares the tape. Present required information to facilitate preparation of the control media in the most economical form and optimize use of existing basic dimensioning practices.

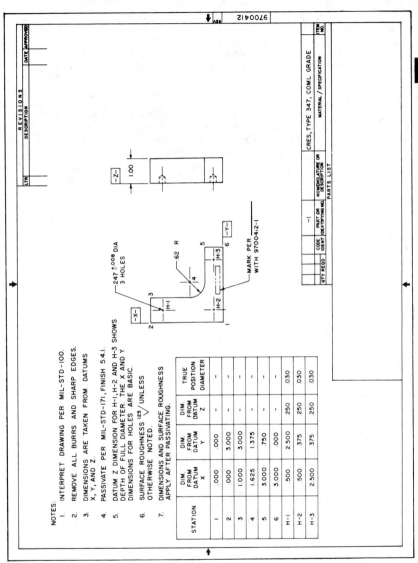

Numerical Control Drawing
Figure 28

OPTICS DRAWING

DESCRIPTION & USE:
   An optics drawing dhows the design and fabrication information of optical lenses, prisms, or mirrors to depict optical elements, parts and systems and to govern the adjustment and alignment of the optical part.

REQUIREMENTS:
1. Prepare optical system drawings to show light source travel from left to right.
2. Number lens surfaces consecutively from left to right.
3. For additional information, see MIL-STD-34.

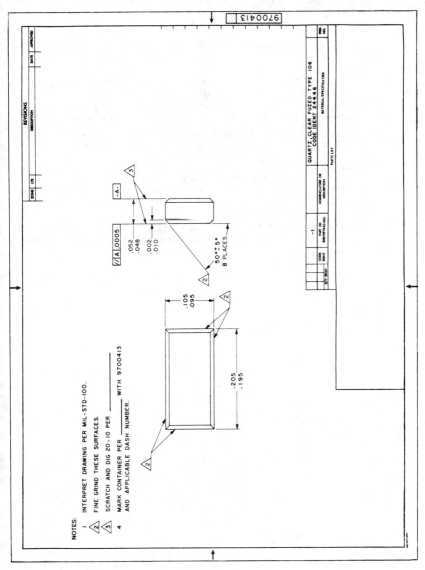

Optics Drawing
Figure 29

PHOTO ASSEMBLY OR INSTALLATION DRAWING

DESCRIPTION AND USE:
    A photo assembly is a drawing which presents pictorial information by means of half-tone photographs of a finished model in place of conventional orthographic line-drawings. Its use is helpful in acquainting a person(s) with or without engineering background, with the item as to its assembly, function, installation, etc.
    NOTE: Microfilming of photo assembly drawings are not acceptable. Delivery of full size photo masters is required and is usually so specified in the contract.

REQUIREMENTS:
1. Carefully compose views to show maximum information. Generally, views should be normal to a major plane of the equipment, using a white background.
2. Paste-up, air brushing, and some superimposed line work may be necessary for clarity and quality of artwork.
3. Screen continuous tone images, using a 60 line, 70 percent density screen on .004 thick matte surface mylar to which the drawing format is added in the printing process.

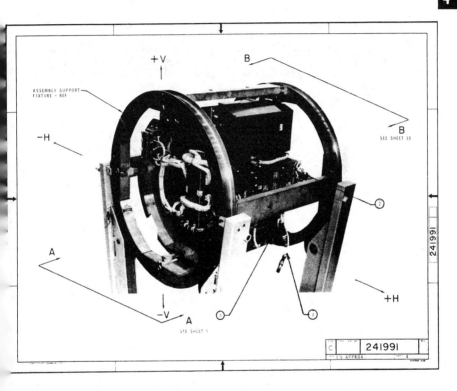

Photo Assembly Drawing
Figure 30

PIPING DIAGRAM

DESCRIPTION & USE:
    A hydraulic, pneumatic or fluid piping diagram which shows by symbols and interconnections the elements, sequence and function of a fluid/air-handling system.

REQUIREMENTS:
    Show sufficient detail to explain (a) the arrangement of the piping, valves, etc., or (b) operational sequence. Symbolic line representation may be used to distinguish functions of various parts. When the objective is to show arrangement the following characteristics may be shown: routing of fluids, physical locations and arrangement of elements, pipe diameters, types and sizes of fittings, flow, pressure, volume, etc.

    Use the following symbology standards:

MIL-STD-17-1	Mechanical Symbols (non-aerospace)
MIL-STD-17-2	Mechanical Symbols for Aeronautical, Aerospace, & Spacecraft Use
MIL-STD-1306	Fluerics, Terminology & Symbols
ANSI Y32.4	Graphic Symbols for Plumbing
ANSI Y32.10	Graphic Symbols for Fluid Power Diagrams
AS 1290	**Graphic Symbols for Aircraft Hydraulic and Pneumatic Systems**
NAS944	Symbols, Hydraulic Test Equipment Drafting

Where a symbol is not standardized by one of these documents, decode it directly on the diagram drawing as part of the general notes.

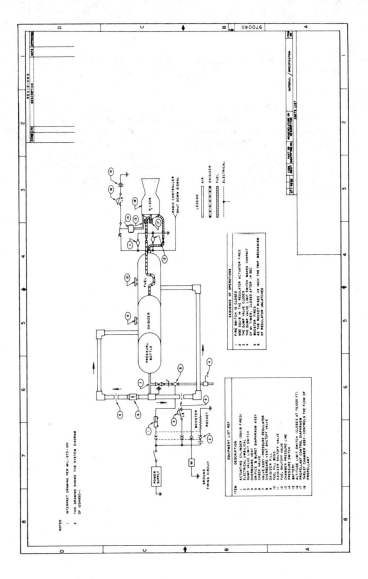

Piping Diagram
Figure 31

PLAN DRAWING

DESCRIPTION & USE:
A plan drawing depicts a horizontal projection of a structure, showing the layout of the foundation, floor, deck, roof, or utility system.

REQUIREMENTS:
As applicable, a plan drawing shall show shapes, sizes and materials of the foundation, its relation to the superstructure and its elevation with reference to fixed datum plane, location of walls, partitions, bulkheads, stanchions, companionways, openings, columns, stairs, shapes and sizes of roofs, parapet walls, drainage, skylights, ventilators, etc. Specify materials of construction and the arrangement of structural framing. The location of equipment or furniture may be indicated. A plan drawing for utilities may have individual layouts for heating, plumbing, air conditioning, electrical or other utility systems.

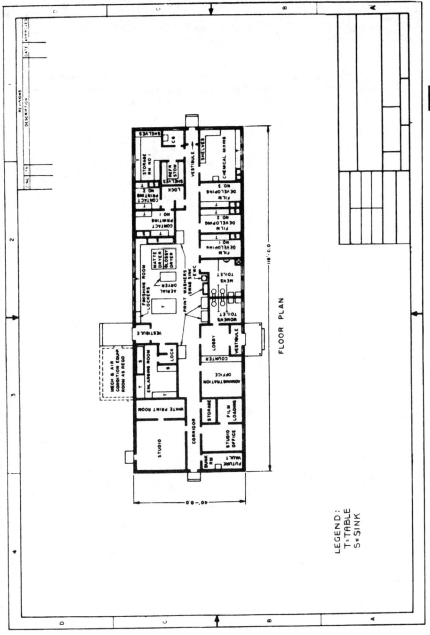

Plan Drawing
Figure 33

PLOT (PLAT) PLAN DRAWING

DESCRIPTION & USE:
    A plot plan drawing depicts areas on which structures are clearly indicated with detailed information regarding their relationship to other structures, existing and proposed utilities, topography, boundary lines, roads, walks, fences, etc.

REQUIREMENTS:
    The plot (plat) plan drawing shall show property lines and locations, contours and profiles, shrubbery, existing and new utilities, sewer and waterlines, building lines, location of proposed structures, existing structures, approaches, finished grades and other pertinent data.

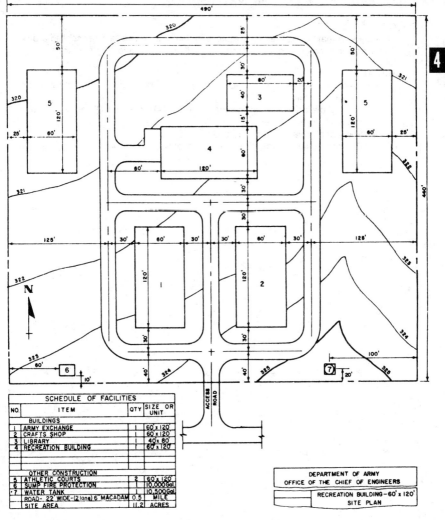

Plot (Plat) Plan Drawing
Figure 34

PROPOSAL DRAWING

DESCRIPTION & USE:
    A drawing that is prepared for the presentation of a new design concept to a prospective customer. It does not contain sufficient information for fabrication.

REQUIREMENTS:
1. Prepare drawings on the design activity standard format and utilize the drawing number system.
2. Identify the drawing with the term "PROPOSAL DRAWING" above the title block in .25 high letters.
3. Show appropriate proprietary notice.
4. Draftsmanship is to be of the highest quality.

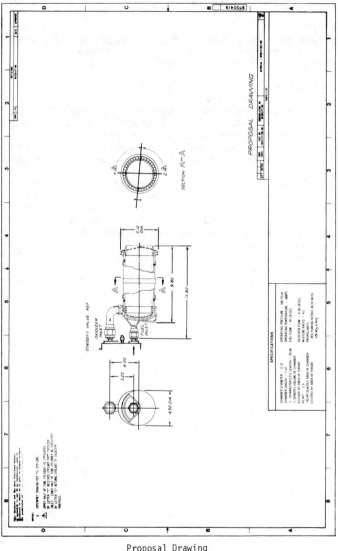

Proposal Drawing
Figure 35

SELECTED ITEM* DRAWING

DESCRIPTION & USE:
A selected item drawing establishes further requirements or restrictions of an existing standard, a design or a vendor's item for fit, tolerance, performance or reliability within the limits prescribed for that item.

REQUIREMENTS:
1. Give detail requirements for selection.
2. Specify in the selected item* parts list the part number* of the unselected (original) item. Also show the code identification number* if the part number of the unselected item was assigned by a vendor.*
3. Marking — specify obliteration of the item's unselected part number. Specify marking with the selected part number, preceded by the design activity* code identification number (See: MIL-STD-130)
4. Drawing submission must include, not only the selected item drawing, but also a document which completely defines the unselected item. If the originating design activity drawing is available, this may be submitted; if not, prepare a Specification Control Drawing. If the unselected item and its selection process are simple, they may be combined on a single drawing.
5. Add the notation "SELECTED ITEM DRAWING" above the title block in .25 high letters.

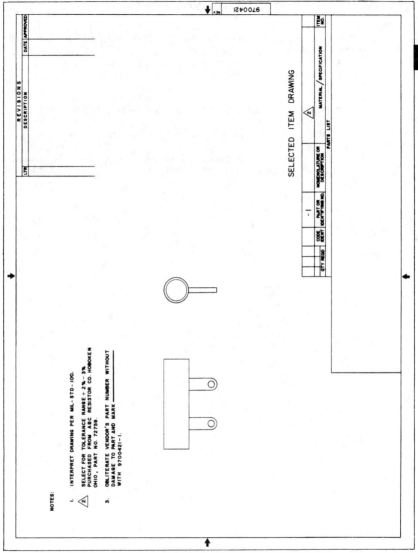

Selected Item Drawing
Figure 36

SEPARABLE ASSEMBLY DRAWING

DESCRIPTION & USE:
    A separable assembly drawing shows the assembled relationship of two or more items where at least one of the items is capable of disassembly for servicing or replacement without causing destruction of any of the items for its designed use. The parts are normally identified and stocked as as assembly.

REQUIREMENTS:
1. Present sufficient views to show the relationship between each part composing the assembly.
2. Cite associated lists, installation drawings, wiring and schematic diagrams, etc., as applicable.
3. When an installation drawing is called out in the associated parts list, include a deltaed note as follows: "FOR INSTALLATION, REFER TO DRAWING XXXX."

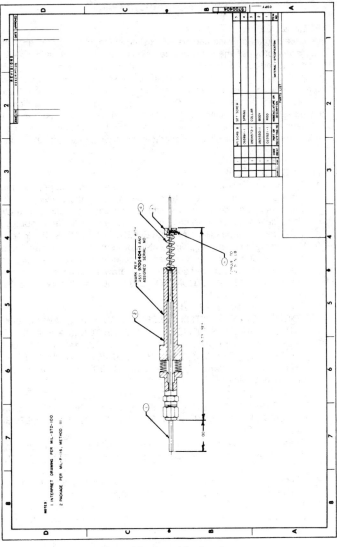

Separable Assembly Drawing
Figure 37

SOURCE CONTROL DRAWING

DESCRIPTION & USE:
A source control drawing depicts an existing commercial* or vendor* item which exclusively provides the performance, installation and interchangeable characteristics required for one or more specific critical applications. If an item can be completely defined for all applications, use a specification control drawing. This type of drawing is used when it is necessary to limit the source(s) of supply for an item to be used in a critical application because a specific supplier's item has proved satisfactory in acceptance tests or in use on an end item and there is not satisfactory experience with similar items from other supplier or it is impractical to isolate the function or physical characteristics which determine the item's suitability.

REQUIREMENTS:
1. Quality conformance and approval procedures shall appear in the general notes or in a specification which is called out on the drawing.
2. The name and address or manufacturer's code identification and the item identification number of each item that has been tested and approved shall be listed under the heading "APPROVED SOURCE(s) OF SUPPLY."
3. When more than one supplier is listed for items that are repairable but repair parts are not interchangeable, each supplier's item shall be assigned a separate dash number of the source control drawing number.
4. The notation "SOURCE CONTROL DRAWING" shall appear above the title block in .25 high letters.
5. Two general notes shall state:
NOTE 1: "ONLY THE ITEM DESCRIBED ON THIS DRAWING WHEN PROCURED FROM THE SUPPLIER(s) LISTED HEREON IS APPROVED BY (name of design activity) FOR USE IN THE APPLICATION(s) SPECIFIED HEREON. A SUBSTITUTE ITEM SHALL NOT BE USED WITHOUT PRIOR TESTING AND APPROVAL BY (name of design activity) OR BY (name of Government procuring activity)."
NOTE 2: "IDENTIFICATION OF THE APPROVED SOURCE(s) HEREON IS NOT TO BE CONSTRUED AS A GUARANTEE OF PRESENT OR CONTINUED AVAILABILITY AS A SOURCE OF SUPPLY FOR THE ITEMS DESCRIBED ON THE DRAWING."
6. On the using assembly, identify the item by its Source Control Drawing and dash number. Also add a general note, "SUPPLIER ITEM — SEE SOURCE CONTROL DRAWING."
7. Altered items, selected items, and Federal, Military and recognized industry standard items shall not be delineated on Source Control Drawings.
8. MARKING: Specify the vendor part number to be enclosed in parentheses, and the design activity code identification number* and source control part number to be marked as the primary identifier (See: MIL-STD-130)
9. When another supplier's item is tested and qualified, it may be added to the drawing provided it is approved for all stated applications.
10. Do not use a Source Control Drawing unless absolutely necessary.
11. Disclosure requirements are otherwise the same as for Specified Control Drawings.

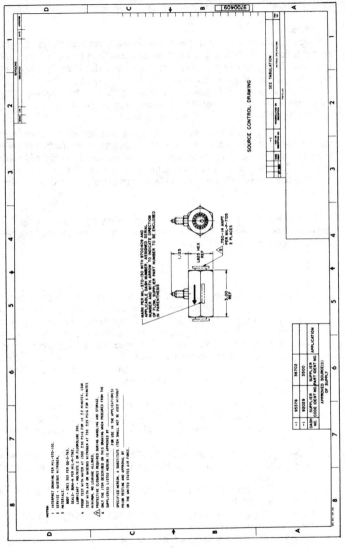

Source Control Drawing
Figure 38

SPECIFICATION CONTROL DRAWING

DESCRIPTION & USE:
A design activity control drawing which depicts and controls the performance and configurations of a commercial item* or a vendor* developed item advertised or cataloged as available on an unrestricted basis on order as an "off-the-shelf" item or an item which, while not commercially available, is procurable on order from a specialized segment of industry. The drawing, under the heading "Suggested Source(s) of Supply" shall list two (2) or more known sources unless, after suitable search, it is determined that there is only one (1) source.

NOTE 1: Vendor* developed items are those products of industries which normally provide customer application engineering services for a commercial product line and their products are commercially available from a specialized segment of industry. Typical examples of such items are: special motors, synchros, transformers, polentiometers, hydraulic valves, carburetors, potted servo-amplifiers, key-boards, tape readers.

NOTE 2: Altered Items, Selected Items and items depicted in Federal, Military and recognized industry association standards or specifications shall not be delineated on specification control drawings.

REQUIREMENTS:
1. Disclose as applicable, configuration, dimensins of envelope, mounting and mating dimensions, interface dimensional characteristics, and limits thereto. In addition, as necessary, inspection and acceptance tests requirements, performance, reliability, maintainability, environmental and other functional requirements, to insure identification and adequate reprocurement of an interchangeable item. If an electrical or electronic (or other engineering) circuit is involved, a schematic and connection or other appropriate diagramatic disclosure shall be included, providing sufficient information for making external connections.
2. Marking: Specify marking with the vendor's code identification number* and part number*. Space permitting, add (in parentheses) the design activity code identification number* and specification control part number (See: MIL-STD-130).
3. Under the heading "SUGGESTED SOURCE(s) OF SUPPLY" list the name, address, code identification number* and item identification number of each known vendor.
4. Add the notation "SPECIFICATION CONTROL DRAWING" above the title block in .25 high letters.
5. Do not specify any requirements other than those normally met by the vendor's item; use altered or selected item drawings instead.
6. On the using assembly parts list, itemize the specification control item by the specification control part number, together with a note, "SUPPLIER ITEM— SEE SPECIFICATION CONTROL DRAWING."

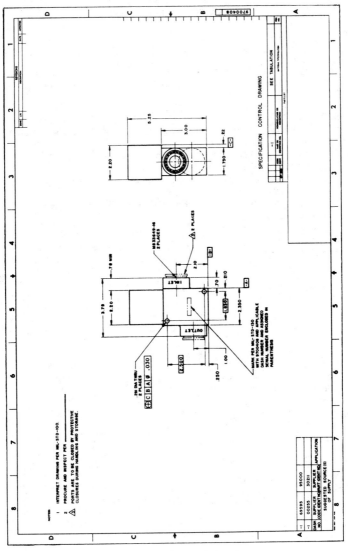

Specification Control Drawing
Figure 39

TABULATED ASSEMBLY DRAWING

DESCRIPTION & USE:
 A tabulated assembly drawing shows how a single assembly drawing can by tabulation of component parts can create additional assemblies without creating separate drawings. The same principle employed by tabulated detail drawings is used and each assembly shall clearly delineate the difference between each tabulated assembly.

REQUIREMENTS:
1. Use parts list to specify the differences in the part & material requirements for each assembly.
2. Assign a uniquely identifying assembly dash (-) number to each assembly. Normally -1 identifies the assembly shown.
3. Each view which shows variable information must be labeled with applicable part number(s) callouts.
4. A General Note, "INSTALL PER (INSTALLATION DRAWING NUMBER), when applicable.

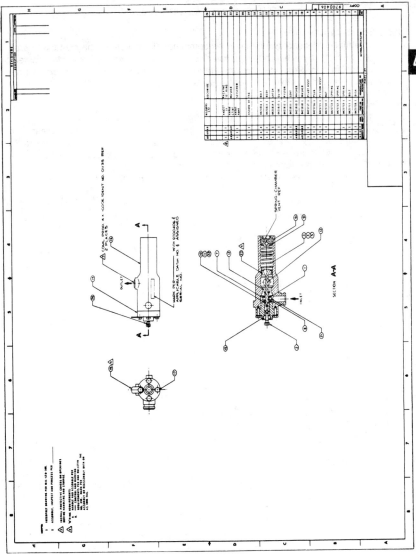

Tabulated Assembly Drawing
Figure

TABULATED DETAIL DRAWING

DESCRIPTION & USE:
A tabulated detail drawing that gives complete information for the manufacture of two or more items which have similar configuration, but vary in dimension, material, finish, etc. Precludes the preparation of a separate drawing for each part by tabulating the variable features or characteristics.

REQUIREMENTS:
1. Draw one item to scale, representing the -1 part if possible.
2. Place a tabulation block on the field of the drawing with column headings appropriate for tabulation of the variables, e.g., "L" for length, "W" for width, "D" for diameter, and other characteristics as applicable.
3. Other letters such as "A", "B", "C", etc., may be used for other variables.
4. Assign a uniquely identifying part dash (-) number to each part.

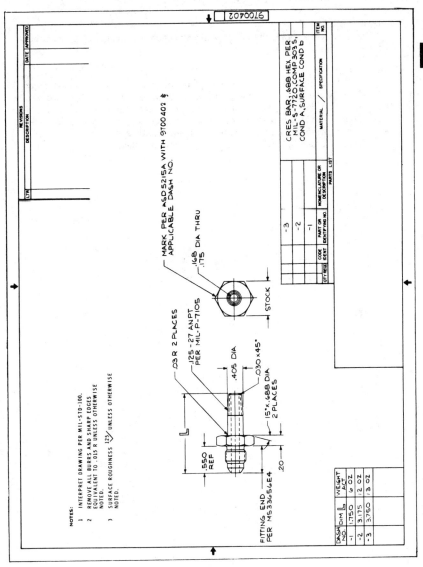

Tabulated Detail Drawing
Figure 41

TEXTILE DRAWING

DESCRIPTION & USE:
A textile drawing shows the design and fabrication information of an item constructed from fabrics.

REQUIREMENTS:
1. Include sufficient information for the fabrication of the item, including dimensions of the finished item and dimensioned flat patterns when possible.
2. Draw the item and flat patterns full size when possible.
3. Seams and stitching should be left to the discretion of the fabricator unless design or contractual requirements require they be specified.
4. Use bulk material callout procedure for fabric articles.

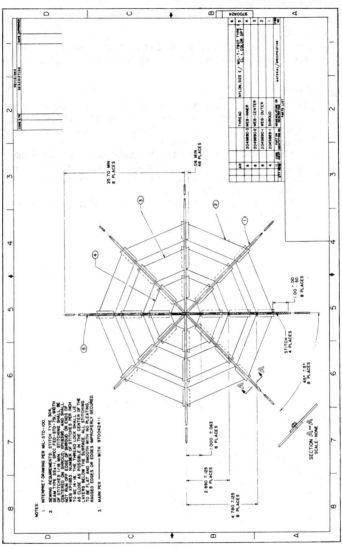

Textile Drawing
Figure 42

TUBE BEND DRAWING

DESCRIPTION & USE:
A tube bend drawing shows, by means of pictorial or tabular delineation, or combinations thereof, complete bend data required for the fabrication of a rigid metal tube.

REQUIREMENTS:
1. Disclose all material, types of ends, identification and quantity of fittings, bend radii and angles, intersection points, intermediate and overall lengths, tolerances, free or restrained state and other data required for the fabrication and acceptance of the item delineated.
2. If it is impractical to determine the bend data before assembly, a tube assembly drawing may be prepared showing a straight tube of approximate developed length and its fittings. A general note shall reference an installation or assembly drawing for forming.
3. Configuration may be shown in pictorial or tabular form depending upon complexity.

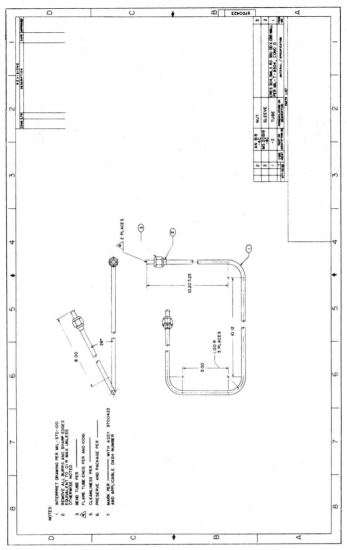

Tube Bend Drawing
Figure 43

UNDIMENSIONED DRAWING

DESCRIPTION & USE:
    An undimensioned master drawing which depicts, to a precise scale, on environmental stable material, loft line information, a template, a pattern, panels, reticles, special scales, sheet metal parts, cable assemblies and other items for which dimensioned detail drawings would be impractical. Drawings shall be accurate to the highest extent possible for photographic generation of parts or templates, without conventional dimensions & tolerances.

REQUIREMENTS:
1. Include information necessary to fabricate parts within necessary manufacturing limits, as for any detail part.
2. Make drawings to high quality standards, with sharp, clear, opaque linework.
3. Draw on dimensionally stable material only.

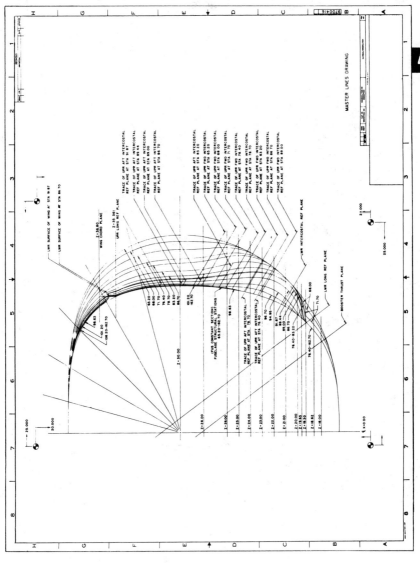

Undimensioned Drawing
Figure 44

NOTES

Section 5 — ENGINEERING DRAWING FORMAT

5.1 SCOPE — This section establishes the standard format for engineering drawings.

5.2 APPLICABLE DOCUMENTS

DOD-STD-100	Engineering Drawing Practices
MIL-M-9868	Microfilming of Engineerng Documents, 35mm, Requirements for
Cataloging HDBK H4	Federal Supply Code for Manufacturers,
H4-1	Name to Code
H4-2	Code to Name
DOD-5220.22-M	Industrial Security Manual for Safeguarding Classified Information
ANSI Y14.1	Drawing Sheet Size and Format

5.3 DEFINITIONS

5.3.1 DRAWING SIZE — A letter designation used to indicate standard format sizes. Refer to page 5-10 for American Drawing Size Letter Designations.

5.3.2 MULTISHEET DRAWING — A drawing consisting of two or more sheets.

5.3.3 BOOK-FORM DRAWING — A type of multisheet drawing used for special applications, e.g., running (wire) list.

5.4 DRAWING ARRANGEMENT

5.4.1 The locations of the title block, supplementary data blocks, parts list, general notes, etc., as shown in Figures 2 through 5.

5.4.2 SUPPLEMENTARY DRAWING NUMBER BLOCKS — On drawing sizes C through J, locate the drawing number in the right hand margin as shown in Figure 1. The block may be subdivided for entering the drawing revision letter. J size drawings shall have additional drawing number and revision letter blocks above the bottom margin and on the reverse side of both ends as shown in Figure 5.

5.4.2.1 ADDITIONAL SUPPLEMENTARY DRAWING NUMBER BLOCKS — Additional drawing number blocks may be added within or adjacent to the margin. For example, a B size drawing may have a drawing number block to the right of the microfilm arrowhead in the top margin. D through J sizes may have an additional block in the right hand margin to facilitate filing microfilm prints, see Figure 5.

5.4.3 MICROFILM ARROWHEADS — Microfilm arrowheads are located as shown in Figure 1 for A through F sizes and Figure 5 for J sizes. Note that on J sizes the arrowheads and match lines are located from the left hand end while the zones are located from the right hand end.

5.4.4 MATCH LINES — Match lines are **dimensionally** located from the left and are used to dimensionally locate microfilm arrowheads, supplemental drawing number blocks and security classification blocks, when required, are located.

5.4.5 ZONING — Zoning lines in the border are dimensionally located from right for each drawing size is shown in Figure 1. A and B sizes are not normally zoned. Zoning on C through F sizes, although optional, is usually part of the preprinted format. Zone J size drawings as shown in Figure 5.

5.4.5.1 ZONES — Determine zones by extending imaginary lines from the marks in the horizontal and vertical margins. These imaginary lines create rectangular zones across the entire drawing and are used for locating sections, views, callouts, changes, etc.

5.4.5.2 ZONE LOCATING SYMBOL FOR MULTIPLE APPLICATIONS — The zone locating symbol for a single cross-reference is:

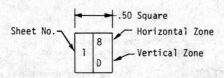

5.4.5.3 ZONE LOCATING SYMBOL FOR MULTIPLICAL APPLICATIONS — The zone locating symbol for multiple cross-references is:

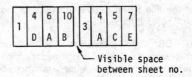

5.5 TITLE BLOCK (See Figures 6 and 7)

5.5.1 TITLE BLOCK FORMAT — The title Block is normally preprinted as part of the drawing format on A through F sizes and in decal form for J sizes. Draw a diagonal line through any block which is not used.

5.5.2 CONTRACT NUMBER BLOCK — Enter the contract number under which the drawing was initially prepared.

5.5.3 DRAWN BLOCK — The draftsman letters his name and the date the drawing was started. Dates shown shall be specified numerically — year, month, day (e.g. 75-04-14).

5.5.4 APPROVAL BLOCKS — The approval blocks are signed only by authorized personnel. The Design Block is signed by the engineer responsible for the design. The Design Activity Block is signed by the program manager or his designee. When additional approvals are required, blocks shall be added as shown in Figure 8 or 9. All approval signatures should be legible; initials will not be accepted.

5.5.5 DESIGN ACTIVITY NAME AND ADDRESS BLOCK — This block contains the Company trademark and name and the design activity address.

5.5.6 DRAWING TITLE BLOCK — Enter the drawing title in accordance with Section 6.

5.5.7 SIZE BLOCK — This block contains the letter designating the drawing size.

5.5.8 Federal Supply Code for Manufacturers (FSCM) NUMBER BLOCK — This block contains the design activity code identification number shown in Handbook H4—1.

5.5.9 DRAWING NUMBER BLOCK — Enter the design activity drawing number. See Section 3.

5.5.10 SCALE BLOCK — Enter drawing scale. See Section 8.

5.5.11 RELEASE DATE BLOCK — The Release Department enters the date the drawing is released. Dates shall be expressed numerically — year, month, date, e.g. 75-10-15.

5.5.12 SHEET BLOCK — A diagonal line is drawn through this block on single sheet drawings. See 5.9.2.2 for instructions on multisheet drawings.

5.6 SUPPLEMENTARY DATA BLOCKS (See Figures 8 and 9)

5.6.1 SUPPLEMENTAL DATA BLOCK FORMAT — The area to the left of the title block is reserved for supplementary data. In addition to the items listed below, any remaining area, up to 22 inches maximum on drawing formats D through J sizes, may be used for tabulation of other supplementary data.

5.6.1.1 The supplementary data blocks are normally part of the preprinted title block format. The blocks are optional, therefore, when a block is omitted the remaining blocks are moved to the right when permitted by use of decals. When the blocks are part of the preprinted format but unused, they are lined out with a diagonal line.

5.6.2 STANDARD TOLERANCE BLOCK — This block contains the standard tolerances that apply to specific dimensions as stated in Section 10. An angular tolerance, if not standardized and preprinted by the design activity, shall be entered when required or a dash entered to indicate the drawing contains no angular dimensions or that each angular dimension is followed by its own tolerance.

5.6.3 TREATMENT BLOCK — Enter the heat treatment requirements by listing the specification number and/or tensile, temper or hardness value. When additional space is needed, enter a delta note as specified in Section 7.

5.6.4 FINISH BLOCK — Enter the coating requirements as specified in Section 19.

5.6.5 SIMILAR TO BLOCK — Enter the drawing number(s) of similar parts where existing manufacturing information such as tooling, planning, etc., could be useful in making the new part.

5.6.6 WEIGHT BLOCKS — Enter the "Actual" or "Calculated" weight of the item shown.

5.6.7 APPLICATION BLOCK — When the application data is maintained on the drawing, the block will be filled in as stated below. On vertical "A" size formats, the block shown in Figure 8 will be inverted and relocated as shown in Figure 3.

5.6.7.1 In the "PART DASH NO." column, enter each parts list dash number that is used on another drawing. The dash number is repeated for each next assembly drawing application.

5.6.7.2 "QTY REQD PER ASSY". For each dash number entry, enter the quantity used for each next assembly in the "NEXT" column and the total quantity used for each final deliverable assembly in the "FINAL" column.

5.6.7.3 In the "NEXT ASSY" column, enter the drawing number of each next assembly. When the drawing depicts the final deliverable item, enter the word "FINAL". When the drawing is used in conjunction with other drawings, enter a delta note number and state the usage in the general notes, e.g. "THIS DRAWING IS USED IN CONJUNCTION WITH _____(enter WIRING DIAGRAM, SCHEMATIC DIAGRAM, ASSEMBLY DRAWING, INTERCONNECTION DIAGRAM, RUNNING LIST, etc. and the drawing number).

5.6.7.4 In the "USED ON" column, enter the model number or other designation assigned to the program.

5.6.8 ADDITIONAL APPROVALS BLOCK — This block is added when approvals other than those contained in the title block are required. On vertical "A" size formats the block as shown in Figure 8 may be inverted as shown in Figure 3.

5.7 PARTS LIST (PL) (See Figures 10 and 11)

5.7.1 PARTS LIST FORMAT — The Parts List Format is normally part of the preprinted title block format and is prepared in accordance with Section 9. Additional columns such as zone, unit weight, symbol, etc., may be added to the left of the item number column when required.

5.8 REVISION BLOCK (See Figures 12 and 13)

5.8.1 REVISION BLOCK FORMAT — The Revision Block Format is normally preprinted as part of the drawing format on A through F sizes and in decal form for J sizes. The entries in this block are made in accordance with Section 18.

5.9 MULTISHEET DRAWINGS

5.9.1 MULTISHEET DRAWINGS FORMAT — A continuation sheet format for sheet two and up is identical to the standard format as far as margins, zones, revision block, etc., except the supplementary data blocks shown in Figures 8 and 9 are omitted and the abbreviated title block shown in Figure 14 is used. The block size dimensions for the continuation sheet title block are the same as those for the corresponding standard format, see Figures 6 and 7.

5.9.2 PREPARING MULTISHEET DRAWING IDENTIFICATION — A multisheet drawing is identified with the same drawing number as sheet number one (1) and is prepared on the same format size. However, continuation sheets prepared by automated techniques need not be the same size as the first sheet.

5.9.2.1 Sheet one (1) is prepared on standard format. The remaining sheets are prepared on standard format or on continuation sheet format. When the standard format is used for the continuation sheets, (Figures 6 or 7) all preprinted supplementary data and title block information is lined out except for the blocks shown in Figure 14. Furthermore, the "RELEASE DATE" block is changed to "REV LTR".

5.9.2.2 SHEET BLOCK — Sheet one (1) records the sheet number and the total number of sheets, e.g. 1 OF 4. The continuation sheets record only the sheet number, e.g. 2, 3, 4, etc.

5.9.2.3 REVISION LETTER BLOCK — The "REV LTR" block in the title block of the continuation sheet(s) is left blank on initial release.

5.9.2.4 REVISION STATUS OF SHEETS BLOCK — The "REV STATUS OF SHEETS" block shown in Figure 15 is added to the left of the title block on sheet 1 for all formats except see 5.10.1.2 for A and B size book-form drawings. Sheet number blocks are added up to and including the last sheet number. The "REV" column carries a dash (-) for each sheet left originally unchanged or which have been added as new drawings.

5.9.3 PARTS LIST AND GENERAL NOTE LOCATION — The parts list and general notes are placed on sheet 1 and continue for as many sheets as necessary. When the parts list extends into the general note area on sheet 1 and any subsequent sheets, the general notes are started on the first sheet following the end of the parts list.

5.10 BOOK-FORM DRAWINGS

5.10.1 BOOK-FORM DRAWING FORMAT — Book-Form Drawing formats are normally preprinted "A" size (8½" x 11") or "B" size (11" x 17") pages for insertion in note books or suitable for binding.

5.10.1.1 SHEET 1 FORMAT — The standard A or B size format is used except the parts list is replaced with the "REV STATUS OF SHEETS" block extending from border to border. See Figure 16.

5.10.1.2 CONTINUATION SHEET FORMAT — The continuation sheet format is made to the requirements of 5.9.1 except the revision block is omitted. See Figure 17.

5.10.2 PREPARING BOOK-FORM DRAWINGS — Book-form drawings are prepared on A or B size format. Each principal section is prepared on one or more sheets and are normally arranged in the following order, as applicable.

    a. Title sheet
    b. Parts list
    c. General notes
    d. Illustrations
    e. Tabulations

5.10.2.1 Book-form drawings are prepared to the requirements of 5.9.2 through 5.9.2.3 as well as the following.

5.10.2.2 REVISION STATUS OF SHEETS BLOCK — When the "REV STATUS OF SHEETS" block is not part of the preprinted format, it is added above the title block and completed as stated in 5.9.2.4. When the block is preprinted, the first unused sheet number block shall be lined out with a diagonal line or additional blocks added up to and including the last sheet number.

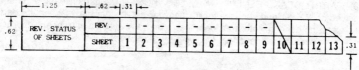

5.11 PRINT FOLDS

5.11.1 Drawing sizes "B" through "J" are normally folded after printing to 8½ x 11 inches to fit standard size file folders and filing cabinets. See Figure 18.

5.12 PROPRIETARY NOTICES

5.12.1 Each drawing or sheet of a multisheet drawing is marked with a proprietary notice as soon as or prior to the time it discloses technical information. The notice is applied in the location(s) shown in Figure 2 or 3 for A sizes, Figure 4 for B through F sizes and Figure 5 for J sizes. Unless otherwise directed by the Contract Administration or the Corporate Patent Department, a proprietary legend is used. It reads:

"THIS DOCUMENT AND THE DATA DISCLOSED HEREIN OR HEREWITH IS NOT TO BE REPRODUCED, USED, OR DISCLOSED IN WHOLE OR IN PART TO ANYONE WITHOUT THE PERMISSION OF (COMPANY NAME)."

5.12.2 When drawing originals or reproductions are submitted to the customer, the Contract Administration will issue instructions on whether the proprietary notice will remain unobscured, be removed, or be covered during reproduction. Normally, drawing originals or reproductions submitted to the Government will not bear proprietary notices.

5.13 SECURITY CLASSIFICATION MARKINGS

5.13.1 TYPES OF MARKINGS — Classified drawings are protected and marked with the applicable security markings as soon as they divulge classified information. Classified drawings contain a minimum of six markings. Two of these, (1) date of origin and (2) the design activity name and address, are part of the standard title block information. The remaining four are (3) security classification, (4) federal statute, (5) group marking and (6) the task code or accountability control number. The markings are applied in the locations shown in Figures 2 and 3 for A size, Figure 4 for B through F sizes and Figure 5 for J sizes. The engineer is responsible for establishing the correct classification and group marking and for informing the draftsman when they will be applied.

5.13.2 SECURITY CLASSIFICATION — The three basic classifications are confidential, secret and top secret. These may be supplemented with one of several prefixes (e.g. SEATO, CENTO, NATO, etc.) or suffixes (e.g. MODIFIED HANDLING AUTHORIZED, RESTRICTED DATA, etc.).

5.13.2.1 DRAWING TITLE LIMITATION — As stated in Section 6, drawing titles can not contain classified information. However, they will still be marked with "(U)" for unclassified.

5.13.3 FEDERAL STATUTE — All classified drawings, except those marked "RESTRICTED DATA", will be marked with the espionage clause, e.g.

CONFIDENTIAL

THIS MATERIAL CONTAINS INFORMATION AFFECTING THE NATIONAL DEFENSE OF THE UNITED STATES WITHIN THE MEANING OF ESPIONAGE LAWS, TITLE 18, U.S.C., SECTIONS 793 AND 794, THE TRANSMISSION OR REVELATION OF WHICH IN ANY MANNER TO AN UNAUTHORIZED PERSON IS PROHIBITED BY LAW.

All drawings marked "RESTRICTED DATA" will be marked with the Atomic Energy Commission restriction, e.g.

CONFIDENTIAL
RESTRICTED DATA
ATOMIC ENERGY ACT OF 1954

5.13.4 GROUP MARKINGS — Classified drawings are marked with one of four group markings, e.g.

GROUP 1
EXCLUDED FROM AUTOMATIC DOWNGRADING
AND DECLASSIFICATION

GROUP 2
EXEMPTED FROM AUTOMATIC DOWNGRADING

GROUP 3
DOWNGRADED AT 12 YEAR INTERVALS; NOT
AUTOMATICALLY DECLASSIFIED

GROUP 4
DOWNGRADED AT 3 YEAR INTERVALS;
DECLASSIFIED AFTER 12 YEARS

Group 3 and 4 may be modified by Government authority to read:

GROUP 3
DOWNGRADE TO:

SECRET ON (effective date)
CONFIDENTIAL ON (effective date)
NOT AUTOMATICALLY DECLASSIFIED

GROUP 4
DOWNGRADE TO:

SECRET ON (effective date)
CONFIDENTIAL ON (effective date)
DECLASSIFY ON (effective date)

5.13.5 TASK CODE OR ACCOUNTABILITY CONTROL NUMBER — The task code number will be marked on all confidential drawings. The accountability control number will be marked on all secret or top secret drawings.

5.13.6 SECURITY MARKINGS ON J SIZE DRAWINGS — J Size drawings are marked as shown in Figure 5. The highest security classification applicable to the entire sheet is repeated in all locations so that it will reproduce on each microfilm frame.

5.13.7 SECURITY MARKINGS ON MULTISHEET DRAWINGS — All sheets of a classified multisheet drawing are marked with the same security classification. Furthermore, all sheets carry the same security markings as sheet 1 except the task code or accountability control number is omitted from the continuation sheets. Any deviation from this procedure must be approved by the program or facility security representative.

5.13.8 RADIOACTIVE MATERIALS — All drawings pertaining to items using radioactive materials shall be suitably marked with a caution note.

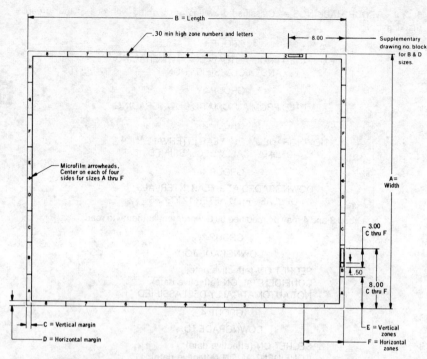

Drawing Size	A Width	B Length	** C Vertical Margin	** D Horizontal Margin	E Vertical Zones	F Horizontal Zones
A Horizontal	8.50	11	.25	.38	2 at 4.25	2 at 5.50
A Vertical	11	8.50	.38	.25	2 at 5.50	2 at 4.25
B	11	17	.62	.38	4 at 2.75	4 at 4.25
C	17	22	.50	.75	4 at 4.25	4 at 5.50
D	22	34	1.00	.50	4 at 5.50	8 at 4.25
E	34	44	.50	1.00	8 at 4.25	8 at 5.50
F	28	40	.50	.50	8 at 3.50	8 at 5.00
J	34	48 to 144	See Fig 5			

Basic Drawing Format
Figure 1

*See Figures 16 and 17 for book-form format requirement.
**The margin sizes in Table 1 and applicable figures have been selected to permit reproduction of drawings made to these standard sizes on sheets which conform to this standard or to international paper sizes. See Appendix A on A5-2.

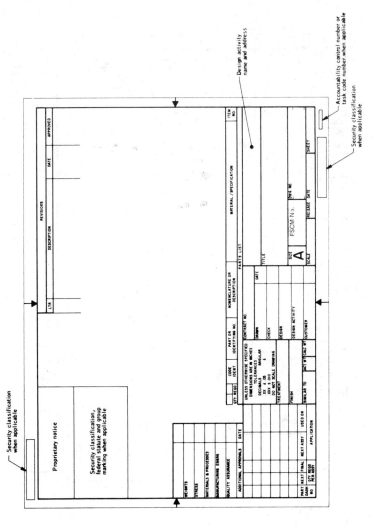

Horizontal A Size Drawing Arrangement
Figure 2

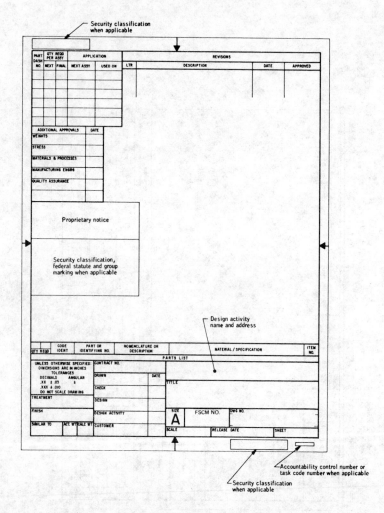

Vertical A Size Drawing Arrangement
Figure 3

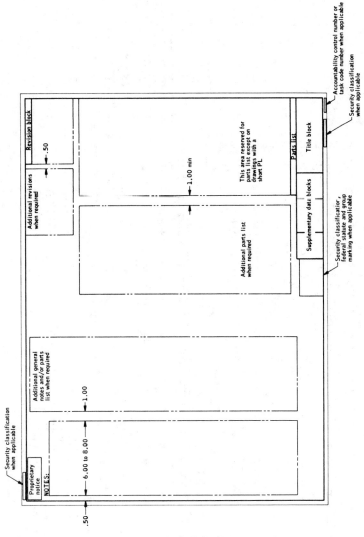

B Through F Size Drawing Arrangement
Figure 4

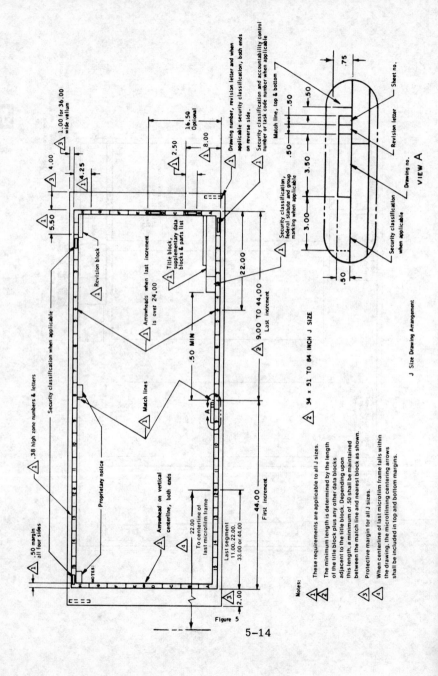

Figure 5 J Size Drawing Arrangement

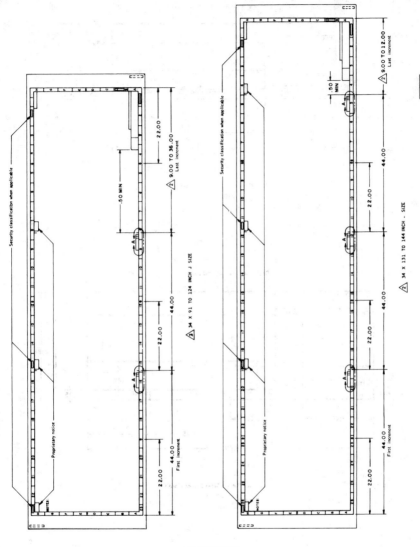

J Size Drawing Arrangement

Figure 5 (continued)

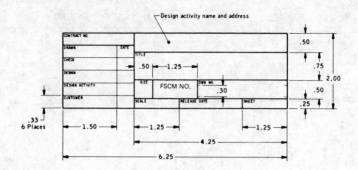

Title Block for A and B Sizes
Figure 6

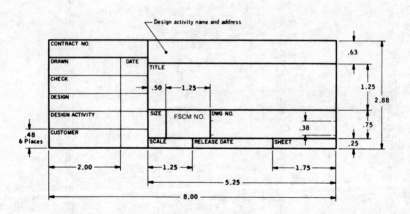

Title Block For All Sizes Except A and B
Figure 7

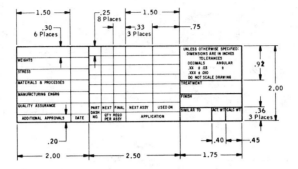

Supplementary Data Blocks For A and B Sizes
Figure 8

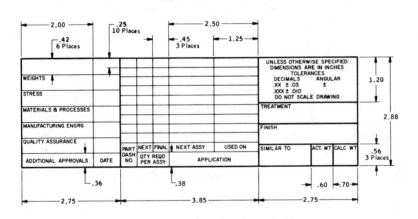

Supplementary Data Blocks For All Sizes Except A and B
Figure 9

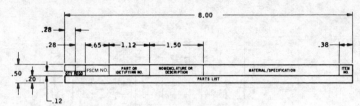

Parts List For A and B Sizes
Figure 10

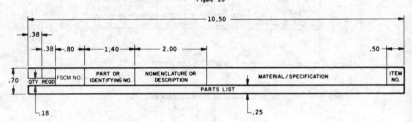

Parts List For All Sizes Except A and B
Figure 11

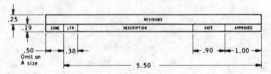

Revision Block For A and B Sizes
Figure 12

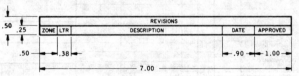

Revision Block For All Sizes Except A and B
Figure 13

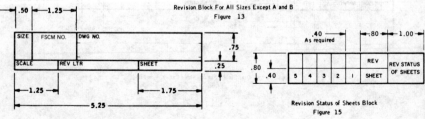

Continuation Sheet Title Block Format
Figure 14

Revision Status of Sheets Block
Figure 15

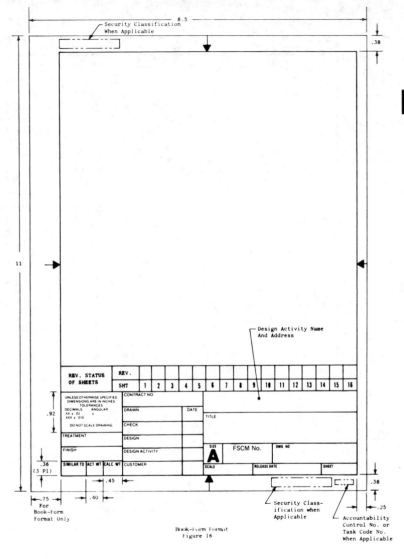

Book-Form Format
Figure 16

5-19

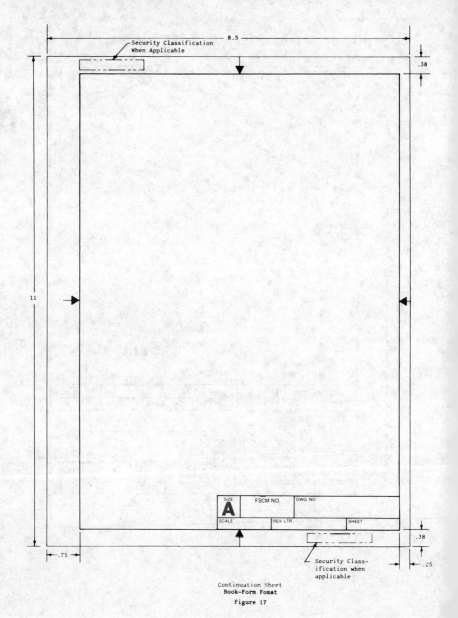

Continuation Sheet
Book-Form Format
Figure 17

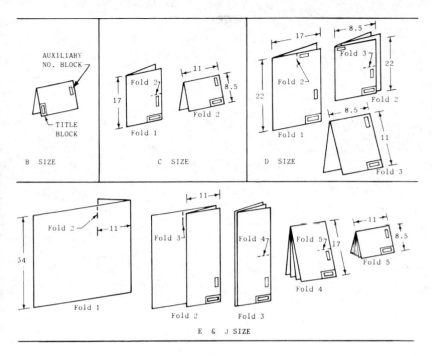

PRINT FOLDS
Figure 18

NOTES

Section 6 — DRAWING TITLES

6.1 SCOPE — This section describes the approved method for developing Drawing Titles.

6.2 APPLICABLE DOCUMENTS
DOD-STD-100   Engineering Drawing Practices
MIL-STD-196   Joint Electronics Type Designation System
Cataloging HDBK H6   Federal Item Identification Guides
                                       for Supply Cataloging

6.3 DEFINITIONS

6.3.1 DRAWING TITLE — The name by which the part or items shall be known and will consist of a basic name, government type designation, if applicable, and sufficient modifiers to differentiate like items in the same major assembly.

6.3.2 APPROVED ITEM NAME — A name approved by the Directorate of Cataloging, Defense Logistics Service Center and published in the Cataloging Handbook H6, Federal Item Identification Guides for Supply Cataloging.

6.3.3 TYPE DESIGNATOR is a combination of letters and/or numbers assigned by the Government to identify a specific item in accordance with MIL-STD-196.

6.3.4 ASSEMBLY — A number of parts and/or subassemblies joined together to perform a specific function.

6.4 TITLE ASSIGNMENT

6.4.1 The first part of the title shall be one of the following in order of preference.

   a. When the part is covered by an approved item name published in H6, the drawing title shall be the approved item name.

   b. When an approved item name is not available, the procedures specified herein shall be used.

6.4.2 When assigning a drawing title, clearly describe the item with the most applicable noun name and, if necessary, additional modifiers.

6.4.3 The first part of the item name shall not contain the method of manufacturing such as Casting, Forging, Machined, etc. The second part may contain the word Casting, Forging, etc., only when there is a machining drawing as well as a casting or forging, etc., drawing.

6.4.4 The following types of drawings shall include the drawing type as the second part of the drawing title.

    ARRANGEMENT DRAWING   SYSTEM DIAGRAM
    BLOCK DIAGRAM   WIRING DIAGRAM
    INTERCONNECTING DIAGRAM   INSTALLATION DRAWING
    SCHEMATIC DIAGRAM

6.4.5 The NOUN NAME is a noun or noun phrase that best establishes a basic concept of the item. It describes what the part is and the usage of the part, and not the material or method of fabrication. A compound noun or noun phrase is used only when a single noun is inadequate. See Figure 1.

    Example:
        Basic Name
            BRACKET   (Noun)
            PISTON   (Noun)
            SLIDE RULE   (Noun Phrase)
            SOLDERING IRON   (Noun Phrase)

        Item Name   Additional Modifiers
        (First Part)   (Second Part)

    GUIDE, SHAFT, BUTTERFLY VALVE-   .010 INCH OVERSIZE, LOWER

Noun Name (REF 6.4.5)
Modifier (REF 6.4.6)
Modifying Phrase (REF 6.4.6)

Modifying Phrase (REF 6.4.7)
Modifier (REF 6.4.7)

Figure 1

6.4.5.1 The singular form of the noun or noun phrase shall be used as the basic name, except as follows:

    a. Where the only form of the noun is plural.

    Example: BELLOWS, TONGS, SCISSORS

    b. Multiple single items appearing on the same drawing (tabulated).

    Example: SPACERS, BOLTS, WASHERS, ETC.
        (In the parts list of next assembly, use singular form of name.)

6.4.5.2 When using the word, "Assembly," "Subassembly," "Installation" in a basic name, it appears as the last word in the noun phrase and may be abbreviated, if space is a problem.

    Examples: ENGINE INSTL   or ENGINE INSTALLATION
                  ROCKET CHAMBER or ROCKET CHAMBER SUBASSEMBLY
                  SUBASSY

6.4.5.3 An ambiguous noun, or one which designates several classes of items, is not used alone as the basic name, but is part of the noun phrase.

Acceptable	Not Acceptable
SHOCK ABSORBER	ABSORBER, SHOCK
TERMINAL BOARD	BOARD, TERMINAL
WIRING HARNESS	HARNESS, WIRING

NOTE: One of the most difficult tasks in naming any item is the determination as to when a noun should be qualified as being ambiguous. The general rule quoted above is amplified to some extent in the succeeding paragraph. When a noun does not expressly fit under any of these rules, a guide in determining whether the selected nouns are or are not ambiguous is to refer to Cataloging Handbook H6 to see if it is listed. For example, if there is a question on the noun "plate" a review of the index will reveal many item names with the noun "plate" used, indicating that the noun is not considered as being ambiguous.

6.4.5.4 A trade-marked or copyrighted name shall not be used as the noun name, except where the technical name is extremely difficult and is not an approved item name in H6.

    Example: "KEL-F" rather than "POLYTRIFLUROMONOCHLOROETHYLENE"

6.4.5.5 When an item is not a container or material, but its basic name involves the use of a noun which ordinarily designates a container or material, use a noun phrase as the basic name.

Acceptable	Not Acceptable
JUNCTION BOX	BOX, JUNCTION
CABLE DRUM	DRUM, CABLE
SOLDERING IRON	IRON, SOLDERING
BUTCHERS STEEL	STEEL, BUTCHERS
GEAR BOX	BOX, GEAR

6.4.5.6 When the name of a container is used as the basic name in a title, it indicates the empty container.

    Examples: BOX, METAL, SHIPPING
                  CONTAINER, SHIPPING AND STORAGE

6.4.6 When the noun name represents an item to which types, grades, variety, etc., are applicable, the remainder of the first part of the title consists of one or two modifiers. A modifier may be a single word or a qualifying phrase. The first modifier serves to narrow the concept established by the basic name. Succeeding modifiers must continue a narrowing of concept by expressing a different type of characteristic. A word directly qualifying a modifier precedes the word it qualifies, thereby forming a modifying phrase.

    Example: BRACKET, PRESSURE TRANSMITTER

The word "PRESSURE" qualifies the word "TRANSMITTER" thus, a qualifying phrase is formed.

6.4.6.1 A modifier is separated from the noun or noun phrase or subsequent modifier by a comma.

    Examples: FRAME, ROCKET ENGINE
               SHAFT, BUTTERFLY VALVE

6.4.7 The second part of the title is separated from the first part by a dash (-) and consists of modifiers or modifying phrases as required. Modifiers indicating what an item is (its shape, structure or form) or what the item does (its function) are preferable to modifiers indicating the application (what it is used for) or location (where it is used) of the item. When two or more drawings in the same major assembly are similar and the parts detailed thereon perform the same general function, they are distinguished by additional modifiers indicating location, position, form, dimensions, etc.

    Examples: GUIDE, SHAFT, BUTTERFLY VALVE-.010 INCH OVERSIZE, LOWER
               GUIDE, SHAFT, BUTTERFLY VALVE-.010 INCH OVERSIZE, UPPER
               TUBE ASSY, METAL - FILTER INLET, TURBOPUMP
               TUBE ASSY, METAL -FILTER DISCHARGE, TURBOPUMP

6.5 GENERAL RULES

6.5.1 Uppercase letters are used. (See Section 8)

6.5.2 The conjunction "or" and the preposition "for" are not used.

6.5.3 Hyphens in compound words are not considered as punctuation marks. Use hyphenated words when a BASIC NAME such as HEATER-GENERATOR UNIT applies to an item with combined functions.

6.5.4 The latest edition of Websters International Dictionary (unabridged) is the authority for spelling and definition of words. For those words with dual or multiple definitions, the Military definition as published in Federal Handbook H6 has precedence.

6.5.5 Abbreviations are not used for any portion of the item name (first part of title) excepting the words ASSEMBLY (ASSY), SUBASSEMBLY (SUBASSY) and INSTALLATION (INSTL).

6.5.6 Abbreviations may be used in the second part of the title providing they conform to those used in Section 20.

6.5.7 Symbols or abbreviations taken from Company standards, electrical symbols, chemical symbols, etc., shall not be used.

6.5.8 Abbreviations which permit more than one interpretation or make the title difficult to read shall not be used. In general, the use of all abbreviations should be avoided.

6.5.9 Disclosure of security categories. No word(s), symbol(s), nor any of their possible combinations which would disclose information in any of the established security categories shall be used in drawing titles.

6.5.10 Parenthetical words or phrases are not used in drawing titles.

    Examples: (CAST), (WELDMENT), (ASSEMBLY OF), (MACHINED), etc.

6.5.11 Titles of detailed parts are consistent with the titles of their next assembly drawings, except where interchangeability of such detail parts between assemblies make consistency impractical, or is prohibited by the procuring agency, or when such use limits application.

    Examples: SHAFT, SHOULDERED - FUEL BUTTERFLY VALVE
                  SHAFT, SHOULDERED - OXIDIZER BUTTERFLY VALVE

6.5.12 Proper tabulation of nomenclature assigned to a drawing should be insured by a quick check in reading the title backward from the last modifier to the basic name.

    Example: (Reference Figure 1)
          "LOWER .010 INCH OVERSIZE BUTTERFLY VALVE SHAFT GUIDE."

6.5.13 The Drawing Title of an item identified in technical reports and/or other documents, published prior to the release of an engineering drawing, shall be in full agreement with the requirements of this DRM. An incorrect ITEM NAME referenced in a technical report and/or other documents, is not used as a drawing title.

6.5.14 New Drawings: When one drawing replaces another, the title is the same provided the old title was prepared in accordance with instructions contained herein. When the title of the drawing being replaced is not in accordance with these instructions, a new drawing title it developed and approved.

NOTES

Section 7 — DRAWING NOTES

7.1 SCOPE — This section standardizes the composition of drawing notes and the method of placing these notes on drawings.

7.2 APPLICABLE DOCUMENTS — None

7.3 DEFINITIONS

7.3.1 LOCAL NOTES — Local notes are those which apply directly to a particular portion of a drawing, indicate local characteristics.

7.3.2 GENERAL NOTES — General notes are those which apply to the drawing in general and would become repetitive if placed at each point of application.

7.4 GENERAL APPLICATION

7.4.1 Use notes to clarify features that are more accurately defined by words than by pictures and dimensions. Notes may also be used to give instructions for the application of special treatments and/or processes or to supplement standard symbols. Any information relating to the drawing or its use, may be placed in the notes.

7.4.2 Notes shall be clear and concise, in the imperative mode, and placed parallel to the bottom edge of the drawing. Do not underline. Use abbreviations only when necessary. (For acceptable abbreviations see Section 20). The note composition should be carefully considered and whenever possible used "as written" in this and other sections of the DRM or in comparable MIL- or Fed- specifications.

7.5 LOCAL NOTES

7.5.1 Place Local Notes on the field of the drawing outside the outline of the object, and as near as practicable to the portion referred to, or to the point where the operation is to be performed.

7.5.2 Extend leader lines from the left or right of a single line local note.

e.g. ⟋—(Note)⟍

7.5.2.1 In the event the local note has multiple lines, the leader line shall be placed to the left of the first line or to the right of the last line.

e.g. ⟋—Rubber Stamp surface after
painting and prior to clear coating —⟍

7.5.3 Punctuate as necessary for clarity; only the longer notes (complete sentences) should terminate with a period.

7.5.4 Do not specify Fabrication operations (e.g. DRILL, REAM, TAP, PUNCH, or BORE). The configuration, surface finish and/or tolerance should permit Manufacturing to establish the type of operation.

7.5.5 SPOTFACE, COUNTERBORE, COUNTERDRILL, COUNTERSINK, THREAD and UNDERCUT, describe features (not fabrication methods) and may be used in notes.

7.5.6   Examples of Local Notes.

(1) .119 DIA THRU 7 PLACES

(2) SF .75 DIA (Near Side Always Understood). -or- SF .75 DIA FAR SIDE -or- SF .75 DIA BOTH SIDES

(3) CSK 100° X .25 DIA

(4) CBORE .875 DIA X .375 DEEP

(5) UNDERCUT .950 DIA X .050 WIDE WITH .030 R IN CORNERS.

7.6   GENERAL NOTES

7.6.1   General Notes are located in the upper left hand corner for drawing sizes "B" through "F". For "J" size drawings locate the "General Notes" column two (2) inches (min.) to the left of the "List of Materials". They are numbered consecutively starting with 1. (See Section 5.)

7.6.2   The General Note column shall not exceed 8 inches in width.

7.6.3   On multisheet drawings, begin the general notes on Sheet 1 whenever possible. (See Section 5.)

7.6.4   General Notes are punctuated according to the rules of English Grammar.

7.6.5   Abbreviations, when used, must conform to Section 20.

7.6.6   Lengthy notes covering complex processes should be avoided. This type of information should be referred to separate process specifications.

7.6.7   SEQUENCE OF NOTES

7.6.7.1   When drawings are prepared to military standards, the first note shall be one of the following depending on contractual requirements and drawing content:
   INTERPRET DRAWING PER MIL-STD-100.
   INTERPRET DIMENSIONS AND TOLERANCES PER ANSI Y14.5

7.6.7.2   The second note, when applicable, is:
   Example 1:
      REMOVE ALL BURRS AND SHARP EDGES
   Example 2:
      REMOVE ALL BURRS AND SHARP EDGES _____ TO _____ .
   Example 3:
      REMOVE ALL BURRS AND SHARP EDGES EQUIVALENT TO _____ R, UNLESS OTHERWISE NOTED.

This note is not required on wiring diagrams, schematics, assembly, or installation drawings, or where no machining is called out.

7.6.7.3   After the above, the notes should be called out in the same sequence as they affect the fabrication and processing of the part or assembly, if possible.

7.6.7.4   REVISION OF A NOTE — Once a note has been assigned a number, the note may be revised, but not revised to a different application or intent. (See Section 18.)

7.6.8   DELTA GENERAL NOTES — Delta are used for information of local significance which is too extensive to be placed in either the limited area of the title block, parts list, or on the field of drawing; or that would require unnecessary repetition in any of these areas. When it becomes necessary to crossreference general notes to items in the title block, parts list, or on the field of the drawing, a delta approximately .50 on each side is used. The applicable general note number appears within the Delta symbol.

Example:

△ SAFETY WIRE PER (Applicable Specification)

7.6.9   COMPOSITION OF NOTES — When composing a note, the general rule to apply is: (process or subject(PER (specification/standard) (additional qualification or acceptance criteria). The specification revision letter or date shall not be called out.

7.6.10   When heat treatment, surface treatment, inspection requirements, etc. affects only a portion of an item, that portion will be noted by a delta and cross-referenced to the applicable general note.

Example, applied in General Notes:

△ HEAT TREAT THIS AREA ONLY PER _____. (Pertinent heat treatment data.)

△ ELECTRO PLATE THIS AREA ONLY PER _____. (Pertinent finish data.)

△ RADIOGRAPHIC INSPECT PER (applicable specification) ACCEPTANCE CRITERIA PER (applicable specification).

Example, applied on Field of Drawing:

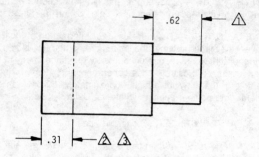

7.6.11 When it is necessary to control dimensional inaccuracies caused by thermal expansion, a general note calling out the applicable standard and referenced to the dimension(s) affected is used.

 Example, applied in General Note:
  ⚠ TEMPERATURE CONTROL DIMENSION PER (applicable specification).

 Example, applied on Field of Drawing:

$$\longmapsto \begin{array}{c} 12.501 \\ 12.503 \end{array} \triangle \longrightarrow$$

Section 8 — DRAFTING PRACTICES

8.1 SCOPE — This section defines various drafting practices such as drawing scale, lettering, line work, sectioning, projection, and descriptive geometry.

8.2 APPLICABLE DOCUMENTS

DOD-STD-100   Engineering Drawing Practice

8.3 SCALE OF DRAWING

8.3.1 The preferred drawing scales and the procedures for specifying on drawings are as follows:

SCALE	DESIGNATION	
Full Size	$\frac{1}{1}$	
Half Size	$\frac{1}{2}$	
Quarter Size	$\frac{1}{4}$	
Tenth Size	$\frac{1}{10}$	(Non Preferred)
Double Size	$\frac{2}{1}$	
Four Times Size	$\frac{4}{1}$	
Ten Times Size	$\frac{10}{1}$	(Non Preferred)

8.3.2 The Division Line of fractional scales shall be horizontal except when included in typewritten notes, tables, lists, and in scale block. In these cases a diagonal line may be used. Example: 1/2 (Typewritten)

8.3.3 All drawings are originally prepared to conform with the scale in the scale block, except schematics, wiring diagrams, etc. (See 8.3.8). Full scale is preferred.

8.3.4 The scale to which most views and sections are drawn shall be that entered in the Scale Block. The scale of each view or section drawn to any other scale shall be entered directly below the title of that view or section. (See Figures 1 & 2)

8.3.5 NEW DRAWINGS are drawn to scale within .03 inches. When dimensional changes are made on released drawings, and it is not practical to change the detail to agree with the new dimension, the dimension is underscored with a straight line to indicate the out of scale condition (see Figure 3). Care must be taken that if this action is taken on an older drawing, other out of scale dimensions shall be changed from the previous wavy line practice to the current straight line practice.

SCALE 1/2

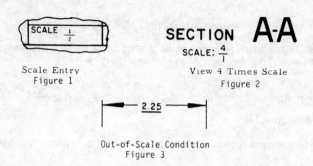

Scale Entry
Figure 1

SECTION A-A
SCALE: 4/1

View 4 Times Scale
Figure 2

2.25

Out-of-Scale Condition
Figure 3

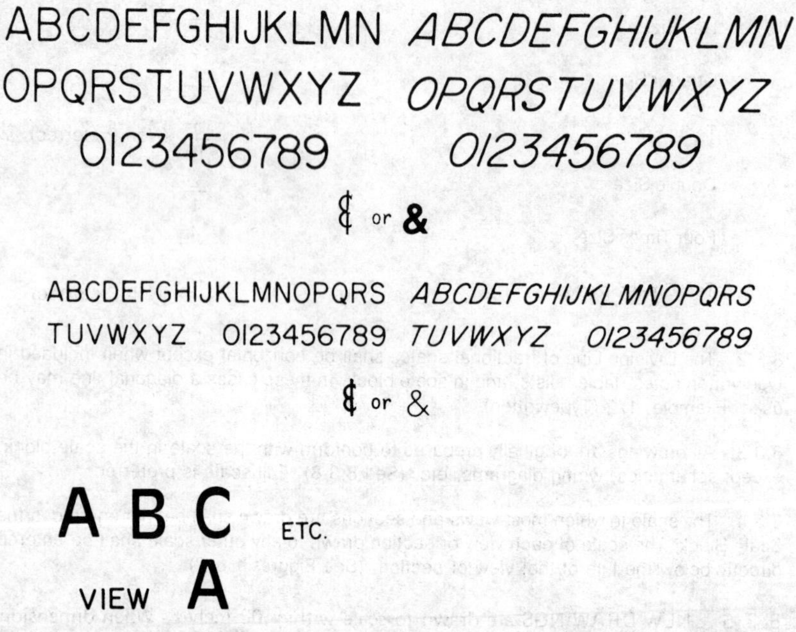

ABCDEFGHIJKLMN ABCDEFGHIJKLMN
OPQRSTUVWXYZ OPQRSTUVWXYZ
0123456789 0123456789
& or &
ABCDEFGHIJKLMNOPQRS ABCDEFGHIJKLMNOPQRS
TUVWXYZ 0123456789 TUVWXYZ 0123456789
& or &
A B C ETC.
VIEW A

Vertical Inclined

Lettering Style
Figure 4

8.3.6 REDUCED SCALE DRAWINGS may be made for large parts whose major view may be clearly shown in the smaller scale to reduce the size of the drawing format. These drawings may contain appropriately labeled larger size views or sections of complicated portions when necessary for clarity.

8.3.7 ENLARGED SCALE DRAWINGS are those of exceptionally small parts or views drawn to an enlarged scale for clarity.

8.3.8 NO SCALE DRAWINGS are certain drawings or figures which cannot be drawn to a specific scale, for example, wiring and schematic diagram drawings, etc. On such drawings the scale block shall read "NONE."

8.4 LETTERING

8.4.1 Lettering (except Section or View "Letters") shall be single stroke upper case, commercial gothic; however, for special design requirements (i.e., name plates, connector pin, etc.) lower case lettering may be used. Inclined or vertical lettering may be used but only one type shall appear on a single drawing. See Figure 4 for lettering styles.

8.4.2 Freehand lettering or typing may be used.

8.4.3 Letters and numerals shall be printed parallel to the bottom of the drawing, regardless of the position of dimension lines.

8.4.4 LETTERING SIZE

USE	SIZE (Min.)
Drawing Number	.38
Drawing Title, FSCM No.	.20
Tabulated & Section Letters	.40
The words "SECTION," "VIEW"	.20
Dimensions & Tolerances	.156
Revision Description	.156
Parts List entries	.156

8.4.5 LETTER SPACING — should be spaced as evenly and consistently as possible. Normally, the space between words and between lines should be the height of one letter. Lettering shall be such that microfilm blowback and reproductions of the drawing shall not show evidence of filled in loops or leaching of character to character.

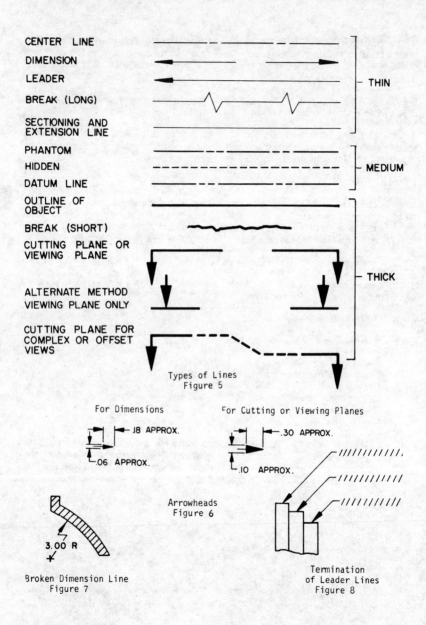

Types of Lines
Figure 5

Arrowheads
Figure 6

Broken Dimension Line
Figure 7

Termination of Leader Lines
Figure 8

8.5 TYPES OF LINES

8.5.1   All lines shall be opaque and of uniform width for each type of line, using the line characteristics recommended in Figure 5, except on diagramatic drawings. All line work shall be of microfilm quality.

8.5.2   CENTER LINES — shall be shown in pertinent views of holes, round shapes on detail drawings, on assembly drawings or to indicate the travel of a center. (See Figure 12) Center lines shall cross without voids (See Figure 10). Very short center lines may be unbroken if there is no confusion with other lines.

8.5.3   DIMENSION LINES — shall terminate in arrowheads and should be unbroken, except for the insertion of the figures expressing the dimension. When dimensioning radii, it is permissible to break the dimension line if true center of the radius falls off the drawing into another view, or interferes with the logical dimensioning procedure. (See Figure 7)

8.5.4   LEADER LINES — shall indicate a part or portion to which a dimension, note, or other reference applies. Leaders terminate in an arrowhead at the end touching the outline of the part or a dot within the outline of an object.

8.5.4.1   Arrowheads are drawn as shown in Figure 6.

8.5.4.2   Leaders terminate at the lettering end with a line approximately .125 inch long, parallel to the lettering of the note and are extended at a constant angle to the part or portion affected by the note. Leader lines are not curved. (See Figure 8)

8.5.4.3   The leader line from a dimension and/or local note to a circular area, or hole, shall be directed toward the center of the area, circle or hole, with the arrowhead terminating at the edge of the circle area or hole. (See Figure 9)

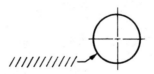

Leader Line Direction
Figure 9

8.5.4.4   Leader lines shall not cross dimension, or other leader extension lines. When it is unavoidable that a leader cross a dimension line, the leader line or dimension line is not broken at the point of intersection.

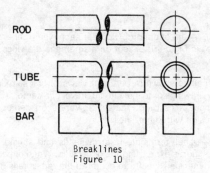

Breaklines
Figure 10

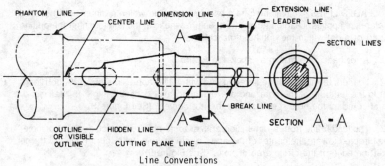

Line Conventions
Figure 11

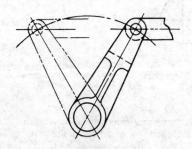

Alternate Position
Figure 12

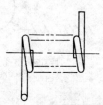

Repeated Detail
Figure 13

8.5.5 BREAK LINES — Short breaks shall be indicated by solid freehand lines; long breaks, by full ruled lines with freehand zig-zags. (See Figure 5)

8.5.5.1 Shafts, rods, tubes, etc., which have a portion of their length broken out, shall indicate the ends of the break as in Figure 10.

8.5.6 PHANTOM LINES — shall be used to indicate an alternate position, repeated detail, or the interfacing position of an absent part. They shall be composed by a series of one long and two short dashes evenly spaced with a long dash at each end. (See Figures 11, 12 and 13)

8.5.7 SECTION LINING — shall be used to indicate the exposed surface of an object in a cross-sectional view. (See Figures 5 and 11)

8.5.8 EXTENSION LINES — shall be used to indicate the extent of a dimension. They shall not touch the object outline. (See Figure 11)

8.5.9 DATUM LINES — Datum lines are used to establish the position of a datum point, line, or plane, and consist of one long dash and two short dashes evenly spaced per Figure 5. (See Section 10 for the use of datums.) Composed the same as Phantom line.

8.5.10 HIDDEN LINES — shall be used only when required for clarity and shall consist of short dashes evenly spaced. They shall begin with a dash in contact with the line from which they start, except when such a dash would form a continuation of a full line. Dashes shall touch at corners and arcs start with dashes on the tangent points. (See Figures 5 and 11)

8.5.10.1 When two or more hidden surfaces intersect, a dash in one line shall cross a dash in the other line.

8.5.11 CUTTING PLANE AND VIEWING LINES — When it is necessary to indicate where a section is taken, a cutting plane line is used. Section letters are placed in front of the directional arrowheads, except for alternate viewing plane in which case they are placed in back of directional arrowheads. (See Figures 5 and 11)

8.5.12 THE OUTLINE OR VISIBLE OUTLINE — shall be used for all lines on the drawing representing the visible outlines of the object. (See Figure 11)

NOT RECOMMENDED
SECTION ROTATED

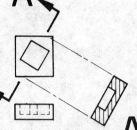

SECTION A-A
CORRECT
BUT SECTION NOT IN
LINE OF PROJECTION

INCORRECT
NOT ALLOWED
SECTION PLACED ON
WRONG SIDE OF
CUTTING
PLANE

SECTION A-A
SECTION PLACED IN
CORRECT POSITION

CORRECT
BUT SECTION NOT IN
LINE OF PROJECTION

INCORRECT
WRONG PROJECTION

SECTION A-A

Placement of Sectional Views
Figure 14

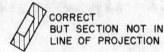

SECTION A-A
ROTATED 90° CLOCKWISE

Rotated Section
Figure 15

SECTION A-A
VIEW B-B

Section Titles
Figure 16

8.6 DRAFTING SYMBOLS

8.6.1 Each drafting symbol shall be defined in a note or legend on the drawing or by reference to an approved source.

See SECTION 5 for ZONE CROSS REFERENCE symbols.
See SECTION 7 for USE & SIZE OF DELTA symbols.
See SECTION 10 for FORM & POSITIONAL TOLERANCE symbols.
See SECTION 11 for SURFACE ROUGHNESS symbols.
See SECTION 15 for WELDING symbols.
See SECTION 17 for ELECTRICAL - ELECTRONIC symbols.

8.7 SECTION OR SECTIONAL VIEWS

8.7.1 A Section or a Sectional View is one obtained by cutting away part of an object to show the shape and construction at the cutting plane. Sectional views are not made from another sectional view, except for clarity.

8.7.2 PLACEMENT OF SECTIONS — Sectional views are placed as closely as practicable behind the arrows showing the direction from which the view was taken. (See Figure 14) When it is absolutely necessary to rotate a sectional view, the degree of rotation and direction are specified beneath the view. (See Figure 15) Cutting planes of a single section of a symmetrical part are not necessary, but when two or more sections are taken, all sections shall be identified.

8.7.2.1 SECTIONAL VIEWS ON SEPARATE SHEET — Sectional views should appear on the same sheet as the subassembly, assembly, or detail drawings from which they are taken. When it is necessary to draw sectional views on a separate sheet, views and cutting plane indications should be suitable cross-referenced by zone designations. When a sectional view appears on a sheet separated from one containing the cutting plane indication, the view will be oriented as it would appear if directly projected from the cutting plane and properly cross-referenced. (See Section 5)

8.7.3 SECTION IDENTIFICATION LETTERS — Section letters are always used in pairs, A-A, B-B, C-C, etc. Should a drawing require so many sections that the single letters of the alphabet are exhausted, further sections may be identified by the use of combinations of different letters such as AA-AA, AB-AB, AC-AC, etc. The letters I, O, and Q are not used as Section letters. Once a letter or combination of letters is established on a released drawing, it is not reused for another section, view, or detail on the same drawing.

8.7.4 SECTION TITLES — The letters which identify a cutting plane are used as part of the section title, which is placed directly under the section view. The word "SECTION" is not abbreviated. (See Figure 16)

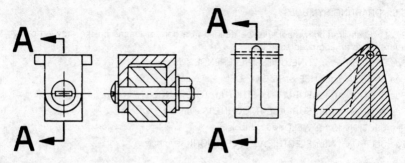

Unlined Sections
Figure 17

Sections Thru Ribs
Figure 18

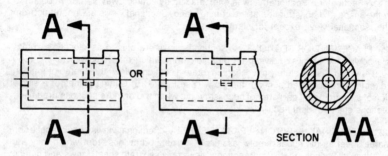

Full Sectional View
Figure 19

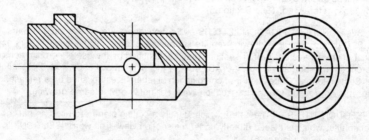

Half Section of Symmetrical Object
Figure 20

8.7.5 TYPES OF SECTIONS AND SECTIONAL VIEWS

8.7.5.1 THIN SECTIONS — Sections of sheet metals, packings, gaskets, etc., which are too thin for section lining, may be shown solid. Where two or more thicknesses are shown, space conducive to microfilming is left between them.

8.7.5.2 UNLINED SECTIONS — Shafts, nuts, bolts, rods, rivets, keys, pins, and similar parts, whose axes lie in the cutting plane, are not cross-sectioned (See Figure 17)

8.7.5.3 SECTIONS THRU RIBS, WEBS, ETC. — When the cutting plane passes thru a rib, web, or similar element, cross-section lines may be omitted from these parts. An acceptable alternate method of representation is to portray the line of intersection of the element (rib, web, etc.) with a hidden line. In cross-sectioning the element, every other section line is omitted to accent the division of the element and the body of the part. (See Figure 18)

8.7.5.4 FULL SECTIONS — A full section is a view which is obtained when the cutting plane extends entirely across the object. (See Figure 19)

8.7.5.5 HALF SECTIONS — A half section is a view of a symmetrical object which shows the internal and external features. This is accomplished by passing two cutting planes at right angles to each other along the center lines of symmetrical axes. Thus, one quarter of the object is considered removed and the interior exposed to view. The cutting plane indications and section titles are omitted. (See Figure 20)

8.7.5.6 REVOLVED SECTIONS — A revolved section drawn directly on an exterior view provides a method of showing the shape of the cross-section of a part such as the spoke of wheel. The cutting plane is passed perpendicular to the center line or axis of the part to be sectioned and the resulting section is rotated in place. The cutting plane indications are omitted for symmetrical sections. (See Figure 21)

8.7.5.7 REMOVED SECTIONS — A removed section may be used to illustrate particular parts of an object. They are drawn like revolved sections, except that they are placed at one side and are often drawn to a larger scale than that of the view on which they are indicated. This is done to show pertinent data. (See Figure 21) Cutting plane indications are omitted.

8.7.5.8 OFFSET SECTIONS — When the cutting plane is not a continuous plane, the resulting section is called an offset section. (See Figure 22) Cutting plane indications and section title are required.

8.7.5.9 BROKEN-OUT SECTION — When a sectional view of only part of the object is needed, broken-out sections may be used. The break line convention is used to separate the sectional view portion from the exterior view. (See Figure 23)

8.7.5.10 SECTION THRU WELDS — When the cutting plane passes thru two or more parts that are being permanently assembled by welding, brazing, etc., each detail part is individually cross-sectioned. When a group of permanently assembled parts is sectioned on its next assembly(s), it is cross-sectioned as one part. (See Section 15)

8.7.6 SCALE OF SECTIONS. Sections should be drawn to the same scale as the views from which they are taken; however, if it is necessary to employ a different scale, it is specified directly below the section title as illustrated in Paragraph 8.3.4 and Figure 2.

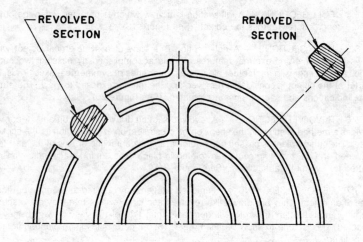

Removed and Revolved Sections
Figure 21

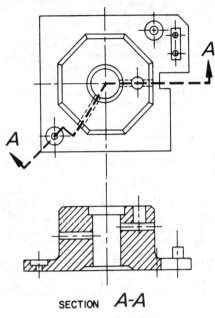

SECTION A-A

Offset Sections
Figure 22

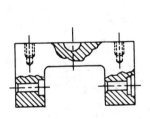

Broken Out Section
Figure 23

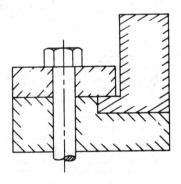

Sectioning Adjacent Parts
Figure 24

8.7.7 SECTIONING CONVENTIONS

8.7.7.1 In addition to showing the shape and construction, sectional views are also used to distinguish the individual components of an assembly or subassembly. This is accomplished by drawing sectional conventions on the exposed surfaces of the sectional view. Sectioning conventions do not cross dimensions or obscure other conventions on drawings. Sectioning conventions may be shown along the borders of the part only when clarity is not sacrificed. (See Figure 24)

8.7.7.2 The cross-sectioning symbol for cast iron is generally used on detail drawings, regardless of material, and on all other sectional views where practicable.

8.7.7.3 When desirable, material from which the parts are made should be indicated by the following symbols:

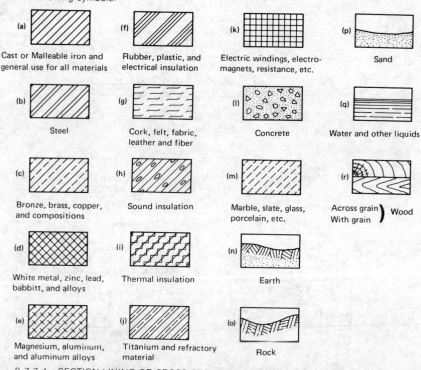

8.7.7.4 SECTION LINING OR CROSS-SECTIONING — They are composed of uniformly spaced lines at an angle of 45 degrees to the base line of the section. On adjacent parts, the 45 degree lines are drawn in the opposite direction. On a third part, adjacent to two other parts, the section lining shall be drawn at an angle of 30 or 60 degrees. (See Figure 24) Where the 45 degree section lining is parallel or nearly parallel, with the outline of the object, another angle is chosen.

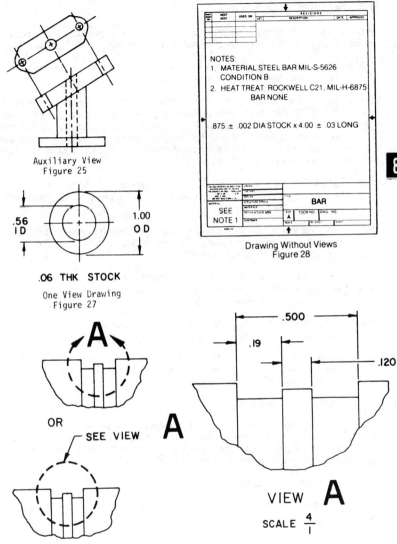

Auxiliary View
Figure 25

One View Drawing
Figure 27

Drawing Without Views
Figure 28

Detail Views
Figure 26

8.8 PROJECTION PROCEDURES

8.8.1 Third angle orthographic projection is used for all drawings. Other types of projection such as isometric, perspective, etc. are not normally used on product drawings.

8.8.2 SELECTION OF VIEWS — Draw only those views necessary to clearly define the object. Avoid views that are meaningless or for picture purpose only.

8.8.3 AUXILIARY VIEWS — Objects having inclined faces or other features which are not parallel to any of the three principal planes of projection may require auxiliary views to show the true shape of such features. In many cases, partial auxiliary views, which show only the pertinent features, may be employed to illustrate features not clearly shown by principal views. The auxiliary view is arranged as though the auxiliary plane were revolved into the plane of the paper. (See Figure 25)

8.8.4 DETAIL VIEWS AND SECTIONS — A detail view or section shows a part of a drawing in the same plane and in the same arrangement, but in greater detail than is shown in the principal view. (See Figure 26)

8.8.5 ONE VIEW DRAWINGS — One view drawings are permissible for simple cylindrical, spherical, square, and thin objects of uniform thickness, if the necessary dimensions can be properly indicated. (See Figure 27)

8.8.6 SPACING AND IDENTIFICATION OF VIEWS — Provide ample space between views to permit the placing of dimensions without crowding and to preclude the possibility of notes pertaining to one view overlapping or crowding the other views.

8.8.7 DRAWINGS WITHOUT VIEWS — In those instances where words will adequately describe the part, no views need be shown. (See Figure 28)

8.8.8 ISOMETRIC AND PICTORIAL VIEWS — These views may be shown on engineering drawings providing there is some particular advantage and clarity is not degraded. See page 4–63 Figure 30.

8.9 DESCRIPTIVE GEOMETRY

8.9.1 This section is included to provide standard practices in the use of descriptive geometry on engineering drawings. The adoption of these methods of constructions and the common use of the various terms, will greatly facilitate the interpretation of drawings, particularly those involving several projections.

8.9.2 FUNDAMENTAL PRINCIPLES

8.9.2.1 In third angle projection, the plane is always between the object and the observer.

8.9.2.2 Projection planes for various views are always rotated into the plane of the drawing through a 90 degree angle. Rotation takes place about lines which may be visualized as hinges between adjacent projection planes. These lines are variously referred to as reference lines, rotation lines, or folding lines, hereafter called reference lines.

8.9.2.3 Considering each view separately, the reference lines represent edge views of other projection planes.

8.9.2.4 The lines of sight from the object to each projection plane are parallel to each other and perpendicular to the projection plane.

8.9.2.5 Perpendicular distances from the projection plane to the object are equal in all views taken perpendicular to that plane. (See Figure 29)

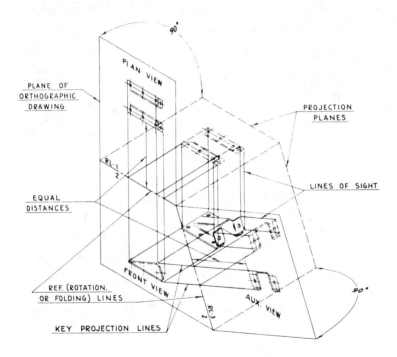

Descriptive Geometry Fundamentals
Figure 29

8-17

8.9.3 USE OF REFERENCE LINES ON DRAWINGS. (See Figure 30)

8.9.3.1 The drawing on the preceding page illustrating the principles of descriptive geometry and, likewise, all the succeeding problems, shows the reference (rotation or folding lines) between all views. The reference lines, if used on releasable drawings, should be temporary construction lines only, and should not appear on the finished drawing. Neither need they remain on layouts for the conventional views (plan, front, side) or where a single auxiliary view is projected from a normal view. The key projection lines for all auxiliary views should, however, remain on both releasable drawings and layouts. On layouts where two or more oblique views are taken in succession, the reference lines should be shown as a phantom line and labeled as shown below.

8.9.3.2 Notice that this manner of labeling the reference lines tells three things.

8.9.3.2.1 Identifies the line as a reference line between views.

8.9.3.2.2 Identifies the views on either side of the line.

8.9.3.2.3 Indicates order in which views are drawn.

8.9.3.3 Where reference lines remain on the layout labeled as above, the views need not be otherwise identified; however, when a view cannot be directly projected due to lack of space and it becomes necessary to place it elsewhere on the drawing, then the conventional method of cutting a view plane and cross-referencing the view shall be used.

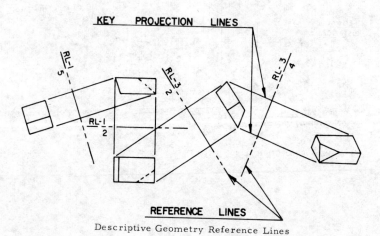

Descriptive Geometry Reference Lines
Figure 30

8-18

8.9.4 PROBLEM 1: To Fine the True Length of a Line. (See Figure 31)

8.9.4.1 A line will show in its true length when projected onto a plane which is parallel to the line in space. When two views of a line are given, neither of which is a true view, a third view may be constructed which will show the true length. This is done by drawing a reference line (edge view of the new projection plane) parallel to one of the given views and projecting the line onto that plane.

8.9.4.2 In the illustration below, draw RL-2/3 parallel to A_2B_2. Draw projections lines through end points A_2 and B_2 at 90° to the reference line. Set off distance from RL-2/3 to A_2 equal to distance from RL-1/2 to A_1. Also, set off distance to B_3 equal to distance to B_1. It should be noted that distances required to construct a view are obtinaed from the second preceeding view. This is a general rule to be followed in all descriptive geometry problems. Draw true length A_3B_3.

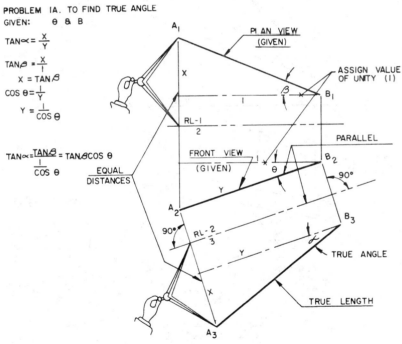

True Length of Line and True Angle
Figure 31

8.9.5 PROBLEM 2: To Find the Perpendicular Distance from a Point to a Line. (See Figure 32)

8.9.5.1 The true perpendicular distance from a point to a line will show in a projection which gives an end view of the line.

8.9.5.2 Construct true view of line A_3B_3, as explained in Problem 1. Draw a reference line RL-3/4 perpendicular to A_3B_3. The line, when projected onto this plane, will show as a point A_4B_4. Project the point P into the 3rd and 4th views, obtaining distances from reference lines each time from the second preceding view. True distance is shown by dotted line $P_4A_4B_4$.

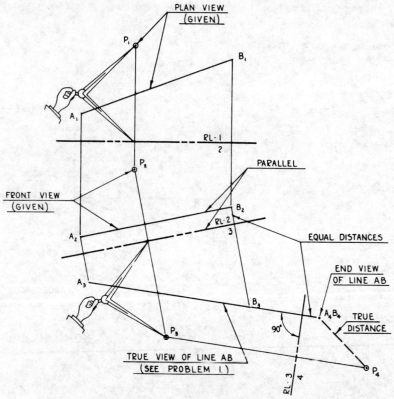

Perpendicular Distance Point to Line
Figure 32

8.9.6 PROBLEM 3: To Fine the Shortest Distance Between Two Non-Intersecting Straight Lines. (See Figure 33)

8.9.6.1 The shortest distance between two non-intersecting straight lines will show in a projection giving the end view of either line.

8.9.6.2 Construct the end view A_4B_4 of line AB, as explained in Problem 2. Project the other line XY into the 3rd and 4th views. The shortest distance will now show as the perpendicular from the point A_4B_4 to the line X_4Y_4. (Note: X_4Y_4 is not a true length.)

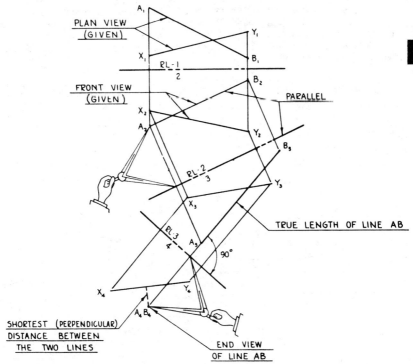

Shortest Distance Non-intersecting Lines
Figure 33

8-21

8.9.7 PROBLEM 4: To Find the True View of a Plane. (See Figure 34)

8.9.7.1 A plane will show in its true size and form when projected onto a plane which is parallel to it. To do this we must first obtain an edge view. The end view of any line in the plane will give an edge view of the plane.

8.9.7.2 This could be done by constructing an end view of any of the three boundary lines AB, AC, or CB. It can be done with one less view, however, by choosing an arbitrary arbitrary line B_1D_1 which lies on the plane ABC and which is parallel to RL-2/1.

8.9.7.3 Since the front view B_1D_1 of the line BD is parallel to the projection plane of the plan view, the plan view B_2D_2 of line BD is a true length. The next projection is onto a plane perpendicular to this true length B_2D_2 and gives an end view of the line and, therefore, an edge view $A_3B_3C_3$ of the plane. The true view of the plane can now be projected onto a plane parallel to A_3C_3.

8.9.7.4 The same results could be obtained by choosing the arbitrary line in the plane view parallel to RL-2/1. The true length would then appear in the front view from which an edge view of the plane could be projected and, hence, a true view.

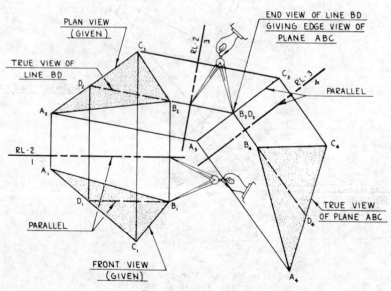

True View of a Plane
Figure 34

8-22

8.9.8 PROBLEM 5: To Find True Distance from a Point to a Plane and Locate Projection of the Point on the Plane. (See Figure 35)

8.9.8.1 Construct edge view and true view of plane as explained in Problem 4. Project the point P into these views. The true distance shows as a perpendicular line from P_3 to the edge view of the plane $A_3B_3C_3$.

8.9.8.2 The intersection of this line with the plane can now be projected back into the plane and front views to locate the projection of the point on the plane in these views. The true position of the point projected on the plane shows in the true view of the plane at P_4.

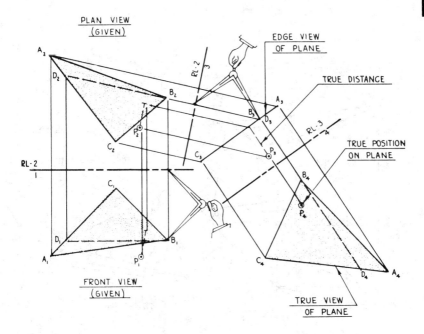

True Distance Point to Plane and Projection of Point
Figure 35

8.9.9 PROBLEM 6: To Find the True Angle Between a Line and a Plane and Locate the Point Where the Line Pierces the Plane. (See Figure 36)

8.9.9.1 The true angle between the line and the plane will show in an edge view of the plane taken in such a manner as to also show the true length of the line. The piercing point may be determined from any edge view.

8.9.9.2 First construct the edge view and true view of the plane as explained in Problem 4 and project the line DE into these views. The piercing point P_3 may now be projected back to the plan and front views. To get the true angle, we need to construct a 5th view. Draw reference line RL-4/5 parallel to D_4E_4. Project the line and plane into this view. This will give another edge view of the plane, which also shows the true length D_5E_5 of the line and, therefore, shows the true angle.

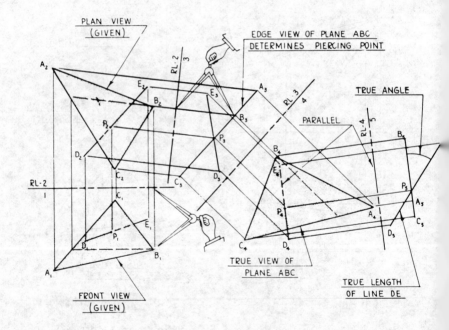

True Angle Line and Plane and Piercing Point
Figure 36

8.9.10 PROBLEM 7: To Find the True Angle Between Two Planes, Intersection Line Known. (See Figure 37)

8.9.10.1 The true angle between the two planes will show in a view which gives the edge view of both planes. Since the intersection line is common to both planes, the end view of the intersection line will show the edge view of both planes.

8.9.10.2 Construct the true view, B_3C_3 and end view, B_4C_4 of line BC as explained in Problems 1 and 2. Project the other points necessary to draw the planes into these views. The true angle may now be measured between the edge views $D_4B_4C_4$ and $A_4B_4C_4$.

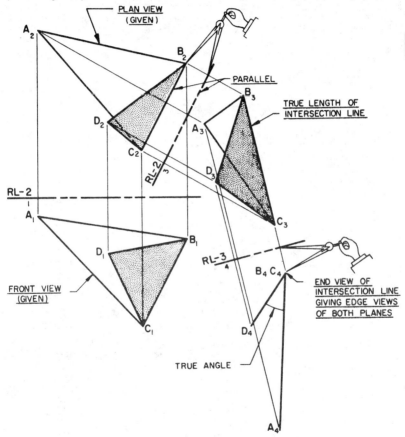

True Angle Between Planes Intersection Known
Figure 37

8-25

8.9.11 PROBLEM 8: To Find the True Angle Between Two Planes, Intersection Line Not Known. (See Figure 38)

8.9.11.1 Since the intersection line is not given, we cannot use the method described in Problem 7.

8.9.11.2 Construct the edge view and true view of plane ABC as explained in Problem 4. Now any view taken off the true view $A_4B_4C_4$ will give an edge view of this plane but we need to select the one which also gives an edge view of plane XYZ. We now apply the same principle to plane XYZ in views 3, 4, and 5 as we did to plane ABC in views 1, 2, and 3, by drawing Q_3X_3 parallel to RL-3/4. This gives a true length Q_4X_4 and, hence, the edge view $Z_5X_5Y_5$. We now construct the edge view $B_5A_5C_5$ of the other plane in this view and obtain the true angle.

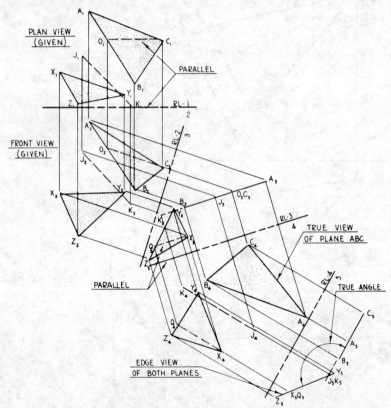

True Angle Between Planes Intersection Not Known
Figure 38

8-26

8.9.11.3 If the intersection line of the two planes is required in the plan and front views, it may be obtained by projecting back from the end view J_5K_5 of the intersection in the last view. This gives a true view of the line in View 4. Any two points J_4K_4 on the line may now be projected into views 3, 2, and 1 to determine the line in the original views.

8.9.12 PULLEY BRACKET LAYOUT. (See Figure 39)

8.9.12.1 The pulley bracket problem below illustrates a common application of some of the foregoing principles of descriptive geometry. We have, given, the plane of the mounting structure 1, 2, 3, 4 and, and the cable plane ABC. It is desired to obtain a view which shows the edge views of both planes and also a true view of the cable plane in order to design the bracket.

8.9.12.2 Notice that since we have the edge view of the mounting structure given in the front view, we can obtain a true view of this plane which also shows the true length J_3K_3 of the intersection line in the first auxiliary view taken (view 3). We can now obtain the edge view (view 4) by projecting onto a plane perpendicular to J_3K_3. The true view of the cable plane is then projected onto a plane parallel to A_4B_4.

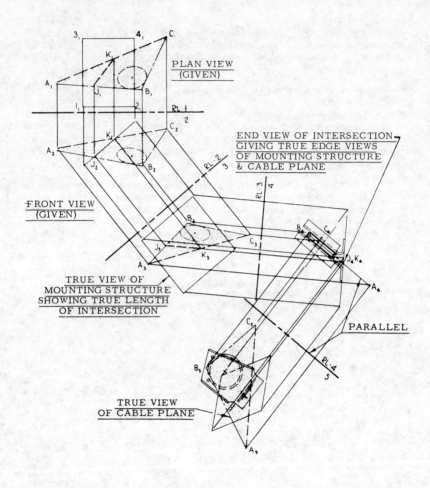

Pulley Bracket Layout
Figure 39

Section 9 — PARTS LIST

9.1 SCOPE — This section establishes the requirements and procedures to be observed in the preparation of a parts list integral with or separate from the drawing.

9.2 APPLICABLE DOCUMENTS

DOD-STD-100	Engineering Drawing Practices
H4	Cataloging Handbook
	Federal Supply Code for Manufacturers
H4-1	Name to Code
H4-2	Code to Name

9.3 DEFINITIONS

9.3.1 PARTS LIST (PL) is a tabulation of parts and materials required to fabricate or procure the item(s) shown on a drawing. Reference documents may also be tabulated on a Parts List.

9.3.2 ITEM NUMBER — Item numbers (find numbers) are assigned to every line entry of an assembly parts list to facilitate the location of that item in the field of the drawing. They are not used for other identification purposes.

9.3.3 SEPARATE PARTS LISTS (SPL) — A Parts List separate from the drawing.

9.3.4 BULK ITEM — "Bulk items" are those necessary constituents of an assembly or part such as oil, wax, solder, cement, ink, damping fluid, grease, powdered graphite, flux, welding rod, thread, twine and chain for which the quantity required is not readily predeterminable or if knowing the quantity, the physical nature of the material is such that it is not adaptable to depiction on a drawing; or which can be cut to finished size by the use of such hand or bench tools as shears, pliers, knives, etc., without any further machining operations and the configuration is such that it can be fully described in writing without the necessity of pictorial presentation.

9.4 GENERAL REQUIREMENTS FOR INTEGRAL PARTS LIST

9.4.1 PARTS LIST FORMAT — when parts lists are prepared integral with the drawing, they shall include, as a minimum, columns shown on Figure 1. Columns may be preprinted to proportions shown in Section 5. Mandatory columns shall appear and additional columns may be added as required.

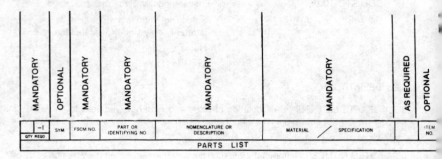

Figure 1

9.4.2 PART AND/OR MATERIAL CALLOUT — Each is entered only once in the parts list. Parts and/or materials referenced from another drawing are not entered.

9.4.3 The parts list entry spaces are normally .50 inch high. When additional space is required, the height of the entry spaces is increased in increments of .50 inch.

9.4.4 BULK ITEMS

9.4.4.1 Bulk items used to produce an inseparable assembly and/or nonpermanent surface protection, e.g., solder, weld rod, paint, primer, etc. are entered in the parts list. The application method is established by a local note, general note, or in the finish block.

9.4.4.2 Materials used to produce permanent surface protection, e.g., electroplating, passivating, anodizing, chromate conversions, etc., are not entered in the parts list. The callout of the appropriate specification will suffice for these materials.

9.4.4.3 When materials used are part of a process or process specification and are called out in the process or process specification, they are not entered in the parts list, e.g., anodize, chromium plate, cadmium plate, etc.

9.4.4.4 Thinning, reducing, cleaning agents, etc., required in conjunction with the application of a protective coating, are not entered in the parts list.

9.4.5 PARTS LIST ENTRIES — Parts entered in the parts list usually follow the order listed below, but this order and the necessity for grouping, through preferred, are not mandatory.
1.) Dash Numbered Parts of the Parent Drawing Number
2.) Subassemblies detailed on the same drawing.
3.) Parts and assemblies detailed on other drawings
4.) Company Numbered parts - Listed in numerically ascending order
5.) Company Standard Parts
6.) Military Standard Parts (AN, MS, etc.)
7.) Industry Standard Parts (AS, NAS, etc.)
8.) Commercial Parts (Suppliers Items)
9.) Bulk Items
10.) Customer Furnished and Controlled Items

9.4.6 ALTERNATE MATERIAL OR ITEM

9.4.6.1 A material or item is an acceptable alternate for another material or item when it possesses interchangeable physical and functional characteristics.

9.6.4.2 The use of alternates shall be kept to a minimum. Alternates are specified on the drawing only upon instruction from the responsible designer. Alternates may require an identity different from the preferred part or material, but do not require a new identity for the assembly in which they are used. Only the alternates listed on the drawing may be used.

9.4.6.3 The following drafting practice is used when alternates are specified on the drawing.

9.4.6.3.1 A single alternate may be specified in a general note and cross-referenced to the preferred item in the parts list. (See 9.11)

Example:
⚠ PERMISSIBLE TO USE (alternate part number and code identification number, when required) IN PLACE OF (parts list part number)

9.4.6.3.2 Two or more alternate items are specified in a tabulation block and qualified by a general note cross-referenced to the parts list entries. On "C" size and smaller drawings, the tabulation may be located wherever convenient on the field of drawing; on "D" size and larger drawing, the tabulation is located adjacent to the lower border and to the left of the title block. (See Figure 2)

9.4.6.3.3 When there is no preference between two or more materials or items, they are entered together in a single space in the parts list. (See Figure 3)

General Notes:

 ALTERNATES PERMISSIBLE, SEE TABULATION.

Field of Drawing:

STANDARD PARTS---------------
SUPPLIER PRODUCTS-----------
COMPANY ITEMS-OTHER PLANTS--
COMPANY ITEMS---------------

14	AN-XXXX	
8	D-XXXX	34567
5	1250046-1	13310
1	1534798-1	
ITEM NO. REF	ALTERNATE PART NO.	FSCM NO.
/X\	ACCEPTABLE ALTERNATES	

Tabulated Alternates
Figure 2

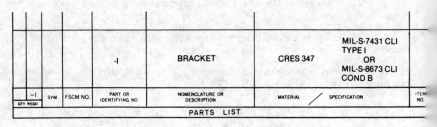

Parts List Entry for Alternates
Figure 3

9.5 PARTS LIST LOCATION

9.5.1 The parts list is located in the lower right hand corner of the drawing, above the title block. (On detail drawings with brief parts lists, this space may be utilized for drawing.) On multiple sheet drawings, the parts list starts on Sheet 1 and continues on as many sheets as necessary.

9.5.2 When the area allotted for the parts list has been fully utilized, (see Section 5), additional blocks may be added to the left of the original block, leaving a space of 1.00 inch minimum between blocks.

9.6 DRAWING APPLICATION

9.6.1 DETAIL DRAWINGS (See Figure 4)

9.6.1.1 "QUANTITY REQUIRED" COLUMNS — These columns are left blank. When a bulk item is required, the letters "AR" (for "As Required") are inserted to signify usage of a quantity that cannot be exactly defined. When explosives are called out in the parts list, the maximum amount is entered in the "quantity required" column. "As required" shall not be used for explosives. (See paragraph 9.6.3.1.4)

9.6.1.2 "CODE IDENTIFICATION NUMBER" COLUMN — No callout required, except when purchased bulk items such as primer, paint, etc. may be listed in the parts list.

9.6.1.3 "PART OR IDENTIFYING NUMBER" COLUMN — Enter the part dash number or bulk item identification number. When a bulk item without an identification number is also required, this column is left blank opposite the bulk item.

9.6.1.4 "NOMENCLATURE OR DESCRIPTION" COLUMN — Leave blank. When bulk items are called out, the name of the item is entered.

9.6.1.5 "MATERIAL" COLUMN — The material name is to be entered and the type designation may be listed when not part of the specification column. List stock size and form only when required by design, e.g.,

    CRES 321
    AL ALLOY 2024-T4 .25 THK PLATE
    CRS - 1020

When bulk items or supplier items are used, this column may be used for trade names of material and/or material identification numbers.

9.6.1.6 "SPECIFICATION" COLUMN — Callout complete material specification information as required. The callout is to include the "grade, condition, class, etc.," as required. (See Section 1)

9.6.1.7 "UNIT WEIGHT" COLUMN — This column contains the actual weight (for calculated weight when the actual weight is not known). This column is to be shown and filled in *only* when directed by the responsible engineer.

9.6.1.8 "ZONE" Column — No callout required.

9.6.1.9 "ITEM NUMBER" COLUMN — No callout required.

QTY REQD	FSCM NO.	PART OR IDENTIFYING NO.	NOMENCLATURE OR DESCRIPTION	MATERIAL	SPECIFICATION	ITEM NO.
AR			SOLDER	SN63 WRAP 2	QQ-S-571	
		-1		COPPER ALLOY NO. 172	QQ-C-530 TEMP A	

PARTS LIST

Parts List for Detail Drawing
Figure 4

9.6.2.1 Column entries are the same as for detail drawings, except for the entering in part number column of all part dash numbers in ascending order, e.g., -1, -2, -3, -4, etc. (See Figures 5 and 6)

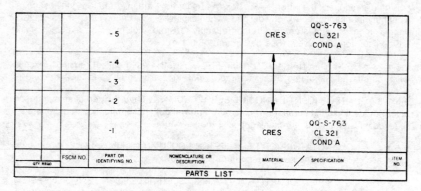

QTY REQD	FSCM NO.	PART OR IDENTIFYING NO.	NOMENCLATURE OR DESCRIPTION	MATERIAL	/	SPECIFICATION	ITEM NO.
		-5		CRES		QQ-S-763 CL 321 COND A	
		-4		↑		↑	
		-3					
		-2		↓		↓	
		-1		CRES		QQ-S-763 CL 321 COND A	

PARTS LIST

Parts List for Tabulated Detail Drawing - Identical Materials
Figure 5

QTY REQD	FSCM NO.	PART OR IDENTIFYING NO.	NOMENCLATURE OR DESCRIPTION	MATERIAL	/	SPECIFICATION	ITEM NO.
		-3		CRES		QQ-S-763 CL 321 COND A	
		-2		STL 4130		MIL-S-6758 COND D4	
		-1		CRES		QQ-S-763 CL 347 COND A	

PARTS LIST

Parts List for Tabulated Detail Drawing - Different Materials
Figure 6

9.6.3 ASSEMBLY DRAWINGS — (See Figures 7, 8, 9, and 10)

9.6.3.1 "QUANTITY REQUIRED" COLUMN — Insert the assembly dash number in right hand column. Each parts list entry reflects the quantity required for one assembly. See Figure 7.

								ITEM NO
BULK ITEMS	AR			COMPOUND, SEALING	MIL-S-8802 CL A2			14
								13
COML PARTS	1	50213	C-7653	VALVE, NEEDLE				12
								11
	2		MS 35334-2	WASHER				10
MIL STD	2		MS 29513-28	PACKING				9
	2		AN 10038	COVER				8
								7
COMPANY STD	1		AS 1012-6	SEAT				6
								5
COMPANY PART NO.	2		1135108	VALVE ASSY				4
								3
	1		-3	FITTING	CRES 316	MIL-S-7720 COND A d		2
DASH NO. OF BASIC DWG NO.	2		-2	BRACKET	CRES 321	MIL-S-6721 COMP Ti		1
	-1 QTY REQD	FSCM NO.	PART OR IDENTIFYING NO	NOMENCLATURE OR DESCRIPTION	MATERIAL	SPECIFICATION	UNIT WT / ZONE	ITEM NO

PARTS LIST

Parts List for Assembly Drawing
Figure 7 (Also see FIG 10)

9–7

9.6.3.1.1 For "Shown" and "Opposite" assembly drawings, insert assembly dash numbers in right and left hand columns. Each column reflects the "quantity required" of each parts list entry to make one assembly. (See Figure 8)

4	4			MS35456-23	SCREW, CAP		1	4 C	11
4	4			MS20365-428A	NUT, SELF LOCKING		1	4 C	10
1	1			MS9014-11	RING		1	4 C	9
									8
1	1			AN12258-4	WASHER		1	4 C	7
									6
2	2			AS1118DZ-118	PACKING		1	4 C	5
									4
1				256781-2	HOUSING		1	3 B	3
	1			256781-1	HOUSING		1	3 B	2
2	2			256780-1	PLUNGER		1	3 A	1
-2	-1	SYM	FSCM NO.	PART OR IDENTIFYING NO.	NOMENCLATURE OR DESCRIPTION	MATERIAL / SPECIFICATION		ZONE	ITEM NO.
QTY REQD					PARTS LIST				

Parts List for Shown and Opposite Assembly
Figure 8

9.6.3.1.2 For multiple assembly drawings, insert the dash number of the lowest numbered assembly in the right hand column and progress to the next numbered assembly in sequence making all entries to the left of the original entry. The entry of "quantity required" of each parts list entry appears in the column headed by its assembly dash number. (See Figure 9)

9.6.3.1.3 When subassemblies are detailed on parent assembly drawings, the dash number of the subassembly is entered in the next open "quantity required" column. The parts list entries required to make one subassembly are reflected in this column. (See Figure 9)

9-8

-4	-3	-2	-1	SYM	FSCM NO.	PART OR IDENTIFYING NO.	NOMENCLATURE OR DESCRIPTION	MATERIAL	SPECIFICATION	ZONE	ITEM NO.
AR				BI	XXXXX	1202	LACQUER	GLYPTAL CLEAR			22
AR				BI	XXXXX		PRIMER COATING	PRIMER PIGMENT SURESEAL			21
	AR	AR	AR	BI	XXXXX	815	ADHESIVE	EPOXY RESIN			20
	AR	AR	AR	BI		MS20995C32	LOCKWIRE			1 / 4C	19
AR				BI			WELD ROD	CRES 347	MIL-R-5031 CLASS 5A		18
	1	1	1	PP	XXXXX	BZ-2RL	SWITCH, FLEXIBLE LEAF	MICROSWITCH CO.		1 / 3C	17
											16
	1	1	1			MS28720-6	FILTER			1 / 3D	15
	1	1	1			AN944-103	FITTING			1 / 3B	14
											13
			1		SCD	408151-1	ACTUATOR			2 / 5C	12
		1			SCD	408150-1	ACTUATOR			2 / 5C	11
	1				SCD	408149-1	ACTUATOR			2 / 5C	10
	1	1	1			408148-1	ELBOW			1 / 3B	9
	1	1	1			408147-1	FLANGE			1 / 3B	8
											7
1						-9	FLANGE	CRES 347	MIL-S-6721 COMP Cb	2 / 6B	6
2						-8	GUSSET			1 / 4C	5
1						-7	SUPPORT	CRES	QQ-S-763 CL 321 COND A	1 / 4C	4
	1	1	1			-6	EXTENSION	CRES	MIL-P-1144 TYPE I, CL3	1 / 3C	3
	1	1	1			-5	ELBOW		COMP 316	1 / 3B	2
	1	1	1			-4	BRACKET ASSY			1 / 2C	1

PARTS LIST

Parts List for Multiple Assembly Drawing
Figure 9

9.6.3.1.4 The quantity for bulk items may be indicated as "AR" (As Required) under the following conditions, provided the item is not normally subject to replacement as a spare part:

    (1) The amount required is less than one of the smallest applicable units of measure.

    (2) The amount required is indeterminate because of variation in manufacturing or tooling.

9.6.3.1.5 Items that are used by weight or volume may be called out as such or "AR" (As Required).

QTY REQD	FSCM NO.	PART OR IDENTIFYING NO.	NOMENCLATURE OR DESCRIPTION	MATERIAL / SPECIFICATION	ITEM NO.
					16
AR			SOLDER	SN63 WRAP QQ-S-571	15
					14
AR			WIRE STRANDED	NYLON JACKETED, WHITE, 22AWG	13
					12
					11
1		8001A-2	TERMINAL BOARD	USECO OR EQUIV	10
					9
1		IN726A	DIODE		8
3		RC07GF100J	RESISTOR		7
					6
8			NUT	#8 CAD PL STL	5
8			SCREW, PAN HD	8-32 x .50 LG CAD PL STL	4
					3
1		-3	PANEL	6061-T6	2
1		-2	PLATE	6061-T6	1

PARTS LIST

NOTE: Minimum Controls are required. Materials and process specifications may be omitted. Hardware is described as commercial quality. Substitutes are left to the discretion of production.

Parts List for Level 1 Drawing Only (Formerly Form 3)
Figure 10

9.6.3.2 "FEDERAL SUPPLY CODE FOR MFGRS" COLUMN–This column lists the design activity's code identification for each item which does not have a code identification common to the parent drawing. It is not a mandatory column except when contractually required. For code identification numbers, see Handbooks H4-1 and H4-2. The supplier's name and address may also be specified in addition to the supplier's code identification number. When no code identification numbers are available, show supplier's name and address in parts list as shown in Figure 11. The code identification number listed in this column applies to the part whose part or identifying number appears in the part number column.

SYM	FSCM NO.	PART OR IDENTIFYING NO.	NOMENCLATURE OR DESCRIPTION	MATERIAL / SPECIFICATION
	21335	7201K	BEARING – BALL	FAFNIR BEARING NEW BRITAIN, CONN.

PARTS LIST

Parts List Entry for Commercial Product
Figure 11

9.6.3.3 "PART OR IDENTIFYING NUMBER" COLUMN

9.6.3.3.1 The identifying number of the item is entered as applicable (see 9.4.5 for types and order of precedence).

9.6.3.3.2 When a standard part number is entered, all identifying letters and/or numbers are called out. (See Figures 7, 8, and 9)

9.6.3.3.3 When entering dash numbered parts of the parent drawing, the parent drawing number is omitted and only the dash number entered. Parts from other than the parent drawing must have the complete part number entered.

9.6.3.3.4 Customer Furnished and Controlled Items. (See 9.8)

9.6.3.3.5 For bulk items, enter applicable manufacturer's number, e.g., formula number, etc.

9–11

9.6.3.4 "NOMENCLATURE OR DESCRIPTION" COLUMN

9.6.3.4.1 DASH NUMBERED PARTS OF PARENT DRAWING — Determine and enter a basic name for each part or subassembly per Section 6, e.g., "BRACKET."

9.6.3.4.2 OTHER COMPANY NUMBERED PARTS — List basic name only of the part as it appears on the drawing where it is detailed. Modifiers are not entered unless necessary for clarity.

9.6.3.4.3 MILITARY STANDARD PARTS (MS, AN) AND OTHER INDUSTRY AND COMPANY STANDARD PARTS — List basic name only of the part as it appears on standard sheet, omitting the modifiers.

9.6.3.4.4 COMMERCIAL PARTS — See 9.7

9.6.3.4.5 CUSTOMER FURNISHED AND CONTROLLED ITEMS — See 9.8

9.6.3.4.6 BULK ITEMS — See 9.4.4

9.6.3.5 "MATERIAL" COLUMN

9.6.3.5.1 DASH NUMBERED PARTS OF PARENT DRAWING — List material name and type designations. (See 9.6.1.5)

9.6.3.5.2 OTHER COMPANY PARTS — No callout required.

9.6.3.5.3 STANDARD PARTS — No callout required. For electronic parts such as resistors, capacitors, etc., this column may be used for helpful information in the form of ratings, values, tolerances, etc.

9.6.3.5.4 COMMERCIAL PARTS — No callout required. Use space if needed for additional description callout, e.g., name of manufacturer, etc.

9.6.3.5.5 BULK ITEM — When bulk items are called out, this column may be used for trade name, if required.

9.6.3.5.6 CUSTOMER FURNISHED AND CONTROLLED ITEMS. (See 9.8)

9.6.3.6 "SPECIFICATION" COLUMN

9.6.3.6.1 DASH NUMBERED PARTS OF PARENT DRAWING — List the material specification information as required per the specification.

9.6.3.6.2 OTHER COMPANY NUMBERED PARTS — No entry required.

9.6.3.6.3 MS, AN, AND OTHER STANDARD PARTS — No entry required.

9.6.3.6.4 COMMERCIAL PARTS. (See 9.7)

9.6.3.6.5 BULK ITEM (BI) — Enter specification, if required.

9.6.3.7 "UNIT WEIGHT" COLUMN — When directed by the responsible engineer, the unit weight is listed in this column. See Figure 7.

9.6.3.8 "ZONE" COLUMN — This column may be used for indicating the zone location of the specific detail or subassembly callout on the field of the drawing. See Figures 7 & 8.

9.6.3.9 "ITEM NUMBER" COLUMN — This column, utilized only on assembly drawings, contains the numbers assigned to the line entries in numerically ascending order, e.g., 1, 2, 3, etc. No number may be omitted or removed.

9.7 COMMERCIAL PRODUCTS

9.7.1 Commercial products are specified on drawings as follows:
   1.) On Assembly Drawings. (See Figure 9, Item number 17)
   2.) On Specification Control Drawings. (See Section 4)
   3.) On Source Control Drawings. (See Section 4)
   4.) On Altered or Selected Item Drawings. (See Section 4)

9.8 CUSTOMER FURNISHED AND CONTROLLED ITEMS

9.8.1 Parts which are furnished by the customer and cannot be controlled by company documentation are subject to the following procedure.

9.8.1.1 Customer Furnished and Controlled Items should appear at the top of the parts list, last in the order of sequence. (See Figure 12)

9.8.2 Parts which are governed by Specification or Source Control Drawings appear in the parts list as shown in Figure 12.

General Note:

⚠ X   CUSTOMER FURNISHED AND CONTROLLED ITEM, AUTONETICS
       DIVISION, NORTH AMERICAN AVIATION, LOS ANGELES, CALIF.

QTY REQD		FSCM NO.	PART OR IDENTIFYING NO.	NOMENCLATURE OR DESCRIPTION	MATERIAL	SPECIFICATION	ITEM NO.
	1	⚠X	27242-507	NOZZLE CONTROL UNIT			
	1		67632-3	SWITCH, THROW TYPE	VENDOR ITEM-SEE SPECIFICATION CONTROL DRAWING		
	–1						

PARTS LIST

Parts List Entry for Customer Furnished and Controlled Items
Figure 12

9-13

9.9 GENERAL REQUIREMENTS FOR PARTS LIST SEPARATE FROM THE DRAWING

9.9.1 A separate parts list is to be made when it is contractually required. In so doing, the following requirements and procedures are to be followed.

General Note:

⚠ CUSTOMER FURNISHED AND CONTROLLED ITEM, AUTONETICS DIVISION, NORTH AMERICAN AVIATION, LOS ANGELES, CALIF.

9.9.2 FORMAT

9.9.2.1 Company Forms are used when preparing a parts list separate from drawings (See Figure 13), except when using automatic data processing techniques.

9.9.2.2 A cover sheet is to be used as Page 1 of all parts lists (see Figure 14). Notes, etc. are to be entered on this sheet.

9.9.3 LIMITATIONS

9.9.3.1 Parts lists separate from the drawing should be limited to end product assemblies.

9.9.3.2 Multi-assemblies are not recommended on a single drawing when a separate parts list is required.

9.9.4 IDENTIFICATION — Parts lists separate from drawings are identified by a document number including the basic drawing number preceded by the prefix "PL." Assemblies requiring separate identification should have different "PL" identifications including "shown" and "opposite" assemblies, e.g., PL1234567-1, PL1234567-2.

9.9.5 CROSS-REFERENCE — When the parts list is separate from the drawing, a reference to the parts list document is made with a note in the parts list area.

Example:

SEE SEPARATE PARTS LIST PL 1234567-1

9.9.6 PARTS LIST ENTRIES

9.9.6.1 ORDER OF PRECEDENCE — Items are entered in separate parts lists starting at the top of the list and proceeding downward. The items shall be listed in the order prescribed by paragraph 9.4.5.

9.9.6.2 ENTRIES — Entries are made by either hand lettering, typing, or automatic data processing techniques. Blank spaces may be left in the parts list between the category groupings to allow for later additions. Item numbers are assigned to each line whether an item is listed on the line or not. Horizontal separating lines between item entries are not mandatory. When separating lines are used, the spacing will be suitable for the height of the entry; no standard increment is required. Entries are made in the blocks and columns of the parts list (Figure 13) and listed as follows:

BLOCK 1, Agency or Contractor Identification (Optional) blocks. The name and address of the Government procurement activity or the name and address of the Contractor (or design activity) whose code identification (FSCM) number appears in Block 2 may be entered.

BLOCK 1(a), Contract Number (Mandatory block). The Government contract number under which the list is initially prepared is entered in this block.

BLOCK 2, Code Identification Number (Mandatory block). Enter the code identification number designation the design activity on each sheet.

BLOCK 3, Parts List Number (Mandatory block). Enter the identifying number of the assembly drawing with the letters PL on each sheet of the list.

BLOCK 4, Revision Letter (Mandatory block). Revisions are recorded on a document Change Notice (DCN). When the list is initially released, enter a dash (-) and release date on each sheet of the list. Each time a sheet of the list is revised, enter the appropriate Revision Letter and revision date on the revised sheet. When the revised list is reissued in its entirety a new revision letter may be entered on all sheets. When an optional cover sheet is used, the release date need be entered on that sheet only. Separate parts lists are changed independently from the applicable drawings. There is no requirement that the drawing and separate PL maintain the same revision letter.

BLOCK 5, List Title (Mandatory block). On sheet 1 enter the basic noun or noun phrase from the title of the drawing to which the list applies.

BLOCK 6, Authentication Signature (Optional). An authentication signature is entered in this block at initial release of manually prepared lists. Signatures are not required on subsequent sheets or when entered on an optional cover sheet. Signatures are not required for machine (ADPA) prepared lists.

BLOCK 7, Revision Authorization Number (Optional). When the list is initially released, enter a dash (-) in this block on each sheet of the list. Each time a sheet of the list is revised, enter the latest revision authorization document number on the revised sheet. When a revised list is reissued in its entirety the latest revision authorization notice may be entered on all sheets. When an optional cover sheet is used, the revision authorization number or description is required on that sheet only.

BLOCK 8, Sheet Number (Mandatory block). Enter the appropriate sheet number on each sheet. The total number of sheets in the list shall be specified.

COLUMN 9, Find Number (Optional column). When find numbers are used to identify items on the field of the drawings, enter the assigned find number for each item. If find numbers are not used on the field of the drawing, this column may be omitted. Reference designations per American National Standard Y32.16 may be used as find numbers.

COLUMN 10, Quantity Required (Mandatory column). Enter the number which denotes the exact quantity or amount of each item required to complete a single article of the item to which the list applies. Additional columns may be added on lists associated with tabulated drawings and, when used, at the head of each column enter a number which identifies the corresponding item. The symbol "AR" (as required) may be used in lieu of the exact quantity of a bulk materials* (such as solder, oil or grease). When the exact quantity of a bulk material is expressed, enter the unit measure in this column or in column 10, if used. Symbols other than "AR" may be used for other conditions, provided that they are explained by an appropriate note or referenced to an explanatory document furnished in the set of engineering data.

COLUMN 11, Unit of Measure (Optional column). When the exact quantity of a bulk material* is expressed in column 10, the unit of measure may be entered in this column in lieu of making the entry in column 10. If this column is not used, it may be omitted.

COLUMN 12, Code Identification Number (Mandatory column). Enter the appropriate code identification number assigned to the design activity whose part or identifying number appears in Column 14. When a Government or Industry standard part or specification number appears in Column 14, no entry need be made in Column 13. When the code identification number for an item is identical to that entered in Block 2, it is not necessary to repeat the number in Column 12.

COLUMN 13, Drawing or Document Number (Optional column). When this column is used, the following may be entered.

    a. The document number applicable to the material from which a listed part delineated on the corresponding drawing is fabricated.

    b. The document number applicable to a listed item for which a type designation type, class condition, etc. has been entered in Column 14.

    c. The drawing number applicable to a listed item in Column 14.

COLUMN 14, Part or Identifying Number (Mandatory column). Enter the part number, or other identifying number, of each item in the list. When the item is controlled by a MILITARY SPECIFICATION, and is individually identified by a designation (such as RC20GF031J, or MR25W010DCUA), enter this designation. When type, grade, class, condition, etc. are required for identification, such information may be entered in this or other appropriate column. When several items from the same tabulated drawing appear consecutively on the list, the complete identifying number shall be recorded for each entry. A line, arrow, or ditto marks may be used between the identical portion of the first and last entries. For items delineated on the drawing to which the list applies, only the dash number associated with each item need be entered.

COLUMN 15, Nomenclature or Description (Mandatory column). The assigned noun or noun phrase shall be entered describing the item whose part or identifying number appears in Column 14.

9.10 PARTS LIST SECURITY CLASSIFICATION — Separate parts lists for classified drawings are not classified unless they also contain classified information.

PARTS LIST

		CONTRACT NO. 1A.	FSCM NO. 2.	3. **PL**		REVISION LTR 4 DATE
	1. Design Activity Identification:					SHEET 8. OF SHEETS
LIST TITLE: 5.	AUTHENTICATION: 6.			REV AUTH NO. 7.		
9. ITEM OR FIND NUMBER	10. QTY RQD	10. QTY RQD	10. QTY RQD	11. UNIT OF MEASURE	12. FSCM	13. DRAWING OR DOCUMENT NUMBER	14. PART OR IDENTIFYING NUMBER	15. NOMENCLATURE OR DESCRIPTION

Separate Parts List
Figure 13

9-17

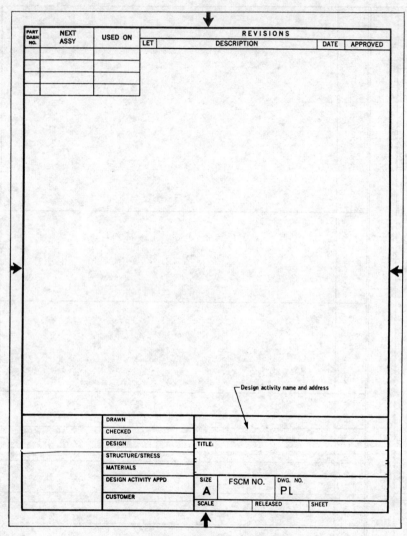

Separate Parts List Cover Sheet
Figure 14

9.11 DRAWING CALLOUT — GENERAL NOTES

9.11.1 A general note is used to establish application requirements that are not otherwise covered by an existing specification, or other acceptable document, or when space will not permit entry of all necessary information.

9.11.2 When requirements are established by a general note, the note is limited to concise statements.

9.11.3 If lengthy or complex data is required to provide adequate information that is not covered by an existing document, the responsible designer initiates action for the preparation of an appropriate company specification.

9.11.4 GENERAL NOTES — The following notes illustrate the type of general note coverage intended to be included on the engineering drawing. They may be varied to suit individual requirements. (See Section 7)

    ABBREVIATIONS NOT LISTED IN MIL-STD-12

        BI   — BULK ITEM
        PP   — PURCHASED PART
        SCD — SPECIFICATION/SOURCE CONTROL DRAWING

⚠ PERMISSIBLE TO USE (alternate part number and code identification number, when required) IN PLACE OF (parts list part number).

⚠ CUSTOMER FURNISHED AND CONTROLLED ITEM (company name and location — list division if applicable).

NOTES

Section 10 — DIMENSIONS AND TOLERANCES
(ALSO SEE SECTION 10M)

10.1 SCOPE — This section establishes and illustrates the methods of specifying dimensions and tolerances on drawings.

10.2 APPLICABLE DOCUMENTS

    ANSI B4.1   Preferred Limits and Fits for Cylindrical Parts

    ANSI Y14.5   Dimensioning and Tolerancing for Engineering Drawings

10.3 DEFINITIONS

10.3.1 ALLOWANCE — The prescribed difference between the maximum material condition of mating parts.

10.3.2 AXIS — The centering of a cylinder, cone, or other surface of revolution as established by the contacting points of the actual surface of the line formed at the intersection of two planes.

10.3.3 CENTER PLANE — The middle or median plane of a feature.

10.3.4 DATUM — Points, lines, planes, cylinders, axes, etc., assumed to be exact for purposes of computation or reference, as established from actual features, and from which the location or form of other features of a part may be established. When datums are implied or specified on the drawing, they exist not in the part itself but in the much more precisely made manufacturing or inspection equipment.

10.3.4.1 DATUM TARGET — A specified point, line, or area on a part which is identified on a drawing as a datum.

10.3.5 DIMENSION — A numerical value expressed in appropriate units of measure and indicated on a drawing along with lines, symbols and notes, to define a geometrical characteristic of an object.

10.3.5.1 BASIC DIMENSION — A numerical value used to describe the theoretically exact size, shape, or location of a feature or datum target. It is the basis from which permissible variations are established by tolerances on other dimensions, in notes or by feature control symbols. Basic dimensions are shown on the drawing as "BASIC", "BSC" or enclosed in a rectangle.

10.3.5.2 LOCATING DIMENSION — One which specifies a position or distance of one feature of an object with respect to another or to a datum.

10.3.5.3 MAXIMUM DIMENSION — A dimension that controls the maximum limit; the minimum limit being controlled by other elements of design.

10.3.5.4 MINIMUM DIMENSION — A dimension that controls the minimum limit; the maximum limit being controlled by other elements of design.

10.3.5.5 REFERENCE DIMENSION

(a) A dimension that has been specified elsewhere on the same drawing or from another drawing or document. These reference callouts state the nominal dimensions, normally without tolerance, or both limits of a limit dimension. The preferred method for indicating reference dimensions on drawings is to enclose the dimensions with parentheses. e.g. (.250). The abbreviation REF may follow or be placed directly under the dimension.

(b) A dimension that is an accumulation of other dimensions. These reference callouts state either the mean or nominal dimension and the tolerance may be stated when necessary. A reference dimension is not used for manufacturing or inspection purposes.

10.3.5.6 SIZE DIMENSION — The specified value of a diameter, length, width, or other geometric characteristic.

10.3.6 DESIGN SIZE — The ideal or preferred size from which variation is allowed by applying tolerances.

10.3.7 FEATURE — A feature is any component portion of a part that can be used as basis for a datum. An individual feature may be:

(a) A plane surface (in which case there is no consideration of feature size).

(b) A single cylindrical or spherical surface, or two plane parallel surfaces (all of which are associated with a size dimension).

Complex features are composed of two or more individual features as defined above.

10.3.8 FIT — A general term used to signify the range of interference or clearance which results from the application of tolerances in the design of mating parts. Fits for cylindrical parts should be established using the preferred limits specified in ANSI B4.1. Fit dimensions shall be specified on the drawing and not called out by the fit designation such as RC1, LC5, etc.

10.3.8.1 CLEARANCE FIT — One having limits so designed that a clearance always results when the mating parts are assembled.

10.3.8.2 INTERFERENCE FIT — One having limits so designed that an interference always results when mating parts are assembled.

10.3.8.3 TRANSITION FIT — One having limits so designed that either a clearance or an interference may result when mating parts are assembled.

10.3.8.4 LINE FIT — One having limits so designed that surface contact or clearance may result when mating parts are assembled.

10.3.9 LIMITS — Maximum and minimum values prescribed for specific dimensions. (See Figure 1.)

10.3.9.1 All limits or tolerances are considered to be absolute. Dimensional limits, regardless of the number of decimal places, are to be used as if they were continued with zeros.

10.3.9.2 For purposes of determining conformance with limits, the measured value is to be compared directly with the specified value and any deviation, however small, outside of the specified limiting values signifies non-conformance with the limits.

Example:

```
1.25 means      1.220--0
                1.280--0

1.250 means     1.240--0
                1.260--0

1.250           1.2500--0
1.252 means     1.2520--0
```

10.3.10 BASIC SIZE — The exact theoretical size from which the limits of size are derived by the application of allowances and tolerances.

10.3.11 MEAN SIZE — The size midway between the limits of size.

10.3.12 NOMINAL SIZE — The designation used for the purpose of general identification. For example, a rod may be referred to as .250 diameter although the actual dimension on the drawing is .249 diameter. In this case, the .250 diameter is the nominal size.

10.3.13 TOLERANCE — The total permissible variation from design size, form, or location.

10.3.13.1 BILATERAL TOLERANCE — A tolerance in which variation is permitted in both directions from the specified dimension. (See Figure 1.)

10.3.13.2 UNILATERAL TOLERANCE — A tolerance in which variation is permitted in only one direction from the specified dimension. (See Figure 1.)

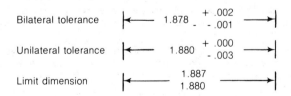

The Application of Tolerances
Figure 1

10.3.14   MAXIMUM MATERIAL CONDITION (MMC or Ⓜ) — The condition where the feature of size contains the maximum amount of material. e.g., minimum hole diameter and maximum shaft diameter.

10.3.15   LEAST MATERIAL CONDITION (LMC or Ⓛ) — The condition where the feature of size contains the least (minimum) amount of material. e.g., maximum hole diameter diameter and minimum shaft diameter. When the symbol Ⓛ is used on drawings, the following note shall also appear:

> SYMBOL Ⓛ INDICATES THAT THE TOLERANCE APPLIES AT LEAST MATERIAL CONDITION AND INCREASES AS THE FEATURE APPROACHES MMC

10.3.16   REGARDLESS OF FEATURE SIZE (RFS or Ⓢ) — The condition where tolerance of position or form must be met irrespective of where the feature lies within its size tolerance.

10.3.17   VIRTUAL CONDITION — The boundary generated by the collective effects of the MMC limit of a feature and any applicable form or positional tolerance.

10.3.18   FULL INDICATOR MOVEMENT (FIM) — The total movement of the indicator when applied to a surface in an appropriate manner. The terms Full Indicator Reading (FIR) and Total Indicator Reading (TIR) were formerly used.

10.4   GENERAL REQUIREMENTS

10.4.1   UNITS OF MEASURE — Unless otherwise specified herein or in other DRM sections, all dimensions and tolerances shall be expressed in inches and decimal parts of an inch or in angular units.

10.4.1.1   RULES APPLICABLE TO INCH UNITS — When specifying decimal dimensions and tolerances the following rules shall apply:

    (a) Zeros shall not be used before the decimal point for values less than one inch.

    (b) All decimals shall have a minimum of two digits following the decimal point.

    (c) A dimension and its tolerance, or both limits of a limit dimension, shall have an equal number of digits following the decimal point.

    (d) Both tolerances shall be specified when using unilateral tolerances. On existing drawings where only one tolerance is shown, the unspecified tolerance shall be interpreted to be zero.

    (e) Unilateral and bilateral tolerances used with dimension lines shall show the tolerances following the dimension. Unilateral or unequal bilateral tolerances shall be shown with the plus tolerance above the minus tolerance. (See Figure 1.)

    (f) When unilateral or unequal bilateral tolerances are specified in general notes, the tolerances may be shown on the same line with the plus tolerance preceding the minus tolerance, e.g. 1.50 + .03 - .00, 1.500 + .003 - .001.

(g) When unilateral tolerances are used, it is preferred that the dimensions specify the maximum position or MMC size with the tolerances applied to the minimum position or LMC size tolerance.

Example:

Position	2.500	+ .000 / − .005
Shaft	1.000	+ .000 / − .002
Hole	.998	+ .002 / − .000

(h) Normally when the tolerance of a dimension is equal to the standard title block tolerance, limit dimensions or dimensions with unilateral or unequal bilateral tolerances are not used.

(i) Limit dimensions shall be shown on drawings as follows:

(1) Position dimensions used with dimension lines shall show the maximum value above the minimum value. e.g. 1.887 / 1.880

(2) When limit dimensions are shown in general notes, the limit dimension that otherwise would be above precedes the other with a dash midway between the digit height, e.g. .990 − 1.000 DIA THRU.

10.4.1.2 RULES APPLICABLE TO ANGULAR UNITS — When specifying angular dimensions and tolerances the following rules shall apply:

(a) Angular dimensions and tolerances shall be expressed in degrees (°) and when necessary, in minutes (') and seconds ('') or in decimal parts of a degree.

(b) An angular dimension and its tolerance or both limits of a limit dimension shall be held to the same units of measure,

e.g. 30° ±5°   (cont.)
      30° 0' 30'' ±0° 0' 15''   30.8° ±.5°
      30° 30' 0''   30° 30.0' ±0° .5'
      30° 30' 45''

An exception to this rule is when the standard title block tolerance applies and is specified in degrees only and the field of the drawing uses degrees and parts of a degree.

(c) The requirements of 10.4.1.1 (d) through (i) also apply to angular dimensions.

10.4.1.3 Tolerances used in geometric characteristic symbols shall be held to a minimum of three decimal places regardless of the number of decimal places in the position or size dimension.

10.4.2 STANDARD TOLERANCES — Dimensions shown without tolerance are controlled by the standard tolerances in the title block, except stock materials, dimensions on welding symbols, undimensioned angles between lines drawn at 90°, dimensions labeled REF, MAX, MIN, BASIC, and similar dimensions that are otherwise controlled. The standard tolerances are .XX = $\pm$ .03 and .XXX = $\pm$ .010.

10.4.2.1 The format tolerances shall not be altered except where a larger tolerance is required for the majority of the dimensions. The revised tolerances cannot be less than the format tolerances and must always be progressively smaller as the number of decimal places increases, e.g., .XXX cannot have a larger tolerance than .XX. When a third tolerance is required, a four place decimal may be added. This is accomplished by adding the following delta note with the applicable correction or addition:

⚠ TOLERANCE ON DECIMALS

.XX $\pm$.-- (Add applicable two place tolerance)

.XXX $\pm$.--- (Add applicable three place tolerance)

.XXXX $\pm$.---- (Add applicable four place tolerance)

The format tolerance(s) will be lined out and the delta note number added in the block, for the revised format tolerance(s).

10.4.3 FUNDAMENTAL RULES OF DIMENSIONING — Dimensioning shall conform to the following rules:

(a) Dimension, extension and leader lines shall not cross each other unless absolutely necessary. When it is unavoidable, a dimension line is never broken except for insertion of the dimension. An extension or leader line shall not run through a dimension nor shall they be broken except where they pass through or adjacent to arrowheads.

(b) Dimensions are shown in the view that most clearly represents the form of the feature.

(c) Sufficient dimensions shall be shown to clearly define size, shape and position of each feature.

(d) A feature shall not be located by more than one toleranced dimension in any one direction.

(e) A dimension shall be enclosed in parentheses or marked "REF" when it is (1) repeated on the same drawing, (2) specified on a subordinate document, (3) an accumulation of other dimensions or (4) shown for information purposes.

(f) Unless clarity is improved, dimensions are shown outside the outline of the part.

(g) Dimensions are selected and arranged to minimize the accumulation of tolerances between related features.

(h) Each dimension shall be expressed clearly so that it can be interpreted in only one way.

(i) Only the end product dimensions and data are shown on drawings unless essential to the definition of engineering requirements. When non-mandatory in-process manufacturing information is shown on the drawing, it shall be marked with a note similar to "NON-MANDATORY, MANUFACTURING DATA".

(j) Chain dimensions should be avoided unless basic dimensions are used.

(k) Center lines, object lines or extension lines should not be used as dimension lines.

(l) Dimensioning to hidden lines shall be avoided.

(m) Maximum and minimum limits must be such that parts will assemble and function under all dimensional conditions that are within limits.

(n) The word "TYPICAL" or the abbreviation "TYP" is not used. Indicate the number of places the dimension applies.

10.5 METHOD OF DIMENSIONING

10.5.1 Parts and features should be dimensioned by the method that most clearly shows the design requirements. Bilateral, unilateral, and limit dimensions may all be used on the same drawing to achieve this requirement.

10.5.2 Figure 2 shows how dimensions are to be spaced.

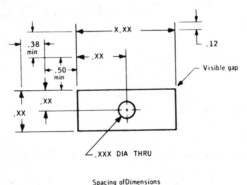

Spacing of Dimensions
Figure 2

10.5.3 Staggered dimensions shall be used to prevent interference with other dimensions. (See Figure 28.)

10.5.4 When shown in profile, diametral dimensions may be specified as DIA or the symbol ⌀ to avoid showing end views of round objects.

10-7

10.5.5 SYMMETRICAL ABOUT ℄ — When the term "Symmetrical About ℄" is used quantities of like features are specified for the entire view.

10.5.6 STOCK SIZE — When the stock size is specified in the parts list and used as furnished, it shall be indicated on the field of the drawing by the word "STOCK". When two stock dimensions are used and the orientation is not obvious from the picture, the dimensions shall be included in the callout (See Figure 3). The foregoing does not apply for shapes such as I-beams, channels, etc., where it is obvious that the shape is used as furnished.

10.5.7 DIMENSIONING OF TUBES — Tubes are dimensioned as illustrated in Section 4. (Tube Drawing)

10.5.8 ROUNDED END PARTS — Parts with rounded ends are dimensioned by giving the overall length and width, and indicating the radius as "R". (See Figure 4.)

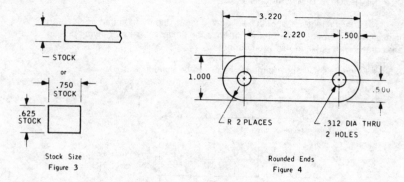

Stock Size
Figure 3

Rounded Ends
Figure 4

10.5.9 COORDINATE DIMENSIONING — Coordinate dimensioning is as shown in Figure 5.

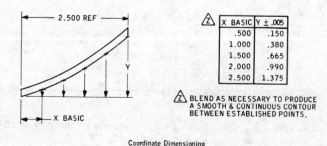

Coordinate Dimensioning
Figure 5

10-8

10.5.10 DIMENSIONING SPHERICAL RADII — Spherical radii are dimensioned as shown in Figure 6.

10.5.11 COUNTERBORES — Counterbores are dimensioned as shown in Figure 7.

Spherical Radius
Figure 6

Counterbore
Figure 7

10.5.12 SPOT FACING — Spot facing is the operation of cleaning up the surface around a feature. The diameter and, when required, the maximum depth or minimum remaining thickness shall be specified. Spot faces are called out by note only, except the depth or remaining thickness may be dimensioned on the drawing. (See Figure 8.)

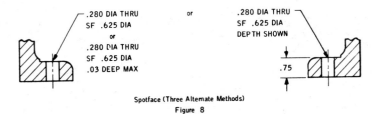

Spotface (Three Alternate Methods)
Figure 8

10.5.13 COUNTERSINKS — Countersinks on a flat surface may be called out by dimensions or notes. (See Figure 9.)

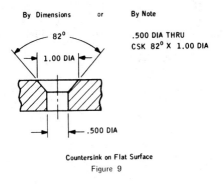

Countersink on Flat Surface
Figure 9

10–9

10.5.13.1 Countersinks on cylindrical surfaces are directly dimensioned as shown in Figure 10.

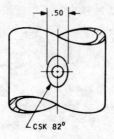

Countersink on Curved Surface
Figure 10

10.5.14 CHAMFERS — Chamfer of fourty-five degree angles are called out by one of the methods shown in Figure 11.

10.5.14.1 Chamfers of other angles are directly dimensioned by one of the methods shown in Figure 12.

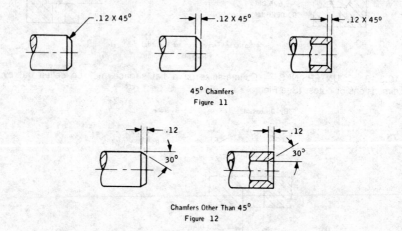

45° Chamfers
Figure 11

Chamfers Other Than 45°
Figure 12

10.5.15 HOLES — Holes are dimensioned in the view where they appear as circles whenever practical.

10.5.15.1 DEPTH OF HOLES — Dimension the depth of holes as shown in Figure 13. Thru holes are either defined by picture or the term "THRU."

10.5.15.2 LOCATING HOLES — Figures 14 through 18 illustrate the positioning of round holes by giving distances, or distances and directions, to the hole centers. These methods can also be used to locate round pins and other features of symmetrical contour. Allowable variations for any of the positioning dimensions may be specified by giving a tolerance with each distance or angle, or by "positional tolerancing dimensioning" explained in 10.10.

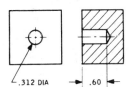 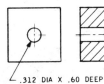

Hole Depth Dimensioning
Figure 13

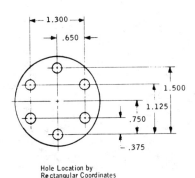

Hole Location by
Rectangular Coordinates
Figure 14

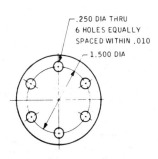

Hole Location by Diameter or
Radius and Equally Spaced
Figure 15

10-11

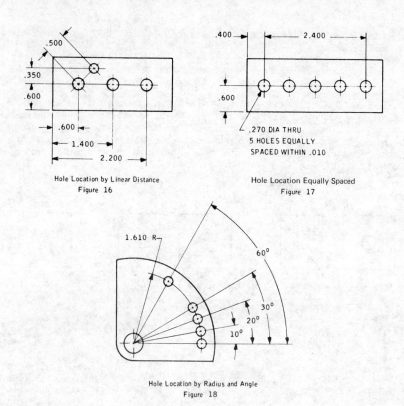

Hole Location by Linear Distance
Figure 16

Hole Location Equally Spaced
Figure 17

Hole Location by Radius and Angle
Figure 18

10.5.16 KEYWAYS — Keyways may be dimensioned as shown in Figures 19 and 20.

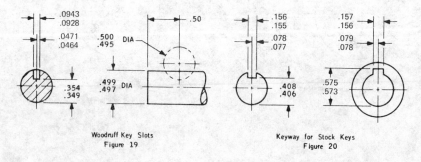

Woodruff Key Slots
Figure 19

Keyway for Stock Keys
Figure 20

10.5.17 TAPERS — Basic diameter method — A diameter on a conical taper labeled "BASIC" is an exact dimension which must be located within specified limits. The BASIC diameter method of dimensioning tapers illustrated in Figure 21 controls the size of the tapered section, as well as its axial position in relation to some other surface. As is shown in the interpretation of Figure 21, the tolerance on the location of the BASIC diameter not only controls the axial position of the tapered section, but also sets up a tolerance zone within which the form of the taper must fall. The word "BASIC" after the taper means that the taper may vary from the value given, but must fall within the zone created by the locational tolerances. If the taper is very slight, it is preferred that the diameter be assigned a tolerance and the dimension locating the diameter be changed to BASIC.

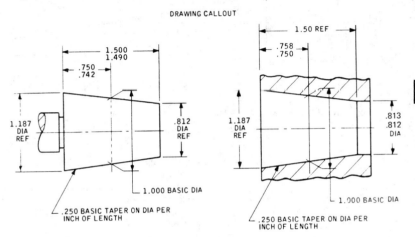

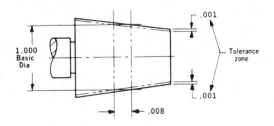

Dimensioning Tapers - Basic Taper & Basic Diameter
Figure 21

10-13

10.5.17.1 If the accuracy of the taper is more important, dimension as shown in Figure 22.

10.5.17.2 FLAT TAPERS — Flat tapers are dimensioned as shown in Figure 23.

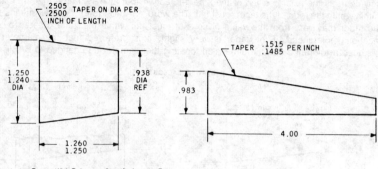

Dimensioning Tapers With Tolerance Specified on the Taper
Figure 22

Flat Taper Dimensioning
Figure 23

10.5.18 KNURLS — Knurls are used to provide a rough surface for gripping, decoration, or for a press fit between mating parts. Knurls for gripping or decorative purposes are called out by type, pitch, and axial length. (See Figure 24.) Knurls for press fits are called out by type, pitch, axial length, diameter before knurling, and should include the minimum diameter after knurling. (See Figure 25.) Types of knurls are diamond, straight and diagonal. Standard pitches are 64, 96, 128 and 164.

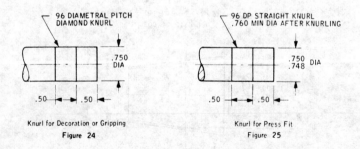

Knurl for Decoration or Gripping
Figure 24

Knurl for Press Fit
Figure 25

10.5.19 TRUE DIMENSIONS — When dimensioning a surface in a plane not perpendicular to the plane of projection, the term "TRUE" may be used as a matter of convenience to avoid showing an auxiliary view. (See Figure 26.)

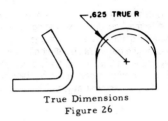

True Dimensions
Figure 26

10.6 GENERAL RULES AND INTERPRETATIONS OF DIMENSIONS AND TOLERANCES

10.6.1 LIMITS OF SIZE AND FORM, RULE 1 — The toleranced dimensions for the size of a feature control the form as well as the size. The basic interpretation of this implied control of form is as follows:

(a) No element of the actual feature (including a datum feature) shall extend beyond the envelope of perfect form at MMC. This envelope is the true form implied by the drawing.

(b) The measured dimensions of the feature at any cross-section shall not be less than the minimum limit of size of an external feature nor greater than the maximum limit of size of an internal feature.

Figure 27 illustrates the extreme variations of form that are permitted by this interpretation.

Note: The above stated interpretation prescribing an envelope of perfect form at MMC applies only to individual features and not to interrelationship of features. such interrelationship should be controlled by form or positional tolerances specified on the drawing.

10.6.1.1 EXCEPTIONS TO RULE 1

(a) The interpretation of Rule 1 does not apply to stock material such as bars, sheets, tubing, etc. These stock materials are controlled by the material specification called out on the drawing or by industry standards for the material when called out commercial grade.

(b) Where it is desired to permit a tolerance of form to exceed the envelope of perfect form at MMC, this may be done by adding to the drawing the suitable form tolerance and a local note or delta note which states "PERFECT FORM NOT REQD AT MMC." The note is applied to the pertinent size dimension(s). When this is done, the form tolerance specified is allowed even through the feature is at its MMC. When this procedure is used, the MMC size of the mating part must be revised (male feature decreased), (female feature increased) by an amount at least equal to the form tolerance.

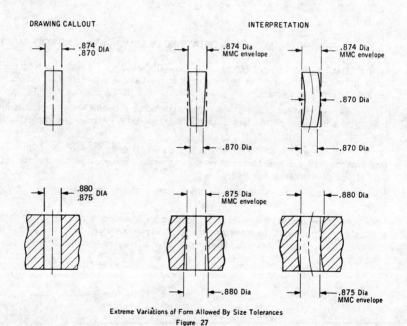

Extreme Variations of Form Allowed By Size Tolerances
Figure 27

10-16

10.6.2 RULES APPLICABLE TO DATUMS — Datums may be implied or specified on a drawing, but in either case the following rules apply:

10.6.2.1 ACCURACY OF DATUM FEATURES — Datum features are only as accurate as stated in Rule 1, therefore, when their deviation of form or position must be controlled more closely than can be assured by Rule 1, the drawing shall include a note or geometric tolerance symbol stating the required accuracy.

10.6.2.2 DIMENSIONS FROM DATUMS — When features are dimensioned from datums, they are located with respect to the datums, not with respect to one another. (See Figure 28.) When a plane surface is used to establish a datum plane, measurements to other features are taken from the datum plane which contacts the high points of the datum surface and not from the actual datum surface. Thus, the specified form or positional tolerance for a feature does not include any variation which may exist in the datum surface.

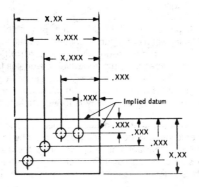

Implied Datums
Figure 28

10.6.2.3 SELECTION OF DATUMS — Features that are selected as datums must be easily recognizable or identified with a datum symbol. A datum on an actual part must be accessible during manufacture so that measurements from it can be readily made. Also, corresponding features on mating parts must be used as datums to insure assembly and facilitate tool and fixture design.

10.6.3 RULES APPLICABLE TO THREADS, GEARS, AND SPLINES — The following rules are applicable to all symbols and notes specifying tolerances of form or position involving screw threads, gears, or splines as toleranced features or datums.

10.6.3.1 SCREW THREADS — Where tolerances of form or position are expressed by symbols and notes, each such tolerance applicable to a screw thread and each datum reference to a screw thread shall be understood to apply to the pitch diameter. If design requirements necessitate an exception to this general rule, a qualifying notation shall supplement the symbol or note, e.g., MAJOR DIA. In the case of symbol applciation, the qualifying notation shall be shown beneath the feature control symbol where applicable to the feature, and beneath the datum identifying symbol where applicable to the datum.

10.6.3.2 GEARS AND SPLINES — For gears and splines, a qualifying notation must be added to the symbol or note, e.g., MAJOR DIA, MINOR DIA, PITCH DIA, or PD.

10.7 GEOMETRIC TOLERANCES

10.7.1 GENERAL — A geometric tolerance is the permissible variation from the specified form of an individual feature of a part. Geometric tolerances control characteristics such as straightness, flatness, roundness, concentricity, perpendicularity, parallelism, angularity, etc. Shapes or forms into which material is fabricated are defined by the use of geometric terms such as a plane, a cylinder, a cone, a square, or a hexagon. The geometric definition assumes a perfect form, but because a perfect form cannot be produced, variation must be controlled if quality is to be maintained.

10.7.2 SYMBOLS — Geometric characteristics are specified on the drawing by the use of symbols. When geometric symbols do not adequately describe the desired condition, a note may be used, either separately or supplementing the symbol, to define the requirement. Figure 29 shows the approved symbol for each characteristic. Figure 30 shows how the symbols are to be drawn with their tolerance, modifiers, and datum references.

10.7.3 DATUM IDENTIFYING SYMBOLS — Each datum (except implied datums) on a drawing is assigned a different identifying reference letter for which letters of the alphabet, except "I", "O", and "Q" are used. Datum assignment begins with the letter "A" and proceeds thorough the alphabet as required. When datum features requiring identification on a drawing are so numerous as to exhaust the single letter series, the double letter series, i.e., "AA" through "AZ" may be used. The datum identifying symbol is formed by a rectangle containing the datum reference letter preceded and followed by a dash. (See Figure 30A.)

10.7.4 SYMBOL FOR BASIC — The symbolic means of labeling a basic is by enclosing each such dimension in a frame. (See Figure 30B.)

10.7.5 SYMBOL PLACEMENT — The feature control symbol shall be associated with the feature(s) being toleranced by one of the methods shown in figure 31.

GEOMETRIC CHARACTERISTIC SYMBOLS

TYPE OF TOLERANCES		SYMBOL	CHARACTERISTIC	FORMER SYMBOL/S AND NOTES
INDIVIDUAL FEATURES	FORM TOLERANCES	▱	FLATNESS	∼
		—	STAIGHTNESS	⌒ or —
		○	ROUNDNESS (CIRCLARITY)	
		⌀	CYLINDRICITY	
INDIVIDUAL OR RELATED FEATURES WHERE DATUMS ARE SPECIFIED		⌒	PROFILE OF ANY LINE	SEE NOTE 1.
		⌒	PROFILE OF ANY SURFACE	SEE NOTE 1.
		//	PARALLELISM	∥
		⊥	PERPENDICULARITY (SQUARENESS)	
		∠	ANGULARITY	
RELATED FEATURES	LOCATION TOLERANCES	◎	CONCENTRICITY	⦿ & SEE NOTE 2.
		═	SYMMETRY	SEE NOTE 3.
		⊕	POSITIONAL	
	RUNOUT TOLERANCES	↗	CIRCULAR	SEE NOTE 4.
		↗	TOTAL	SEE NOTE 5.

NOTES:

1.) ALTHOUGTH INCLUDED UNDER "FORM TOLERANCE", PROFILE TOLERANCE CONTROL SIZE AS WELL AS FORM.
2.) WHERE CONCENTRICITY RFS APPLIES, IT IS PREFERRED THAT THE RUNOUT SYMBOL BE USED. WHERE CONCENTRICITY AT MMC APPLIES, IT IS PREFERRED THAT THE POSITIONAL SYMBOL BE USED. FOR AN APPLICATION OF CONCENTRICITY SEE FIGURE 62. THE INNER CIRCLE OF THE SYMBOL IS NOT FILLED SOLID.
3.) WHERE SYMMETRY APPLIES, IT IS PREFERRED THAT THE POSITIONAL SYMBOL BE USED.
4.) THE SYMBOL ↗ WITHOUT THE QUALIFIER "CIRCULAR" FORMERLY DENOTED TOTAL RUNOUT.
5.) "TOTAL" MUST BE SPECIFIED UNDER THE FEATURE CONTROL SYMBOL.

Figure 29

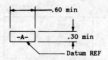
A. Datum Identifying Symbol

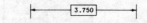
B. Symbol for Basic or True Position Dimension

symbol	Modifier
Ⓜ	Maximum Material Condition (MMC)
Ⓢ	Regardless of Feature Size (RFS)
*Ⓛ	Least Material Condition (LMC)
⌀	Diameter (DIA)
Ⓟ	Projected Tolerance Zone (TOL ZONE PROJ)
(1.250)	Reference (REF)
3.875	Basic (BSC)
*	See Paragraph 10.3.15

C. Symbols for MMC, RFS and LMC & Others

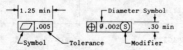
D. Feature Control Symbols

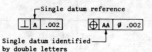
E. Feature Control Symbols Incorporating Datum References

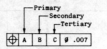
(a) Usual Sequence

(b) International Sequence

F. Datum Reference Showing Order Of Precedence

G. Multiple Datum Features Establishing Single Datum Reference

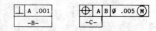
H. Combined Feature Control Symbol

J. Composite Feature Control Symbol

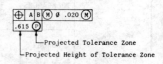
K. Feature Control Symbol With A Projected Tolerance Zone

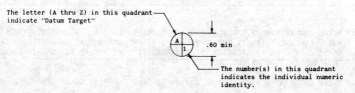
L. Datum Target Symbol

Geometric Tolerance Symbols
Figure 30

10.7.6 MULTIPLE DATUMS — Where reference is made to more than one datum used to control the positional or form tolerance of a feature, the first datum reference letter in the sequence shall be the primary datum, the second letter the secondary datum, etc. The datum reference letters shall be placed in order of importance with respect to establishing the location, form, or position of other features. Thus, the datum reference letters will not necessarily be in alphabetical order. An example of the use of primary, secondary, and tertiary datums would be when it is desired to place a part in a three axis coordinate reference frame so that the dimensions may be given in the three planes. Since an actual part will not have perfect squareness between any of the three datum feature surfaces used for dimensioning, it is necessary for the drawing to show how the datums are applied to the part. This is accomplished as follows: The primary datum is a plane established by any three or more points not in a line that contact the surface identified as the primary datum; the secondary datum plane is established as perpendicular to the primary datum and contacting the part at two or more points on the surface identified as the secondary datum; and the tertiary datum is established as perpendicular to both the primary and secondary datums and contacting the part at one or more points on the surface identified as the tertiary datum. (See Figure 30F.) In some designs the datums obviously are of equal importance. For example, a runout tolerance related to two equally important bearing surfaces. (See Figure 30G.)

10.7.7 COMBINED SYMBOLS — When a feature serves as a datum and is also controlled by a positional or form tolerance, the geometric characteristic symbol and the datum identifying symbol are combined as shown in Figure 30H. In such cases, the length of the box for the datum symbol may be the same as that of the geometric characteristic symbol or .60 inch.

10.7.8 COMPOSITE FEATURE CONTROL SYMBOL — Composite feature control symbol is used where more than one tolerance of a given geometric characteristic applies to the same feature. It is also used to locate a pattern of holes, then refine the locations of the holes within the pattern. The upper portion of the composite feature control symbol specifies the datums related to the pattern (usually three for non-cylindrical datums) and the lower portion indicates the datum which controls the perpendicularity. The tolerance for the interrelated holes in the pattern must be less than the tolerance for the pattern location. A single entry of the geometric characteristic symbol is followed by each tolerance requirement, one above the other, separated by a horizontal line. (See Figure 30J.)

10.7.9 COMBINED FEATURE CONTROL AND PROJECTED TOLERANCE ZONE SYMBOL — Where a positional or perpendicularity tolerance is specified as a projected tolerance zone, a frame containing the projected height followed by the appropriate symbol is placed beneath the feature control symbol. (See Figure 30K.)

10.7.10 RULES APPLICABLE TO USE OF GEOMETRIC FORM TOLERANCE SYMBOLS.

(a) Rule 1 applies even when geometric form tolerances are specified.

(b) Geometric form tolerance control is applied to features only when it is necessary to control form more precisely than the limits established by Rule 1 or its exceptions.

(c) Profile tolerances, in themselves, establish an envelope of perfect form at MMC and are therefore not subject to usage limitations in (b) above.

(d) Runout tolerances are considered form tolerances but they also control position. These tolerance symbols may be used when it is necessary to control the interrelationship between features even when size tolerances adequately control the form of each individual feature.

(e) Form tolerances in feature control symbols are not modified by such terms as DIA, TIR, FIR, or R.

(f) Form tolerances which always require a datum reference are parallelism, perpendicularity, angularity, and runout.

(g) Form tolerances which never utilize a datum reference are flatness, straightness, roundness and cylindricity.

(h) Profile tolerances may or may not utilize datum references depending on the design requirements.

10.7.11 FORMER SYMBOLS — The symbol ⌒ was formerly used for expressing a tolerance on flatness. The symbols ⌒ and — were formerly used for expressing tolerances on straightness and flatness. Whenever these symbols appear on existing drawings, they may be interpreted as specified herein for the specific case shown. The symbol ⌖ A.XXX TIR was formerly used for expressing a tolerance on concentricity. Whenever this symbol appears on existing drawings it shall be interpreted the same as runout on a diameter.

10.7.12 DIAMETER SYMBOL — The symbol used to designate a diameter is as shown in Figure 30C. It precedes the specified tolerance in a feature control symbol. The symbol may be used elsewhere on a drawing in place of the word DIAMETER or the abbreviation DIA.

10.8 DATUM TARGETS

10.8.1 QUALIFICATION OF DATUM SURFACES — The datum features of certain parts frequently require further qualification before they can be related to the three-plane framework. Examples are surfaces produced by casting, forging, and molding; surfaces adjacent to welds; and thin sheet metal. All of these parts are subject to bowing, warping, and distortion, therefore, it is not recommended that an entire surface be designated as a datum. For example, a cast surface may actually rock or "Teeter-Totter" when placed in contact with a datum plane such as a machine table or surface plate, thereby making accurate and repeatable measurements very difficult. To overcome this problem, the datum target method (formerly referred to as tooling points) should be used.

10.8.2 DATUM TARGET METHOD — The datum target method is a useful technique for relating the above mentioned parts to the three-plane framework. Normally, three datum targets are required to restablish the first or primary plane, two datum targets to prevent rotation of the secondary plane and a single datum target to position the third plane in relation to the first and second plane. Additional datum targets may be indicated when necessary. It is at these points or areas that contact is made with the processing and inspection equipment. Drawings utilizing datum targets shall include the following general note: " ⊕ Symbol Designates Datum Targets". (See Figure 30L.)

10.8.3 LOCATING DATUM TARGETS — Datum targets are separated as far apart as possible, dimensioned relative to each other, and located on surfaces that will not be machined. On castings and forgings they should be located on one side of the parting line, not too close to a fillet or corner, and not on the parting line or on a gate. If a separate machining drawing is made of the casting or forging, or if a separate machining view is made on the casting or forging drawing, the datum target points shall be shown in the same location as on the casting or forging but shall not be dimensioned. When a separate machining drawing is made, modify the general note to read: " ⊕ Symbol Designates Datum Targets. See Drawing XXXXXX". For an example of locating and dimensioning datum targets see Section 12.

10.9 DRAWING APPLICATION AND INTERPRETATION — The following illustrations show the correct method of applying geometric tolerances to drawings and the correct interpretation of the tolerance zone provided by each.

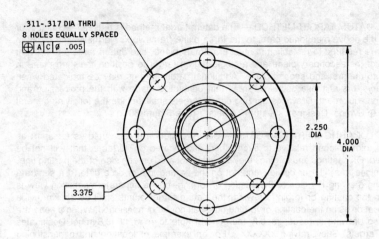

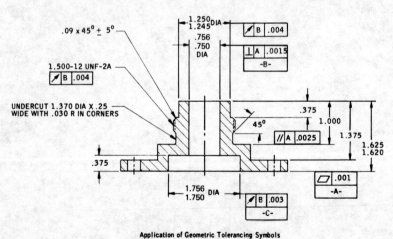

Application of Geometric Tolerancing Symbols
Figure 31

10.9.1 FLATNESS — Flatness is the condition of a surface having all elements in one plane.

10.9.1.1 FLATNESS TOLERANCE — A flatness tolerance specifies a tolerance zone confined by two parallel planes within which the entire surface must lie. (See Figure 32.) The expression "MUST BE CONCAVE" or "MUST BE CONVEX" may be added if necessary.

DRAWING CALLOUT

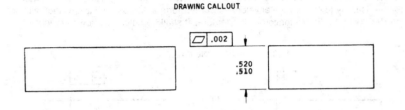

INTERPRETATION

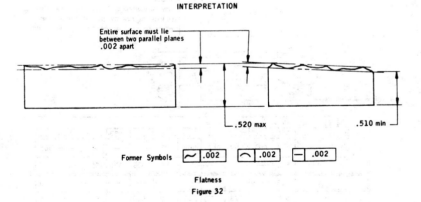

Flatness
Figure 32

10.9.2 STRAIGHTNESS — Straightness is a condition where an element of a surface is a straight line.

10.9.2.1 STRAIGHTNESS TOLERANCE — A straightness tolerance specifies a tolerance zone of uniform width along a straight line, within which all points of the considered line must lie. Straightness tolerance symbols are attached to leader lines only and are applied in a view where the surface elements to be controlled are represented as a straight line. (See Figure 33.) This symbol is not used to specify straightness of axis. Existing drawings which specify straightness of an axis are to be interpreted to mean that the axis must lie within a diameter or width zone equal to the specified tolerance.

10.9.2.2 STRAIGHTNESS TOLERANCE EXCEEDING MMC ENVELOPE — When a feature is likely to exceed the envelope of perfect form at MMC and this deviation will not interfere with the fit or function of the feature, it shall be noted as shown in Figure 33.

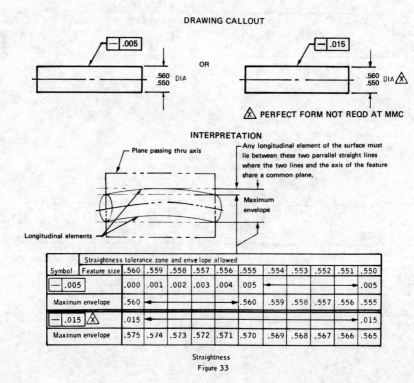

Straightness
Figure 33

10.9.3 ROUNDNESS — Roundness is a condition of a surface of revolution such as a cylinder, cone, or sphere, where all points of the surface intersected by any plane, (1) perpendicular to a common axis (cylinder, cone), or (2) passing through a common center (sphere), are equidistant from the axis.

10.9.3.1 ROUNDNESS TOLERANCE — A roundness tolerance specifies a tolerance zone bounded by two concentric circles in that plane within which the periphery must lie. Figures 34, 35, and 36 show how roundness is specified and interpreted for rigid parts. Note that specified tolerances are always on a radial basis. Existed drawings showing roundness tolerances on a diameter basis are interpreted to mean a radial tolerance of one/half the tolerance shown.

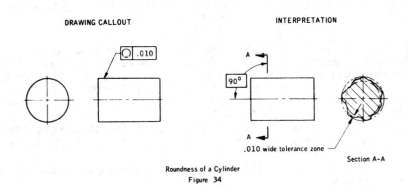

Roundness of a Cylinder
Figure 34

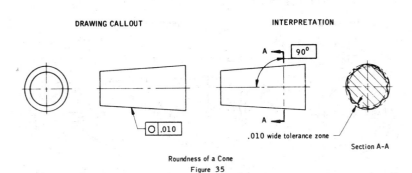

Roundness of a Cone
Figure 35

10-27

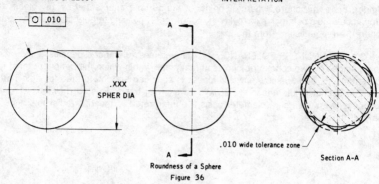

Roundness of a Sphere
Figure 36

10.9.3.2 ROUNDNESS, FREE-STATE VARIATION — Free-state variation can exist in two ways: (1) distortion due to the weight or flexibility of the part, or (2) distortion due to internal stresses set up in fabrication. Parts that are subject to such distortion are referred to as "nonrigid" parts and such distortion is referred to as "free-state variation." The above distortions are accounted for on drawings only when the feature(s) may fall outside the drawing limits and are controlled as follows:

(a) By adding a note stating "DIMENSIONS APPLY IN RESTRAINED CONDITION" (specify the amount of restraining force allowable to bring feature(s) within drawing limits when necessary).

(b) State the allowable free-state variation and show average diameter as shown in Figure 37.

Note: The term "average diameter" or "AVG DIA" is the mean of several diameters (not less than four) used to determine conformance of the diameter tolerance only. Average diameter may also be measured with a periphery tape.

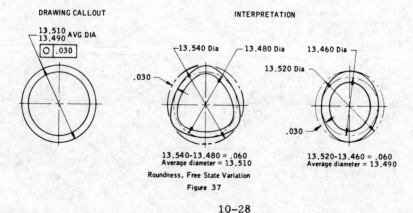

Roundness, Free State Variation
Figure 37

10.9.4 CYLINDRICITY — Cylindricity is a condition of a surface of revolution in which all elements form a cylinder.

10.9.4.1 CYLINDRICITY TOLERANCE — A cylindricity tolerance specifies a tolerance zone confined to the annular space between two concentric cylinders within which the surface must lie. (See Figure 38.)

Note that the cylindricity tolerance controls roundness and straightness, as well as parallelism of the elements and that the specified tolerance is always on a radial basis.

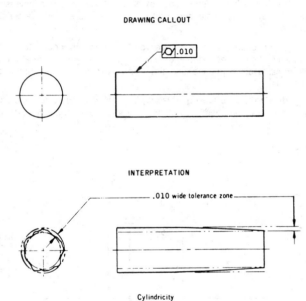

Cylindricity
Figure 38

10.9.5 PROFILE TOLERANCING — Profile tolerancing is a method used where a uniform amount of variation may be permitted along a line or surface. The line or surface may consist of straight lines or curved lines, the latter being either arcs or irregular curves.

10.9.5.1 PROFILE TOLERANCE — A profile tolerance (either bilateral or unilateral) specifies a tolerance zone, always measured normal to the profile at all points of the profile, within which the specified line or surface must lie.

10.9.5.2 APPLICATION OF PROFILE TOLERANCES — Figures 39 and 40 illustrate methods of dimensioning profiles and comply with the following requirements:

(a) A view or section is drawn which shows the desired basic profile.

(b) The profile is dimensioned by basic dimensions. This dimensioning may be in the form of located radii and angles, or it may consist of coordinate dimensions to points on the profile.

(c) An exaggerated tolerance zone is shown by one or two phantom lines drawn parallel to the profile. The tolerance zone may be shown unilaterally to either side of the profile or bilaterally with a locating dimension (normally one half the profile tolerance) from the profile.

(d) Line and surface controls may be applied to the same feature when the line elements in one direction need to be controlled more closely than the surface as a whole.

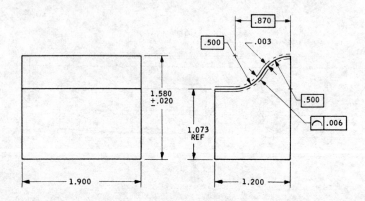

Profile of a Line
Figure 39

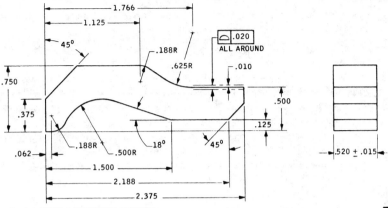

X. ALL PROFILE DIMENSIONS ARE BASIC.

A. Bilateral Tolerance Zone

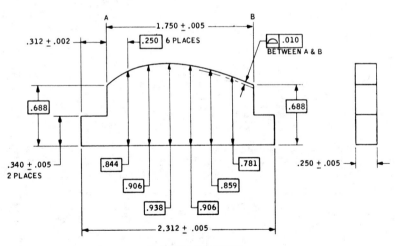

B. Unilateral Tolerance Zone
Profile of a Surface
Figure 40

10-31

10.9.6 PARALLELISM — Parallelism is the condition of a surface, axis, or line which is equidistant at all points from a datum plane or axis.

10.9.6.1 PARALLELISM TOLERANCE — A parallelism tolerance specifies one of the following:

    (a) A tolerance zone confined by two planes parallel to a datum plane within which the considered feature (surface or axis) must lie. (See Figures 41 and 42.)

    (b) A cylindrical tolerance zone parallel to a datum feature axis and within which the axis of a feature must lie. (See Figure 43.)

    (c) A tolerance zone confined by two straight lines parallel to a datum plane or datum axis within which an element of the surface must lie (See Figure 44.)

Note that the parallelism tolerance when applied to a plane surface controls flatness if a flatness tolerance is not specified.

DRAWING CALLOUT

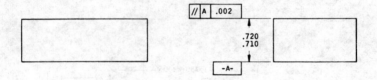

INTERPRETATION

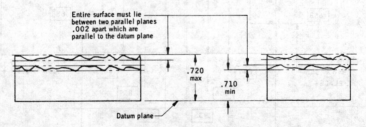

Parallelism of a Feature Surface to a Datum Plane
Figure 41

10-32

DRAWING CALLOUT   INTERPRETATION

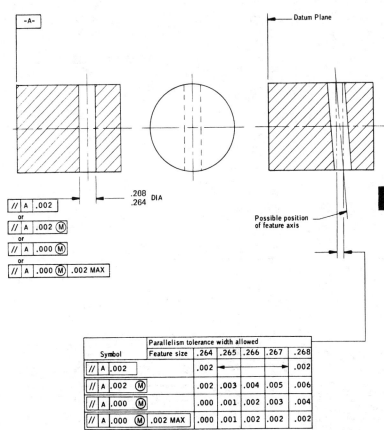

Parallelism of a Feature Axis to a Datum Plane
Figure 42

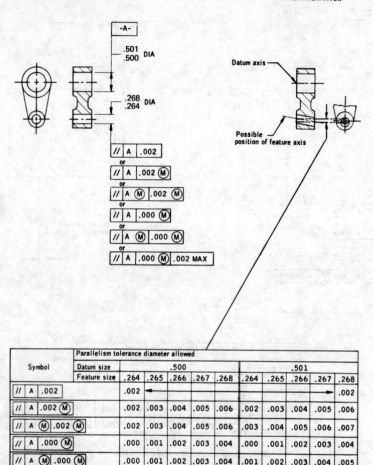

Parallelism of a Feature Axis to a Datum Axis
Figure 43

DRAWING CALLOUT

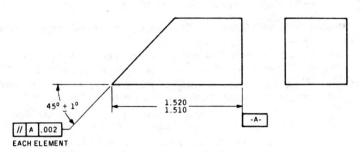

INTERPRETATION

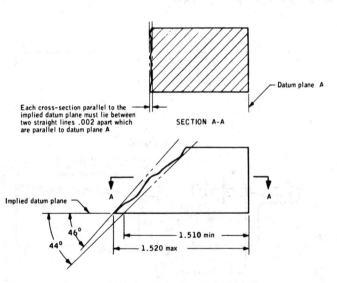

Parallelism of Feature Elements to a Datum Plane
Figure 44

10.9.7 PERPENDICULARITY — Perpendicularity is the condition of a surface, axis, or line which is at right angles to a datum plane or axis.

10.9.7.1 PERPENDICULARITY TOLERANCE — A perpendicularity tolerance specifies one of the following:

(a) A tolerance zone confined by two parallel planes perpendicular to a datum plane within which the surface of a feature must lie. (See Figures 45 and 46.)

(b) A tolerance zone confined by two parallel planes perpendicular to a datum plane within which the centerplane of a feature must lie. (See Figure 47.)

(c) A cylindrical tolerance zone perpendicular to a datum plane within which the axis of the feature must lie. (See Figures 48, 49, 50, and 51.)

(d) A tolerance zone confined by two parallel planes perpendicular to a datum axis within which the axis of a feature must lie. (See Figure 52.)

(e) A tolerance zone confined by two parallel straight lines perpendicular to a datum plane or datum axis within which an element of the surface must lie. (See Figure 53.)

Note that the perpendicularity tolerance when applied to a plane surface controls flatness if a flatness tolerance is not specified.

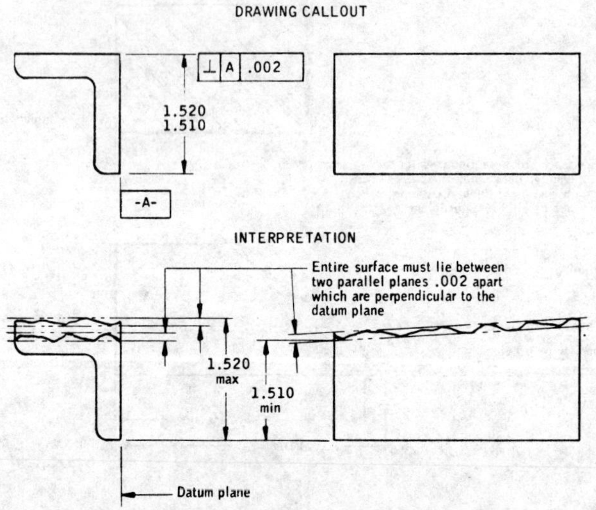

DRAWING CALLOUT

INTERPRETATION

Entire surface must lie between two parallel planes .002 apart which are perpendicular to the datum plane

Perpendicularity of a Feature Surface to a Datum Plane
Figure 45

10-36

DRAWING CALLOUT

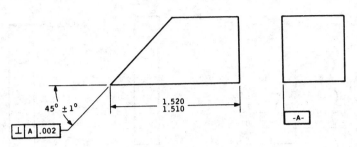

INTERPRETATION

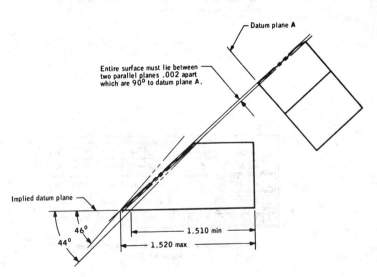

Perpendicularity of a Feature Surface to a Datum Plane
Figure 46

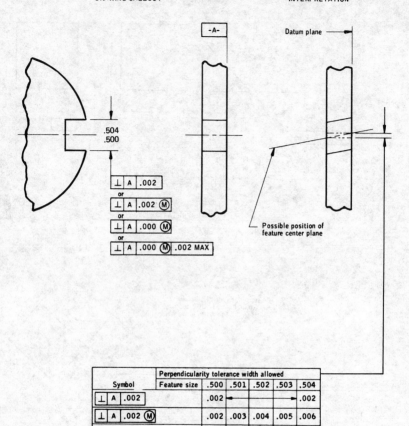

Perpendicularity of a Feature Center Plane to a Datum Plane
Figure 47

DRAWING CALLOUT INTERPRETATION

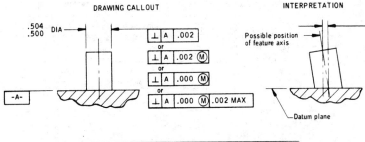

Symbol	Feature size	Perpendicularity tolerance diameter allowed				
		.504	.503	.502	.501	.500
⊥ A .002		.002				.002
⊥ A .002 Ⓜ		.002	.003	.004	.005	.006
⊥ A .000 Ⓜ		.000	.001	.002	.003	.004
⊥ A .000 Ⓜ .002 MAX		.000	.001	.002	.002	.002

Perpendicularity of a Feature Axis to a Datum Plane, Fixed Pin
Figure 48

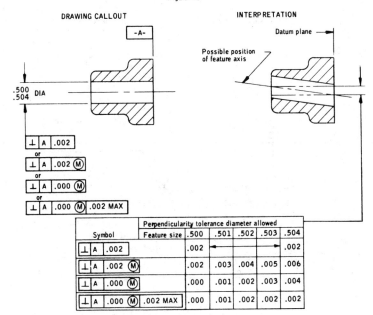

Symbol	Feature size	Perpendicularity tolerance diameter allowed				
		.500	.501	.502	.503	.504
⊥ A .002		.002				.002
⊥ A .002 Ⓜ		.002	.003	.004	.005	.006
⊥ A .000 Ⓜ		.000	.001	.002	.003	.004
⊥ A .000 Ⓜ .002 MAX		.000	.001	.002	.002	.002

Perpendicularity of a Feature Axis to a Datum Plane
Figure 49

10-39

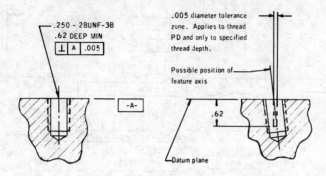

Perpendicularity of Threaded Holes and/or Inserts
Figure 50

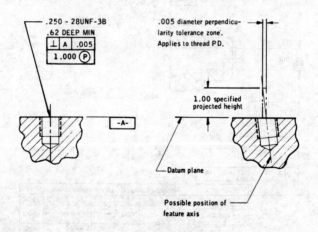

Perpendicularity of Threaded Holes and/or Inserts
Tolerance Zone Projected
Figure 51

DRAWING CALLOUT   INTERPRETATION

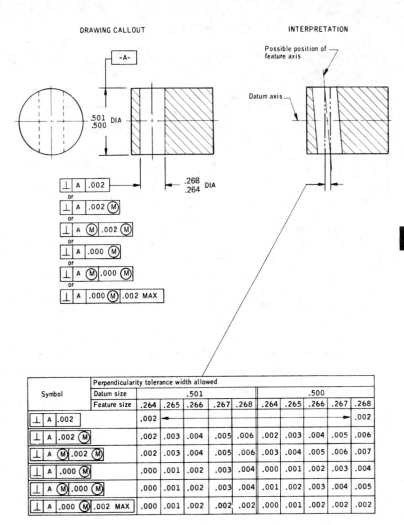

Perpendicularity of a Feature Axis to a Datum Axis
Figure 52

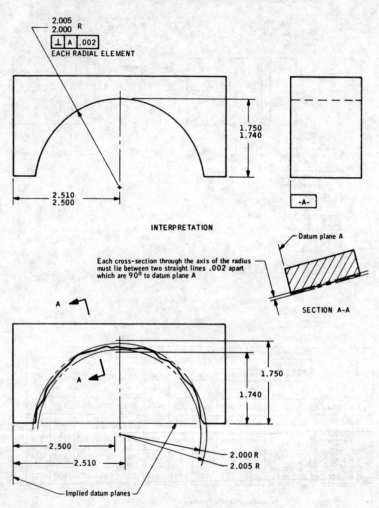

Perpendicularity of Feature Elements to a Datum Plane
Figure 53

10.9.8 ANGULARITY — Angularity is the condition of a surface or line which is at the specified angle (other than 90°) from a datum plane or axis.

10.9.8.1 ANGULARITY TOLERANCE — An angularity tolerance for a surface specifies a tolerance zone confined by two parallel planes, inclined at the specified angle to a datum plane, within which the toleranced surface must lie. (See Figure 54.) Note that the angularity tolerance when applied to a plane surface controls flatness if a flatness tolerance is not specified.

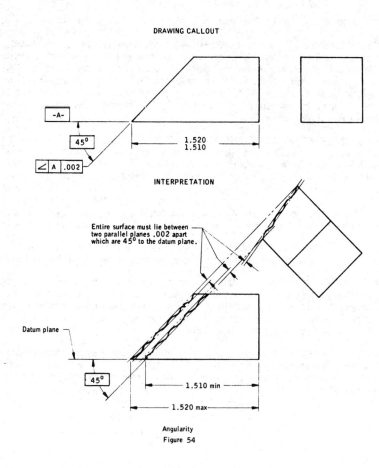

Angularity
Figure 54

10.9.9 RUNOUT — Runout is the condition of perfect form and axial alignment of two or more surfaces of revolution such as cylinders, cones, or contours and may include plane surfaces perpendicular to and generated about a common axis.

10.9.9.1 RUNOUT TOLERANCE — A runout tolerance controls the relationship of two or more features within the allowable errors of concentricity, perpendicularity, and alignment of the features. It also controls variations in roundness, straightness, flatness, angularity, and parallelism of individual surfaces. In essence, runout establishes composite form control of those features of a part having a common axis.

Note: Runout always applies RFS, therefore, the symbol for MMC shall not be used.

10.9.9.2 SELECTION OF RUNOUT DATUMS — To control the relationship of features, it is necessary to establish a datum axis about which the features are to be related. This axis may be established by a diameter of considerable length; two diameters having considerable axial separation; or a diameter and a surface which is at right angles to it. Insofar as possible, surfaces used as datums for establishing axes should be functional and must be accessible during manufacturing and inspection. Pitch diameters of features should be avoided as datums for runout.

10.9.9.3 INTERPRETATION OF RUNOUT TOLERANCES — Figure 55 illustrates the interpretation of runout tolerances. Measurements are taken under a single setup for all runout tolerances related to a common axis. However, features that are functionally related to each other and not to the common axis may be toleranced to reflect this requirement. (See Figure 57). Any two features on a common axis which are individually within their specified runout tolerance are related to each other within the sum of their runout tolerance. Therefore, to ensure 100% interchangeability, the sum of the runout tolerances of two mating diameters shall not exceed the clearance of the diameters at MMC.

10.9.9.4 APPLICATION OF RUNOUT TOLERANCES — Figures 55 through 60 illustrate various methods of specifying datum axes and applying the feature control symbols.

10.9.9.5 CIRCULAR RUNOUT — Circular runout is the maximum permissible surface variation at any *fixed point* during one complete rotation of the part about the datum axis. Control is applied on an individual basis to each circular element rather than total form control of the surface area. (See Figure 55). When circular runout is to be applied at a specific location, it is so stated on the drawing.

10.9.9.6 TOTAL RUNOUT — Total runout is the maximum permissible surface variation at *all* surface elements during one complete rotation of the part about the datum axis. Total runout is indicated by the addition of the word TOTAL beneath the feature control symbol. (See Figure 61.)

DRAWING CALLOUT

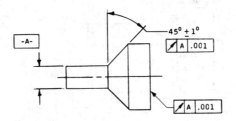

INTERPRETATIONS

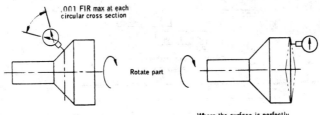

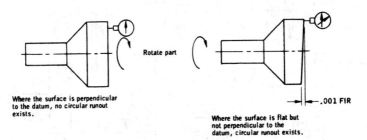

Circular Runout
Figure 55

10-45

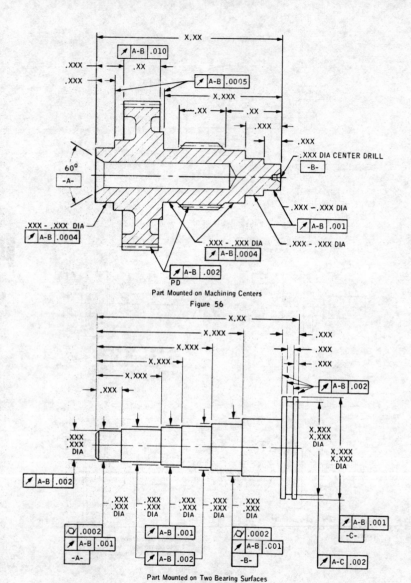

Part Mounted on Machining Centers
Figure 56

Part Mounted on Two Bearing Surfaces
Figure 57

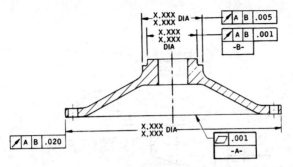

Part Mounted on Large Flat Surface with Narrow Finished Diameter
Figure 58

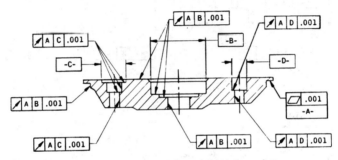

Part Mounted on Large Flat Surface with Multiple Common Axes
Figure 59

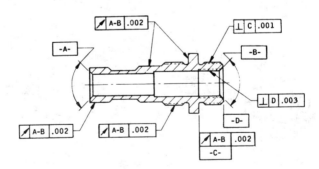

Part Mounted on Tapered Surfaces
Figure 60

10–47

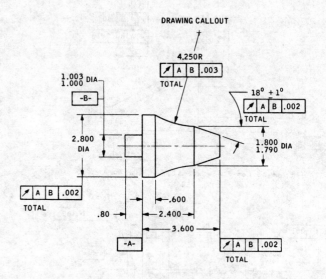

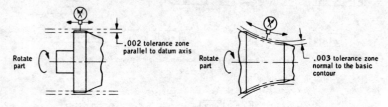

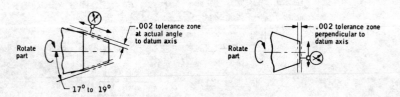

Note: A tolerance may be assigned to the 4.250 R or 18° angle when required, however, basic dimensions are preferred.

Interpretation of Runout Tolerance Zones
Figure 61

10.9.10 CONCENTRICITY — Concentricity is the condition of surfaces of revolution, such as cylinders, cones, or spheres, wherein they have a common axis.

10.9.10.1 CONCENTRICITY TOLERANCE — A concentricity tolerance is the diameter of the cylindrical tolerance zone within which the axis of the feature(s) so toleranced must lie. (See Figure 62). The positional relationship of features on a common axis is not controlled unless concentricity, runout, or true position is specified on the drawing.

Note: Irregularities in the form of the feature to be inspected may make it difficult to actually establish the axis of the feature. For instance, a nominally cylindrical surface may be bowed or out-of-round in addition to being offset from its datum surface; in such cases, finding the axis of the feature may entail a time-consuming analysis of the surface. Therefore, unless there is a definite need for the control of axes (as in the case shown in Figure 62), it is recommended that the control be specified in terms of runout or positional tolerance. (See Figures 55 through 61 and 63.)

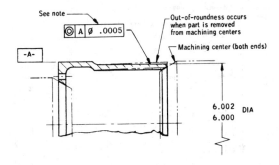

Note: A concentricity tolerance (rather than a runout tolerance, or a true position tolerance) has been applied to the item shown above because of the following supposed conditions:

1. A precise degree of coaxiality is required when part is assembled with mating parts.

2. The toleranced diameter, when removed from supporting tooling (machining centers), is likely to go out-of-round to the full amount permitted by the limits of size. This would preclude verification by either a runout inspection or a true position inspection, and would require a careful "analysis" of the surface; see Paragraph 10.9.10.1

Example Where Concentricity Tolerance is Required

Figure 62

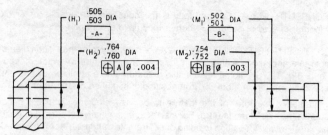

True Position Formula

Where Y = Total positional tolerance for both parts at MMC parts at MMC
H_1 = MMC of female datum
H_2 = MMC of related female diameter
M_1 = MMC of male datum
M_2 = MMC of related male diameter

$Y = (H_1 - M_1) + (H_2 - M_2) = (.503 - .502) + (.760 - .754) = .001 + .006 = .007$
Must be divided between mating diameters, e.g. .004 and .003

A. Parts with Two Mating Diameters

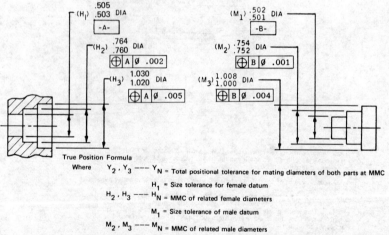

True Position Formula

Where $Y_2, Y_3 --- Y_N$ = Total positional tolerance for mating diameters of both parts at MMC
H_1 = Size tolerance for female datum
$H_2, H_3 --- H_N$ = MMC of related female diameters
M_1 = Size tolerance of male datum
$M_2, M_3 --- M_N$ = MMC of related male diameters

$Y_2 = (H_2 - M_2) - (H_1 + M_1) \qquad Y_3 = (H_3 - M_3) - (H_1 + M_1) \qquad Y_N = (H_N - M_N) - (H_1 + M_1)$

Example: $Y_2 = (H_2 - M_2) - (H_1 + M_1) - (.760 - .754) - [(.505 - .503) + (.502 - .501)] =$
$.006 - (.002 + .001 - .003$ Must be divided between mating diameters, e.g. .002 - .001
$Y_3 = (H_3 - M_3) - (H_1 + M_1) - (1.020 - 1.008) - .003$ from First example) $-.012 - .003 =$
.009 Must be divided between mating diameters, e.g. .005 and .004

B. Parts with Three or More Mating Diameters with MMC Datum References

Concentricity Controlled by True Position Tolerances
Figure 63

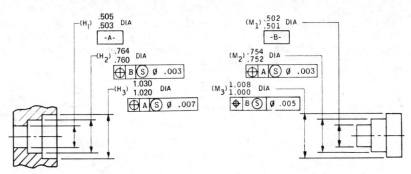

True Position Formula

Where: $Y_2, Y_3 \text{---} Y_N$ = Total true position tolerance for mating diameters of both parts at MMC
H_1 = Female datum
$H_2, H_3 \text{---} H_N$ = MMC of related female diameters
M_1 = Male datum
$M_2, M_3 \text{---} M_N$ = MMC of related male diameters

$$Y_2 = H_2 - M_2 \qquad Y_3 = H_3 - M_3 \qquad Y_N = H_N - M_N$$

Example: $Y_2 = H_2 - M_2 = .760 - .754 = .006$ Must be divided between mating diameters, e.g. .003 and .003
$Y_3 = H_3 - M_3 = 1.020 - 1.008 = .012$ Must be divided between mating diameters, e.g. .007 and .005

C. Parts with Three or More Diameters with RFS Datum References

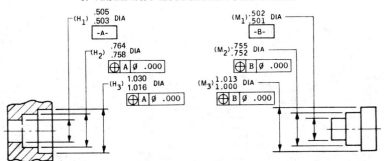

True Position Formula

When using this method, the tolerance is normally preset at .000, therefore, the formula becomes a simple rule as follows:

The clearance between mating diameters at MMC must be equal to or greater than the sum of the size tolerances of both datums.

Example: $(.505 - .503) + (.502 - .501) = .002 + .001 = .003$
$.758 - .755 = .003$
$1.016 - 1.013 = .003$

D. Parts with Three or More Diameters with Zero Tolerance at MMC

Figure 63 (Contd)

10.9.11 SYMMETRY — Symmetry is a condition wherein a part or a feature has the same contour and size on opposite sides of a central plane, or a condition in which a feature is symetrically disposed about the central plane of a datum feature.

10.9.11.1 SYMMETRY TOLERANCE — Where it is required that a feature be located symmetrically with respect to a datum feature, positional tolerancing should be used. This permits the tolerance to be expressed on an MMC basis or on an RFS basis. (See Figure 64.)

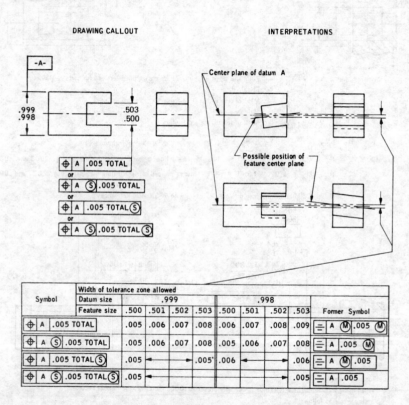

Symmetry Controlled by True Position
Figure 64

10.10 POSITIONAL TOLERANCE — Positional tolerance is a term used to describe the perfect or exact locatin of a point, line, or plane of a feature in relationship with a datum or other feature.

10.10.1 POSITIONAL TOLERANCE — A positional tolerance is the total permissible variation in the location of a feature about its true position. For cylindrical features (holes and bosses), the positional tolerance is the diameter (cylinder) of the tolerance zone within which the axis of the feature must lie, the center of the tolerance zone being at the true position. For other features (tabs, slots, etc.), the positional tolerance is the total width of the tolerance zone within which the center plane of the feature must lie, the center plane of the zone being at true position.

10.10.2 APPLICATION OF MMC, LMC AND RFS

(a) Unless the symbol for LMC or RFS is shown on the drawing, true position applies at MMC of both the feature and any datum reference, except when the datum is one plane surface. To apply this technique, the statement 'true position applies at MMC of both the feature and any datum reference' will be added to the general notes thereby eliminating the addition of 'M' to the symbol block.

(b) The symbol for RFS should be applied to all positional tolerances used with screw threads. Specifying MMC is not practical because of the centralizing effect of threads when tightened and because of the difficulty in determining the actual deviation from MMC. When MMC is applied, the additional tolerance available is the tolerance of the thread PD.

(c) The LMC symbol can be used to advantage on drawings of castings, forgings, molded parts, etc., where the positional tolerances are normally calculated and based on the minimum material condition. When the LMC symbol is used, the general note in 10.3.15 must also be included.

10.10.3 FORMULAS FOR POSITIONAL TOLERANCING — The formulas shown below may be used for determining the positional tolerance of round or threaded holes of mating parts. These formulas will result in a "no-interference, no-clearance" fit at maximum material condition of the mating features. They are based on equal positional tolerances for each part; however, the tolerances may be divided unequally when required. For example, it is normally more practicable to assign a larger tolerance to the threaded holes in one part, and a smaller tolerance to the corresponding clearance holes in the mating part. The threaded holes or holes for tight-fitting members such as dowels should be specified as "projected tolerance zone XXX" (See Figure 101), otherwise fastener interference may occur. The assembly conditions are commonly referred to as "floating fasteners" and "fixed fasteners." The "floating fastener formula" is used where two or more mating parts contain clearance holes and the "fixed fastener formula" is used where one part contains threaded holes, or holes for tight-fitting dowels, and the mating part has clearance holes.

Floating Fastener Formula: Fixed Fastener Formula:
$$Y = H - F, \quad H = F + Y \qquad Y = \frac{H - F}{2}, \quad H = F + 2Y$$

Where: Y = Diameter of positional tolerance zone
           H = Minimum diameter of clearance hole
           F = Maximum diameter of fastener

Note: When using solid dowel pins, the size tolerance of the pin and retaining hole are usually very small (± .0005 max), therefore, the "fixed fastener formula" may be used and the tolerances applies at MMC. However, when using rolled spring pins an additional tolerance must be considered since the pin will conform to the actual hole size. To accommodate this additional tolerance, the "fixed fastener formula" may be used with one of the following changes.

LMC Formula:
$$Y = \frac{H - F}{2}, \quad H = F + 2Y$$

Where: F = Maximum diameter of retaining hole.
           Y of retaining hole applies at LMC
           Y of clearance hole applies at MMC

See Figure 94A

MMC Formula:
$$Y = \frac{H - F - S}{2}, \quad H = F + 2Y + S$$

Where: F = Maximum diameter of retaining hole
           S = Size tolerance of retaining hole.

10.10.4 DATUMS FOR POSITIONAL TOLERANCE — Datums, either implied or specified, are always used with positional dimensioning. When two or more circular features could be used as a datum, one feature must be selected and identified with a datum symbol.

10.10.5 APPLICATION AND INTERPRETATION OF POSITIONAL TOLERANCES — Figures 63 through 103 illustrate various methods of applying positional tolerances and their interpretations.

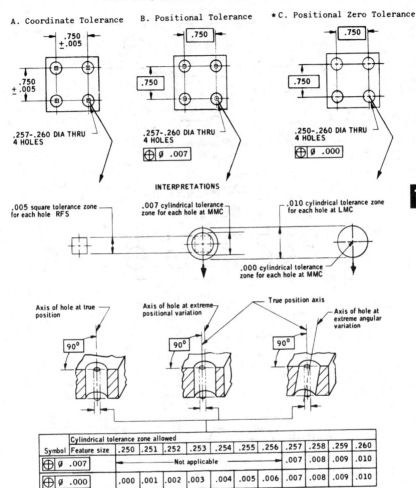

Advantages of True Position Tolerances
Figure 65

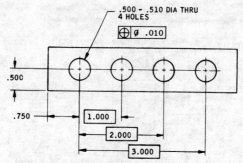

Base Line Dimensioning of Aligned Holes (Datums implied)
Figure 66

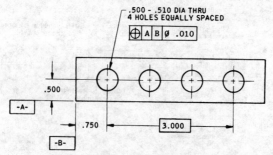

Dimensioning Equally Spaced Aligned Holes (Datums Specified)
Figure 67

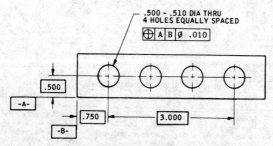

Locating Aligned Holes with Basic Dimensions (Datums Specified)
Figure 68

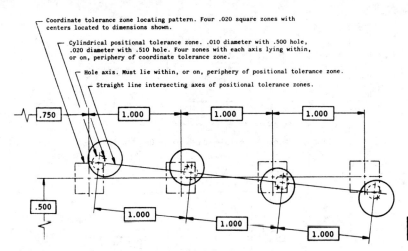

Interpretation of Figure 66 and 67
Figure 69

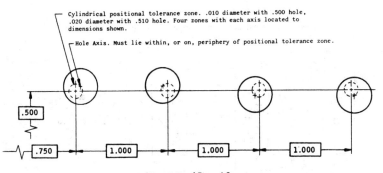

Interpretation of Figure 68
Figure 70

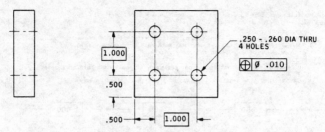

Positional & Coordinate Tolerances From Implied Datums
Figure 71

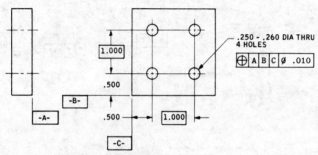

Positional & Coordinate Tolerances From Specified Datums.
Figure 72

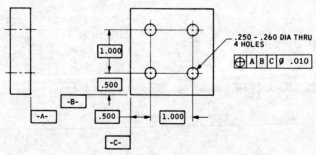

Positional With All Dimensions Basic From Specified Datums
Figure 73

Coordinate tolerance zone locating pattern. Four .020 square zones with centers located to dimensions shown.

Cylindrical positional tolerance zone. .010 diameter with .250 hole, .020 diameter with .260 hole. Four zones with each axis lying within, or on, periphery of coordinate tolerance zone.

Hole axis. Must lie within, or on, periphery of positional tolerance zone.

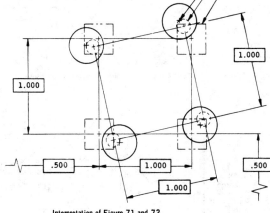

Interpretation of Figure 71 and 72
Figure 74

Cylindrical positional tolerance zone. .010 diameter with .250 hole .020 diameter with .260 hole. Four zones with each axis located to dimensions shown.

Hole axis. Must lie within, or on, periphery of positional tolerance zone.

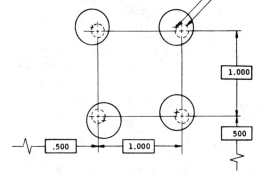

Interpretation of Figure 73
Figure 75

10-59

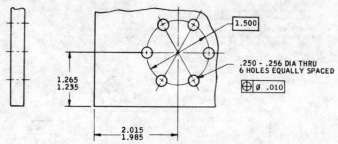

Hole Pattern Located by Coordinate Dimensions from Implied Datums
Figure 76

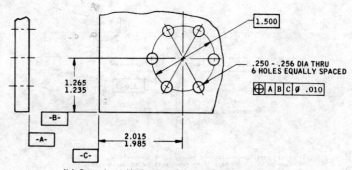

Hole Pattern Located by Coordinate Dimensions from Specified Datums
Figure 77

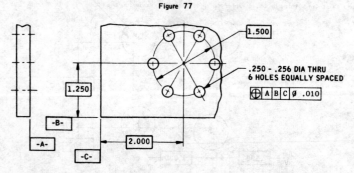

Hole Pattern Located by Basic Dimensions from Specified Datums
Figure 78

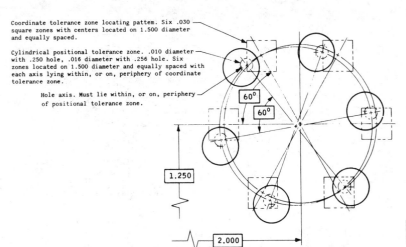

Note: If the coordinate tolerances were unequal, the square zones would be changed to rectangular zones of the specified size, but all other conditions would remain the same.

Interpretation of Figure 76 and 77
Figure 79

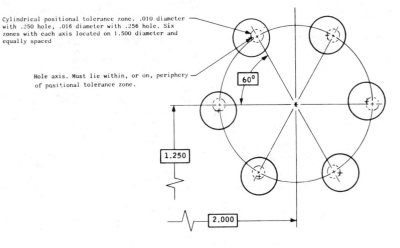

Interpretation of Figure 78
Figure 80

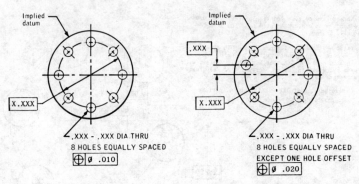

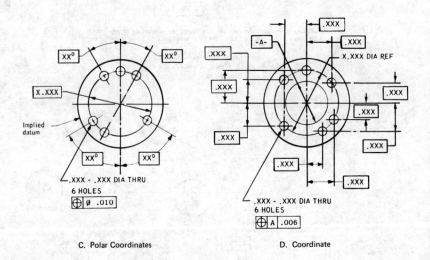

Methods of Dimensioning Hole Patterns in Relation to a Datum Axis
Figure 81

DRAWING CALLOUT

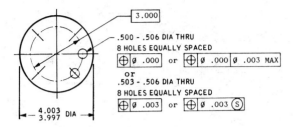

INTERPRETATION

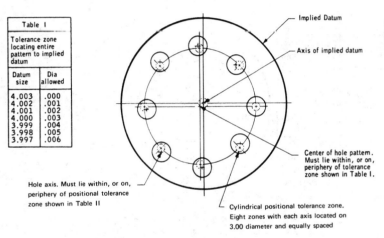

Table I	
Tolerance zone locating entire pattern to implied datum	
Datum size	Dia allowed
4.003	.000
4.002	.001
4.001	.002
4.000	.003
3.999	.004
3.998	.005
3.997	.006

Hole axis. Must lie within, or on, periphery of positional tolerance zone shown in Table II

Implied Datum

Axis of implied datum

Center of hole pattern. Must lie within, or on, periphery of tolerance zone shown in Table I.

Cylindrical positional tolerance zone. Eight zones with each axis located on 3.00 diameter and equally spaced

Table II								
	Cylindrical tolerance zone allowed for each hole							
Symbol	Feature size	.500	.501	.502	.503	.504	.505	.506
⊕ ⌀ .000		.000	.001	.002	.003	.004	.005	.006
⊕ ⌀ .000 ⌀ .003 MAX		.000	.001	.002	.003	←	→	.003
⊕ ⌀ .003		Not Applicable			.003	.004	.005	.006
⊕ ⌀ .003 Ⓢ		Not Applicable			.003	←	→	.003

Hole Pattern Located to an Implied Datum Axis
Figure 82

10-63

DRAWING CALLOUT

4.003 / 4.000 DIA
2.002 / 2.000 DIA
-A-
-B-
⊕ | A | ⌀ .005

3.000

.500 - .506 DIA THRU
8 HOLES EQUALLY SPACED
⊕ | B | ⌀ .000   or   ⊕ | B | .000 | ⌀ .003 MAX
or
.503 - .506 DIA THRU
8 HOLES EQUALLY SPACED
⊕ | B | ⌀ .003   or   ⊕ | B (S) | ⌀ .003 (S)

INTERPRETATION

Table I
Tolerance zone locating datum B to datum A

Datum A size	Datum B size	Dia allowed
4.003	2.000	.005
	2.001	.006
	2.002	.007
4.002	2.000	.006
	2.001	.007
	2.002	.008
4.001	2.000	.007
	2.001	.008
	2.002	.009
4.000	2.000	.008
	2.001	.009
	2.002	.010

Table II
Tolerance zone locating entire pattern to datum B

Datum B size	Dia allowed
2.000	.000
2.001	.001
2.002	.002

Axis of Datum A

Axis of datum B. Must lie within, or on, periphery of tolerance zone shown in Table I. This tolerance is not allowed when datum reference is RFS.

Center of hole pattern. Must lie within, or on, periphery of tolerance zone shown in Table II. This tolerance is not allowed when datum reference is RFS.

See note

Hole axis. Must lie within, or on, periphery of positional tolerance zone. See Figure 82 Table II.

Cylindrical positional tolerance zone. Eight zones with each axis located on 3.00 diameter and equally spaced.

Note: In some designs the remaining wall thickness may be critical. The calculations for determining this are:

(a) Subtract the basic hole pattern diameter from the O.D. at LMC.
(b) From (a), subtract the LMC hole diameter.
(c) Find the sum of the positional tolerances and related size tolerances.
(d) Subtract (c) from (b) and divide by 2.

Example using ⊕ | B | ⌀ .003

(a) 4.000 datum A
    -3.000 hole pattern dia
     1.000
(b) -.506 LMC hole
      .494
(d) -.018
     2).476 = .238 remaining wall

(c) TP of B to A = .005 + size tol of A (.003) + size tol of B (.002) = .010
    TP of hole pattern to B = .003 + size tol of B (.002) = .005
    Size tol of hole = .003
                      .018

Hole Pattern Located to a Specified Datum
Figure 83

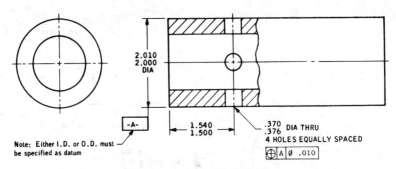

Radial Holes Located by Coordinate Dimension from Implied Datum
Figure 84

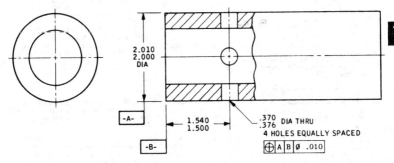

Radial Holes Located by Coordinate Dimension from Specified Datum
Figure 85

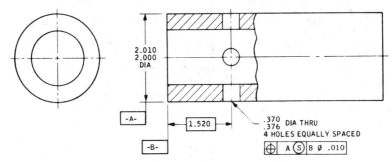

Radial Holes Located by Basic Dimension from Specified Datum
Figure 86

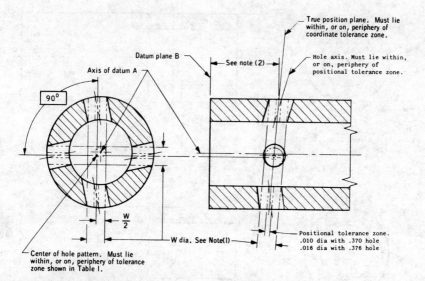

Table I	
Tolerance zone locating entire pattern of holes in relation to datum A axis.	
Datum A size	Diameter allowed
2.010	.000
2.009	.001
2.008	.002
2.007	.003
2.006	.004
2.005	.005
2.004	.006
2.003	.007
2.002	.008
2.001	.009
2.000	.010

Note: (1) W diameter equals the MMC hole minus the positional tolerance, e.g. .370 – .010 = .360. The holes may deviate from true position or true direction provided W diameter is not violated and the holes are within limits of size.

(2) The distance from datum plane B (or the implied datum plane in Figure 84) to the true position plane may vary anywhere within the coordinate dimension tolerance, e.g. 1.500 min to 1.540 max. Each hole axis may deviate from this plane by an amount equal to 1/2 the positional tolerance. Therefore, the distance from datum plane B to the actual hole axis may be 1.495 to 1.545 with .370 hole (1.500 – .005 and 1.540 + .005) or 1.492 to 1.548 with .376 hole (1.500 – .008 and 1.540 + .008).

Interpretation of Figure 84 and 85
Figure 87

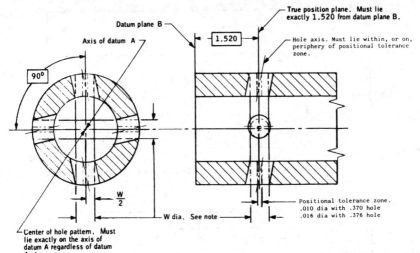

Note: W diameter equals the MMC hole minus the positional tolerances, e.g. .370-.010=.360. The holes may deviate from true position or true direction provided W diameter is not violated and the holes are within limits of size.

Interpretation of Figure 86
Figure 88

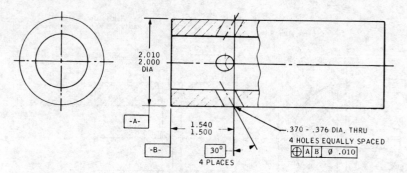

Angular Radial Holes Located by Coordinate Dimension at O.D.
Figure 89

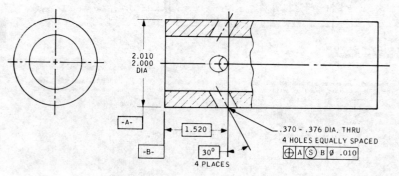

Angular Radial Holes Located by Basic Dimension at O.D.
Figure 90

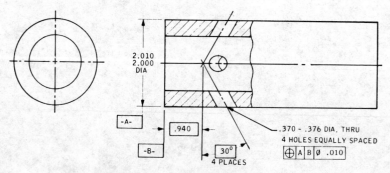

Angular Radial Holes Located by Basic Dimension at Axis of O.D.
Figure 91

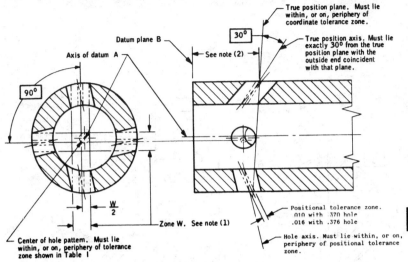

Table I	
Tolerance zone locating entire pattern of holes in relation to datum A axis	
Datum A size	Diameter allowed
2.010	.000
2.009	.001
2.008	.002
2.007	.003
2.006	.004
2.005	.005
2.004	.006
2.003	.007
2.002	.008
2.001	.009
2.000	.010

Note: (1) Zone W equals the MMC hole minus the positional tolerance, e.g. .370 - .010=.360. The holes may deviate from true position or true direction provided zone W is not violated and the holes are within limits of size.

(2) The distance from datum plane B to the true position plane may vary anywhere within the coordinate dimension tolerance, e.g. 1.500 min to 1.540 max. Each hole axis may deviate from this plane by an amount equal to 1/2 the true position tolerance. Therefore, the distance from datum plane B to the actual hole axis may be 1.495 to 1.545 with .370 hole(1.500 - .005 and 1.540 + .005) or 1.492 to 1.548 with .376 hole(1.500 - .008 and 1.540 + .008).

Interpretation of Figure 89
Figure 92

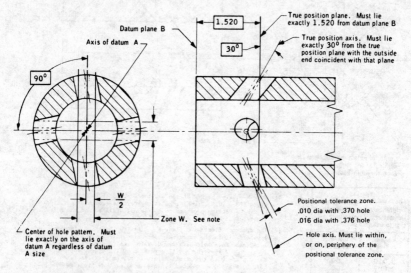

Note: Zone W equals the MMC hole minus the positional tolerance, e.g. .370-.010 = .360
The holes may deviate from true position or true direction provided zone W is not violated and the holes are within limits of size.

Interpretation of Figure 90
Figure 93

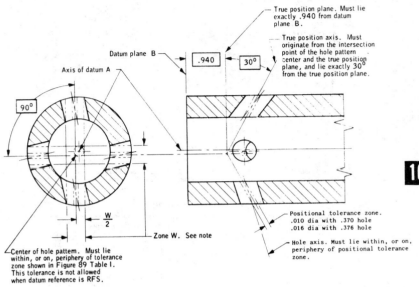

Interpretation of Figure 91
Figure 94

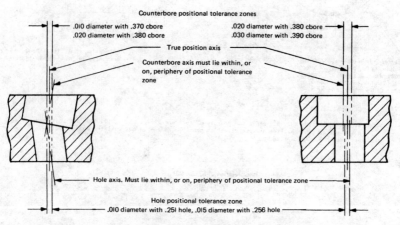

Position of Multifeatures in a Single Callout
Figure 95

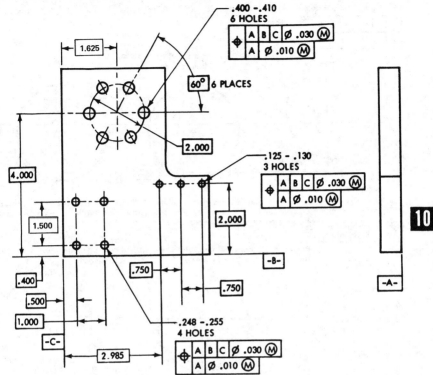

HOLE PATTERNS LOCATED BY COMPOSITE POSITIONAL TOLERANCING, SPECIFIED DATUMS

Figure 96

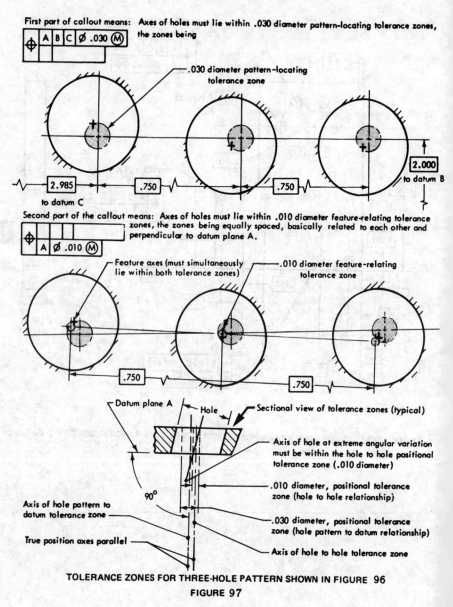

TOLERANCE ZONES FOR THREE-HOLE PATTERN SHOWN IN FIGURE 96
FIGURE 97

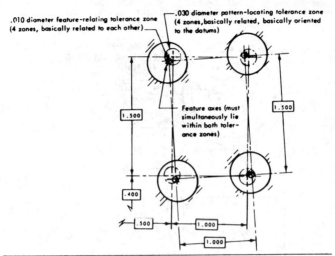

TOLERANCE ZONES FOR FOUR-HOLE PATTERN SHOWN IN FIGURE 96
FIGURE 98

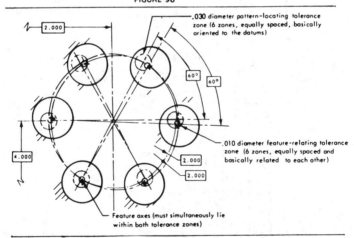

TOLERANCE ZONES FOR SIX-HOLE PATTERN SHOWN IN FIGURE 96
FIGURE 99

10-75

DRAWING CALLOUT

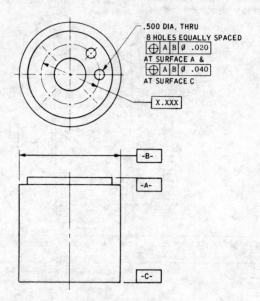

INTERPRETATION

Larger Positional Tolerance at One End of Hole Than at Other End

Figure 100

ASSEMBLY CONDITIONS

A. Spring Pin B. Bolt C. Stud

DRAWING CALLOUTS

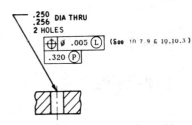 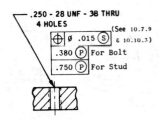

Note: Projected tolerance height should equal maximum pin height above surface.

Note: Projected tolerance height for bolt should equal maximum thickness of mating part; height for stud should equal maximum stud height above surface.

INTERPRETATIONS

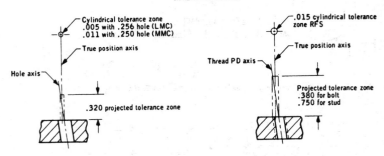

Projected Tolerance Zone and LMC Application
Figure 101

ASSEMBLY CONDITION

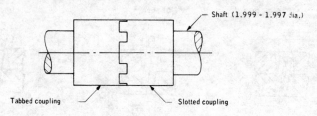

Tabbed coupling — Slotted coupling
Shaft (1.999 - 1.997 dia.)

DRAWING OF TABBED PART

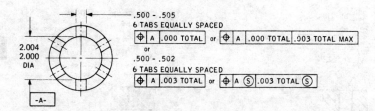

DRAWING OF SLOTTED PART

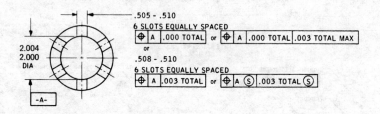

Note: The formula for fixed fastners may be used for determining the tolerance of a pattern of mating tabs or slots.

Example:

$$Y = \frac{H - F}{2} \qquad Y = \frac{.505 - .505}{2} = .000 \quad \text{or} \quad Y = \frac{.508 - .502}{2} = .003$$

Where: Y = permissible positional tolerance
H = minimum slot width
F = maximum tab width

Positional Tolerancing of Tabs and Slots
Figure 102

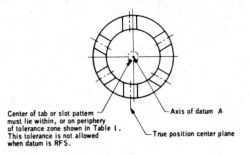

Table I

Tolerance zone locating entire pattern of tabs or slots in relation to datum A axis	
Datum A size	Diameter allowed
2.000	.000
2.001	.001
2.002	.002
2.003	.003
2.004	.004

Center of tab or slot pattern must lie within, or on periphery of tolerance zone shown in Table I. This tolerance is not allowed when datum is RFS.

Axis of datum A
True position center plane

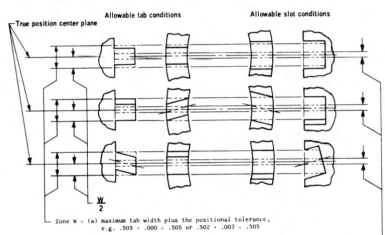

Zone W = (a) maximum tab width plus the positional tolerance,
e.g. .505 + .000 = .505 or .502 + .003 = .505

(b) minimum slot width minus the positional tolerance,
e.g. .505 - .000 = .505 or .508 - .003 = .505

Side surfaces of each tab or slot may deviate from true position or true direction, provided zone W is not violated and the tab or slot is within limits of size.

Symbol	Table II Width of tolerance zone allowed						
	Tab size	.505	.504	.503	.502	.501	.500
	Slot size	.505	.506	.507	.508	.509	.510
⊕ A .000 TOTAL		.000	.001	.002	.003	.004	.005
⊕ A .000 TOTAL .003 TOTAL MAX		.000	.001	.002	.003	.003	.003
⊕ A .003 TOTAL		Not applicable		.003	.004	.005
⊕ A(S) .003 TOTAL (S)		Not applicable		.003	.003	.003

Interpretation of Figure 102

Figure 103

NOTES

NOTES

NOTES

10.11 With the signing of H.R. 8674, the Metric Conversion Act of 1975 into Public Law 94-168 by President Ford on December 23, 1975, government contracts as well as international trade with other countries using the SI Metric system will continue to grow at an increasing rate. The following conversion tables are provided, as well as the appendix in the rear of this publication to aid the changeover from one system to the other. The information will also assist in reading metric drawings for comparison with inch drawings now in use.

    1.) See section 16.5 for Metric Screw Threads
    2.) See appendix, Section A2 and Table 3, for additional conversion measurements and for exact conversion factors.

DECIMAL EQUIVALENTS
Fractions of an Inch Expressed as Decimals and Millimetres

Fractional inch	Decimal inch	mm	Fractional inch	Decimal inch	mm
1/64	0.015625	0.396 9	33/64	0.515625	13.096 9
1/32	0.03125	0.793 8	17/32	0.53125	13.493 8
3/64	0.046875	1.190 6	35/64	0.546875	13.890 6
1/16	0.0625	1.587 5	9/16	0.5625	14.287 5
5/64	0.078125	1.984 4	37/64	0.578125	14.684 4
3/32	0.09375	2.381 3	19/32	0.59375	15.081 3
7/64	0.109375	2.778 1	39/64	0.609375	15.478 1
1/8	0.125	3.175	5/8	0.625	15.875
9/64	0.140625	3.571 9	41/64	0.640625	16.271 9
5/32	0.15625	3.968 8	21/32	0.65625	16.668 8
11/64	0.171875	4.365 6	43/64	0.671875	17.065 6
3/16	0.1875	4.762 5	11/16	0.6875	17.462 5
13/64	0.203125	5.159 4	45/64	0.703125	17.859 4
7/32	0.21875	5.556 3	23/32	0.71875	18.256 3
15/64	0.234375	5.953 1	47/64	0.734375	18.653 1
1/4	0.25	6.35	3/4	0.75	19.05
17/64	0.265625	6.746 9	49/64	0.765625	19.446 9
9/32	0.28125	7.143 8	25/32	0.78125	19.843 8
19/64	0.296875	7.540 6	51/64	0.796875	20.240 6
5/16	0.3125	7.937 5	13/16	0.8125	20.637 5
21/64	0.328125	8.334 4	53/64	0.828125	21.034 4
11/32	0.34375	8.731 3	27/32	0.84375	21.431 3
23/64	0.359375	9.128 1	55/64	0.859375	21.828 1
3/8	0.375	9.525	7/8	0.875	22.225
25/64	0.390625	9.921 9	57/64	0.890625	22.621 9
13/32	0.40625	10.318 8	29/32	0.90625	23.018 8
27/64	0.421875	10.715 6	59/64	0.921875	23.415 6
7/16	0.4375	11.112 5	15/16	0.9375	23.812 5
29/64	0.453125	11.509 4	61/64	0.953125	24.209 4
15/32	0.46875	11.906 3	31/32	0.96875	24.606 3
31/64	0.484375	12.303 1	63/64	0.984375	25.003 1
1/2	0.5	12.7	1	1	25.4

DECIMAL EQUIVALENTS
Inches to Millimetres

Inch	mm	Inch	mm	Inch	mm
.001	0.0254	.027	0.6858	.062	1.5748
.0015	0.0381	.028	0.7112	.063	1.6002
.002	0.0508	.029	0.7366	.064	1.6256
.0025	0.0635	.030	0.7620	.065	1.6510
.003	0.0762	.031	0.7874	.066	1.6764
.0035	0.0889	.032	0.8128	.067	1.7018
.004	0.1016	.033	0.8382	.068	1.7272
.0045	0.1143	.034	0.8636	.069	1.7526
.005	0.1270	.035	0.8890	.070	1.7780
.0055	0.1397	.036	0.9144	.071	1.8034
.006	0.1524	.037	0.9398	.072	1.8288
.0065	0.1651	.038	0.9652	.073	1.8542
.007	0.1778	.039	0.9906	.074	1.8796
.0075	0.1905	.040	1.0160	.075	1.9050
.008	0.2032	.041	1.0414	.076	1.9304
.0085	0.2159	.042	1.0668	.077	1.9558
.009	0.2286	.043	1.0922	.078	1.9812
.0095	0.2413	.044	1.1176	.079	2.0066
.010	0.2540	.045	1.1430	.080	2.0320
.011	0.2794	.046	1.1684	.081	2.0574
.012	0.3048	.047	1.1938	.082	2.0828
.013	0.3302	.048	1.2192	.083	2.1082
.014	0.3556	.049	1.2446	.084	2.1336
.015	0.3810	.050	1.2700	.085	2.1590
.016	0.4064	.051	1.2954	.086	2.1844
.017	0.4318	.052	1.3208	.087	2.2098
.018	0.4572	.053	1.3462	.088	2.2352
.019	0.4826	.054	1.3716	.089	2.2606
.020	0.5080	.055	1.3970	.090	2.2860
.021	0.5334	.056	1.4224	.091	2.3114
.022	0.5588	.057	1.4478	.092	2.3368
.023	0.5842	.058	1.4732	.093	2.3622
.024	0.6096	.059	1.4986	.094	2.3876
.025	0.6350	.060	1.5240	.095	2.4130
.026	0.6604	.061	1.5494	.096	2.4384

METRIC CONVERSION PLANNER

in./mm

Inch	mm	Inch	mm	Inch	mm
.097	2.4638	.132	3.3528	.167	4.2418
.098	2.4892	.133	3.3782	.168	4.2672
.099	2.5146	.134	3.4036	.169	4.2926
.100	2.5400	.135	3.4290	.170	4.3180
.101	2.5654	.136	3.4544	.171	4.3434
.102	2.5908	.137	3.4798	.172	4.3688
.103	2.6162	.138	3.5052	.173	4.3942
.104	2.6416	.139	3.5306	.174	4.4196
.105	2.6670	.140	3.5560	.175	4.4450
.106	2.6924	.141	3.5814	.176	4.4704
.107	2.7178	.142	3.6068	.177	4.4958
.108	2.7432	.143	3.6322	.178	4.5212
.109	2.7686	.144	3.6576	.179	4.5466
.110	2.7940	.145	3.6830	.180	4.5720
.111	2.8194	.146	3.7084	.181	4.5974
.112	2.8448	.147	3.7338	.182	4.6228
.113	2.8702	.148	3.7592	.183	4.6482
.114	2.8956	.149	3.7846	.184	4.6736
.115	2.9210	.150	3.8100	.185	4.6990
.116	2.9464	.151	3.8354	.186	4.7244
.117	2.9718	.152	3.8608	.187	4.7498
.118	2.9972	.153	3.8862	.188	4.7752
.119	3.0226	.154	3.9116	.189	4.8006
.120	3.0480	.155	3.9370	.190	4.8260
.121	3.0734	.156	3.9624	.191	4.8514
.122	3.0988	.157	3.9878	.192	4.8768
.123	3.1242	.158	4.0132	.193	4.9022
.124	3.1496	.159	4.0386	.194	4.9276
.125	3.1750	.160	4.0640	.195	4.9530
.126	3.2004	.161	4.0894	.196	4.9784
.127	3.2258	.162	4.1148	.197	5.0038
.128	3.2512	.163	4.1402	.198	5.0292
.129	3.2766	.164	4.1656	.199	5.0546
.130	3.3020	.165	4.1910	.200	5.0800
.131	3.3274	.166	4.2164	.201	5.1054

in./mm

METRIC CONVERSION PLANNER

Inch	mm	Inch	mm	Inch	mm
.202	5,130.8	.237	6,019.8	.272	6,908.8
.203	5,156.2	.238	6,045.2	.273	6,934.2
.204	5,181.6	.239	6,070.6	.274	6,959.6
.205	5,207.0	.240	6,096.0	.275	6,985.0
.206	5,232.4	.241	6,121.4	.276	7,010.4
.207	5,257.8	.242	6,146.8	.277	7,035.8
.208	5,283.2	.243	6,172.2	.278	7,061.2
.209	5,308.6	.244	6,197.6	.279	7,086.6
.210	5,334.0	.245	6,223.0	.280	7,112.0
.211	5,359.4	.246	6,248.4	.281	7,137.4
.212	5,384.8	.247	6,273.8	.282	7,162.8
.213	5,410.2	.248	6,299.2	.283	7,188.2
.214	5,435.6	.249	6,324.6	.284	7,213.6
.215	5,461.0	.250	6,350.0	.285	7,239.0
.216	5,486.4	.251	6,375.4	.286	7,264.4
.217	5,511.8	.252	6,400.8	.287	7,289.8
.218	5,537.2	.253	6,426.2	.288	7,315.2
.219	5,562.6	.254	6,451.6	.289	7,340.6
.220	5,588.0	.255	6,477.0	.290	7,366.0
.221	5,613.4	.256	6,502.4	.291	7,391.4
.222	5,638.8	.257	6,527.8	.292	7,416.8
.223	5,664.2	.258	6,553.2	.293	7,442.2
.224	5,689.6	.259	6,578.6	.294	7,467.6
.225	5,715.0	.260	6,604.0	.295	7,493.0
.226	5,740.4	.261	6,629.4	.296	7,518.4
.227	5,765.8	.262	6,654.8	.297	7,543.8
.228	5,791.2	.263	6,680.2	.298	7,569.2
.229	5,816.6	.264	6,705.6	.299	7,594.6
.230	5,842.0	.265	6,731.0	.300	7,620.0
.231	5,867.4	.266	6,756.4	.301	7,645.4
.232	5,892.8	.267	6,781.8	.302	7,670.8
.233	5,918.2	.268	6,807.2	.303	7,696.2
.234	5,943.6	.269	6,832.6	.304	7,721.6
.235	5,969.0	.270	6,858.0	.305	7,747.0
.236	5,994.4	.271	6,883.4	.306	7,772.4

in./mm

METRIC CONVERSION PLANNER

Inch	mm	Inch	mm	Inch	mm
.307	7,797.8	.342	8,686.8	.377	9,575.8
.308	7,823.2	.343	8,712.2	.378	9,601.2
.309	7,848.6	.344	8,737.6	.379	9,626.6
.310	7,874.0	.345	8,763.0	.380	9,652.0
.311	7,899.4	.346	8,788.4	.381	9,677.4
.312	7,924.8	.347	8,813.8	.382	9,702.8
.313	7,950.2	.348	8,839.2	.383	9,728.2
.314	7,975.6	.349	8,864.6	.384	9,753.6
.315	8,001.0	.350	8,890.0	.385	9,779.0
.316	8,026.4	.351	8,915.4	.386	9,804.4
.317	8,051.8	.352	8,940.8	.387	9,829.8
.318	8,077.2	.353	8,966.2	.388	9,855.2
.319	8,102.6	.354	8,991.6	.389	9,880.6
.320	8,128.0	.355	9,017.0	.390	9,906.0
.321	8,153.4	.356	9,042.4	.391	9,931.4
.322	8,178.8	.357	9,067.8	.392	9,956.8
.323	8,204.2	.358	9,093.2	.393	9,982.2
.324	8,229.6	.359	9,118.6	.394	10,007.6
.325	8,255.0	.360	9,144.0	.395	10,033.0
.326	8,280.4	.361	9,169.4	.396	10,058.4
.327	8,305.8	.362	9,194.8	.397	10,083.8
.328	8,331.2	.363	9,220.2	.398	10,109.2
.329	8,356.6	.364	9,245.6	.399	10,134.6
.330	8,382.0	.365	9,271.0	.400	10,160.0
.331	8,407.4	.366	9,296.4	.401	10,185.4
.332	8,432.8	.367	9,321.8	.402	10,210.8
.333	8,458.2	.368	9,347.2	.403	10,236.2
.334	8,483.6	.369	9,372.6	.404	10,261.6
.335	8,509.0	.370	9,398.0	.405	10,287.0
.336	8,534.4	.371	9,423.4	.406	10,312.4
.337	8,559.8	.372	9,448.8	.407	10,337.8
.338	8,585.2	.373	9,474.2	.408	10,363.2
.339	8,610.6	.374	9,499.6	.409	10,388.6
.340	8,636.0	.375	9,525.0	.410	10,414.0
.341	8,661.4	.376	9,550.4	.411	10,439.4

in./mm

METRIC CONVERSION PLANNER

Inch	mm	Inch	mm	Inch	mm
.412	10,464 8	.447	11,353 8	.482	12,242 8
.413	10,490 2	.448	11,379 2	.483	12,268 2
.414	10,515 6	.449	11,404 6	.484	12,293 6
.415	10,541 0	.450	11,430 0	.485	12,319 0
.416	10,566 4	.451	11,455 4	.486	12,344 4
.417	10,591 8	.452	11,480 8	.487	12,369 8
.418	10,617 2	.453	11,506 2	.488	12,395 2
.419	10,642 6	.454	11,531 6	.489	12,420 6
.420	10,668 0	.455	11,557 0	.490	12,446 0
.421	10,693 4	.456	11,582 4	.491	12,471 4
.422	10,718 8	.457	11,607 8	.492	12,496 8
.423	10,744 2	.458	11,633 2	.493	12,522 2
.424	10,769 6	.459	11,658 6	.494	12,547 6
.425	10,795 0	.460	11,684 0	.495	12,573 0
.426	10,820 4	.461	11,709 4	.496	12,598 4
.427	10,845 8	.462	11,734 8	.497	12,623 8
.428	10,871 2	.463	11,760 2	.498	12,649 2
.429	10,896 6	.464	11,785 6	.499	12,674 6
.430	10,922 0	.465	11,811 0	.500	12,700 0
.431	10,947 4	.466	11,836 4	.501	12,725 4
.432	10,972 8	.467	11,861 8	.502	12,750 8
.433	10,998 2	.468	11,887 2	.503	12,776 2
.434	11,023 6	.469	11,912 6	.504	12,801 6
.435	11,049 0	.470	11,938 0	.505	12,827 0
.436	11,074 4	.471	11,963 4	.506	12,852 4
.437	11,099 8	.472	11,988 8	.507	12,877 8
.438	11,125 2	.473	12,014 2	.508	12,903 2
.439	11,150 6	.474	12,039 6	.509	12,928 6
.440	11,176 0	.475	12,065 0	.510	12,954 0
.441	11,201 4	.476	12,090 4	.511	12,979 4
.442	11,226 8	.477	12,115 8	.512	13,004 8
.443	11,252 2	.478	12,141 2	.513	13,030 2
.444	11,277 6	.479	12,166 6	.514	13,055 6
.445	11,303 0	.480	12,192 0	.515	13,081 0
.446	11,328 4	.481	12,217 4	.516	13,106 4

in./mm

METRIC CONVERSION PLANNER

Inch	mm	Inch	mm	Inch	mm
.517	13,131 8	.552	14,020 8	.587	14,909 8
.518	13,157 2	.553	14,046 2	.588	14,935 2
.519	13,182 6	.554	14,071 6	.589	14,960 6
.520	13,208 0	.555	14,097 0	.590	14,986 0
.521	13,233 4	.556	14,122 4	.591	15,011 4
.522	13,258 8	.557	14,147 8	.592	15,036 8
.523	13,284 2	.558	14,173 2	.593	15,062 2
.524	13,309 6	.559	14,198 6	.594	15,087 6
.525	13,335 0	.560	14,224 0	.595	15,113 0
.526	13,360 4	.561	14,249 4	.596	15,138 4
.527	13,385 8	.562	14,274 8	.597	15,163 8
.528	13,411 2	.563	14,300 2	.598	15,189 2
.529	13,436 6	.564	14,325 6	.599	15,214 6
.530	13,462 0	.565	14,351 0	.600	15,240 0
.531	13,487 4	.566	14,376 4	.601	15,265 4
.532	13,512 8	.567	14,401 8	.602	15,290 8
.533	13,538 2	.568	14,427 2	.603	15,316 2
.534	13,563 6	.569	14,452 6	.604	15,341 6
.535	13,589 0	.570	14,478 0	.605	15,367 0
.536	13,614 4	.571	14,503 4	.606	15,392 4
.537	13,639 8	.572	14,528 8	.607	15,417 8
.538	13,665 2	.573	14,554 2	.608	15,443 2
.539	13,690 6	.574	14,579 6	.609	15,468 6
.540	13,716 0	.575	14,605 0	.610	15,494 0
.541	13,741 4	.576	14,630 4	.611	15,519 4
.542	13,766 8	.577	14,655 8	.612	15,544 8
.543	13,792 2	.578	14,681 2	.613	15,570 2
.544	13,817 6	.579	14,706 6	.614	15,595 6
.545	13,843 0	.580	14,732 0	.615	15,621 0
.546	13,868 4	.581	14,757 4	.616	15,646 4
.547	13,893 8	.582	14,782 8	.617	15,671 8
.548	13,919 2	.583	14,808 2	.618	15,697 2
.549	13,944 6	.584	14,833 6	.619	15,722 6
.550	13,970 0	.585	14,859 0	.620	15,748 0
.551	13,995 4	.586	14,884 4	.621	15,773 4

in./mm

METRIC CONVERSION PLANNER

Inch	mm	Inch	mm	Inch	mm
.622	15.798 8	.657	16.687 8	.692	17.576 8
.623	15.824 2	.658	16.713 2	.693	17.602 2
.624	15.849 6	.659	16.738 6	.694	17.627 6
.625	15.875 0	.660	16.764 0	.695	17.653 0
.626	15.900 4	.661	16.789 4	.696	17.678 4
.627	15.925 8	.662	16.814 8	.697	17.703 8
.628	15.951 2	.663	16.840 2	.698	17.729 2
.629	15.976 6	.664	16.865 6	.699	17.754 6
.630	16.002 0	.665	16.891 0	.700	17.780 0
.631	16.027 4	.666	16.916 4	.701	17.805 4
.632	16.052 8	.667	16.941 8	.702	17.830 8
.633	16.078 2	.668	16.967 2	.703	17.856 2
.634	16.103 6	.669	16.992 6	.704	17.881 6
.635	16.129 0	.670	17.018 0	.705	17.907 0
.636	16.154 4	.671	17.043 4	.706	17.932 4
.637	16.179 8	.672	17.068 8	.707	17.957 8
.638	16.205 2	.673	17.094 2	.708	17.983 2
.639	16.230 6	.674	17.119 6	.709	18.008 6
.640	16.256 0	.675	17.145 0	.710	18.034 0
.641	16.281 4	.676	17.170 4	.711	18.059 4
.642	16.306 8	.677	17.195 8	.712	18.084 8
.643	16.332 2	.678	17.221 2	.713	18.110 2
.644	16.357 6	.679	17.246 6	.714	18.135 6
.645	16.383 0	.680	17.272 0	.715	18.161 0
.646	16.408 4	.681	17.297 4	.716	18.186 4
.647	16.433 8	.682	17.322 8	.717	18.211 8
.648	16.459 2	.683	17.348 2	.718	18.237 2
.649	16.484 6	.684	17.373 6	.719	18.262 6
.650	16.510 0	.685	17.399 0	.720	18.288 0
.651	16.535 4	.686	17.424 4	.721	18.313 4
.652	16.560 8	.687	17.449 8	.722	18.338 8
.653	16.586 2	.688	17.475 2	.723	18.364 2
.654	16.611 6	.689	17.500 6	.724	18.389 6
.655	16.637 0	.690	17.526 0	.725	18.415 0
.656	16.662 4	.691	17.551 4	.726	18.440 4

in./mm

METRIC CONVERSION PLANNER

Inch	mm	Inch	mm	Inch	mm
.727	18.465 8	.762	19.354 8	.797	20.243 8
.728	18.491 2	.763	19.380 2	.798	20.269 2
.729	18.516 6	.764	19.405 6	.799	20.294 6
.730	18.542 0	.765	19.431 0	.800	20.320 0
.731	18.567 4	.766	19.456 4	.801	20.345 4
.732	18.592 8	.767	19.481 8	.802	20.370 8
.733	18.618 2	.768	19.507 2	.803	20.396 2
.734	18.643 6	.769	19.532 6	.804	20.421 6
.735	18.669 0	.770	19.558 0	.805	20.447 0
.736	18.694 4	.771	19.583 4	.806	20.472 4
.737	18.719 8	.772	19.608 8	.807	20.497 8
.738	18.745 2	.773	19.634 2	.808	20.523 2
.739	18.770 6	.774	19.659 6	.809	20.548 6
.740	18.796 0	.775	19.685 0	.810	20.574 0
.741	18.821 4	.776	19.710 4	.811	20.599 4
.742	18.846 8	.777	19.735 8	.812	20.624 8
.743	18.872 2	.778	19.761 2	.813	20.650 2
.744	18.897 6	.779	19.786 6	.814	20.675 6
.745	18.923 0	.780	19.812 0	.815	20.701 0
.746	18.948 4	.781	19.837 4	.816	20.726 4
.747	18.973 8	.782	19.862 8	.817	20.751 8
.748	18.999 2	.783	19.888 2	.818	20.777 2
.749	19.024 6	.784	19.913 6	.819	20.802 6
.750	19.050 0	.785	19.939 0	.820	20.828 0
.751	19.075 4	.786	19.964 4	.821	20.853 4
.752	19.100 8	.787	19.989 8	.822	20.878 8
.753	19.126 2	.788	20.015 2	.823	20.904 2
.754	19.151 6	.789	20.040 6	.824	20.929 6
.755	19.177 0	.790	20.066 0	.825	20.955 0
.756	19.202 4	.791	20.091 4	.826	20.980 4
.757	19.227 8	.792	20.116 8	.827	21.005 8
.758	19.253 2	.793	20.142 2	.828	21.031 2
.759	19.278 6	.794	20.167 6	.829	21.056 6
.760	19.304 0	.795	20.193 0	.830	21.082 0
.761	19.329 4	.796	20.218 4	.831	21.107 4

in./mm

METRIC CONVERSION PLANNER

Inch	mm	Inch	mm	Inch	mm
.832	21.132 8	.867	22.021 8	.902	22.910 8
.833	21.158 2	.868	22.047 2	.903	22.936 2
.834	21.183 6	.869	22.072 6	.904	22.961 6
.835	21.209 0	.870	22.098 0	.905	22.987 0
.836	21.234 4	.871	22.123 4	.906	23.012 4
.837	21.259 8	.872	22.148 8	.907	23.037 8
.838	21.285 2	.873	22.174 2	.908	23.063 2
.839	21.310 6	.874	22.199 6	.909	23.088 6
.840	21.336 0	.875	22.225 0	.910	23.114 0
.841	21.361 4	.876	22.250 4	.911	23.139 4
.842	21.386 8	.877	22.275 8	.912	23.164 8
.843	21.412 2	.878	22.301 2	.913	23.190 2
.844	21.437 6	.879	22.326 6	.914	23.215 6
.845	21.463 0	.880	22.352 0	.915	23.241 0
.846	21.488 4	.881	22.377 4	.916	23.266 4
.847	21.513 8	.882	22.402 8	.917	23.291 8
.848	21.539 2	.883	22.428 2	.918	23.317 2
.849	21.564 6	.884	22.453 6	.919	23.342 6
.850	21.590 0	.885	22.479 0	.920	23.368 0
.851	21.615 4	.886	22.504 4	.921	23.393 4
.852	21.640 8	.887	22.529 8	.922	23.418 8
.853	21.666 2	.888	22.555 2	.923	23.444 2
.854	21.691 6	.889	22.580 6	.924	23.469 6
.855	21.717 0	.890	22.606 0	.925	23.495 0
.856	21.742 4	.891	22.631 4	.926	23.520 4
.857	21.767 8	.892	22.656 8	.927	23.545 8
.858	21.793 2	.893	22.682 2	.928	23.571 2
.859	21.818 6	.894	22.707 6	.929	23.596 6
.860	21.844 0	.895	22.733 0	.930	23.622 0
.861	21.869 4	.896	22.758 4	.931	23.647 4
.862	21.894 8	.897	22.783 8	.932	23.672 8
.863	21.920 2	.898	22.809 2	.933	23.698 2
.864	21.945 6	.899	22.834 6	.934	23.723 6
.865	21.971 0	.900	22.860 0	.935	23.749 0
.866	21.996 4	.901	22.885 4	.936	23.774 4

in./mm

METRIC CONVERSION PLANNER

Inch	mm	Inch	mm	Inch	mm
.937	23.799 8	.972	24.688 8	8.0	203.200 0
.938	23.825 2	.973	24.714 2	9.0	228.600 0
.939	23.850 6	.974	24.739 6	10.0	254.000 0
.940	23.876 0	.975	24.765 0	11.0	279.400 0
.941	23.901 4	.976	24.790 4	12.0	304.800 0
.942	23.926 8	.977	24.815 8	13.0	330.200 0
.943	23.952 2	.978	24.841 2	14.0	355.600 0
.944	23.977 6	.979	24.866 6	15.0	381.000 0
.945	24.003 0	.980	24.892 0	16.0	406.400 0
.946	24.028 4	.981	24.917 4	17.0	431.800 0
.947	24.053 8	.982	24.942 8	18.0	457.200 0
.948	24.079 2	.983	24.968 2	19.0	482.600 0
.949	24.104 6	.984	24.993 6	20.0	508.000 0
.950	24.130 0	.985	25.019 0	21.0	533.400 0
.951	24.155 4	.986	25.044 4	22.0	558.800 0
.952	24.180 8	.987	25.069 8	23.0	584.200 0
.953	24.206 2	.988	25.095 2	24.0	609.600 0
.954	24.231 6	.989	25.120 6	25.0	635.000 0
.955	24.257 0	.990	25.146 0	26.0	660.400 0
.956	24.282 4	.991	25.171 4	27.0	685.800 0
.957	24.307 8	.992	25.196 8	28.0	711.200 0
.958	24.333 2	.993	25.222 2	29.0	736.600 0
.959	24.358 6	.994	25.247 6	30.0	762.000 0
.960	24.384 0	.995	25.273 0	31.0	787.400 0
.961	24.409 4	.996	25.298 4	32.0	812.800 0
.962	24.434 8	.997	25.323 8	33.0	838.200 0
.963	24.460 2	.998	25.349 2	34.0	863.600 0
.964	24.485 6	.999	25.374 6	35.0	889.000 0
.965	24.511 0	1.0	25.400 0	36.0	914.400
.966	24.536 4	2.0	50.800 0	48.0	1 219.200
.967	24.561 8	3.0	76.200 0	60.0	1 524.000
.968	24.587 2	4.0	101.600 0	72.0	1 828.800
.969	24.612 6	5.0	127.000 0	84.0	2 133.600
.970	24.638 0	6.0	152.400 0	96.0	2 438.400
.971	24.663 4	7.0	177.800 0	108.0	2 743.200
				120.0	3 048.000

10-88

METRIC CONVERSION PLANNER
DECIMAL EQUIVALENTS
Millimetres to Inches

mm	Inch	mm	Inch	mm	Inch
0.01	.000 39	0.34	.013 39	0.67	.026 38
0.02	.000 79	0.35	.013 78	0.68	.026 77
0.03	.001 18	0.36	.014 17	0.69	.027 17
0.04	.001 57	0.37	.014 57	0.70	.027 56
0.05	.001 97	0.38	.014 96	0.71	.027 95
0.06	.002 36	0.39	.015 35	0.72	.028 35
0.07	.002 76	0.40	.015 75	0.73	.028 74
0.08	.003 15	0.41	.016 14	0.74	.029 13
0.09	.003 54	0.42	.016 54	0.75	.029 53
0.10	.003 94	0.43	.016 93	0.76	.029 92
0.11	.004 33	0.44	.017 32	0.77	.030 31
0.12	.004 72	0.45	.017 72	0.78	.030 71
0.13	.005 12	0.46	.018 11	0.79	.031 10
0.14	.005 51	0.47	.018 50	0.80	.031 50
0.15	.005 91	0.48	.018 90	0.81	.031 89
0.16	.006 30	0.49	.019 29	0.82	.032 28
0.17	.006 69	0.50	.019 69	0.83	.032 68
0.18	.007 09	0.51	.020 08	0.84	.033 07
0.19	.007 48	0.52	.020 47	0.85	.033 46
0.20	.007 87	0.53	.020 87	0.86	.033 86
0.21	.008 27	0.54	.021 26	0.87	.034 25
0.22	.008 66	0.55	.021 65	0.88	.034 65
0.23	.009 06	0.56	.002 05	0.89	.035 04
0.24	.009 45	0.57	.022 44	0.90	.035 43
0.25	.009 84	0.58	.022 83	0.91	.035 83
0.26	.010 24	0.59	.023 23	0.92	.036 22
0.27	.010 63	0.60	.023 62	0.93	.036 61
0.28	.011 02	0.61	.024 02	0.94	.037 01
0.29	.011 42	0.62	.024 41	0.95	.037 40
0.30	.011 81	0.63	.024 80	0.96	.037 80
0.31	.012 20	0.64	.025 20	0.97	.038 19
0.32	.012 60	0.65	.025 59	0.98	.038 58
0.33	.012 99	0.66	.025 98	0.99	.038 98
				1.00	.039 37

METRIC CONVERSION PLANNER
mm/in.

mm	Inch	mm	Inch	mm	Inch
1.01	.039 76	1.34	.052 76	1.67	.065 75
1.02	.040 16	1.35	.053 15	1.68	.066 14
1.03	.040 55	1.36	.053 54	1.69	.066 54
1.04	.040 94	1.37	.053 94	1.70	.066 93
1.05	.041 34	1.38	.054 33	1.71	.067 32
1.06	.041 73	1.39	.054 72	1.72	.067 72
1.07	.042 13	1.40	.055 12	1.73	.068 11
1.08	.042 52	1.41	.055 51	1.74	.068 50
1.09	.042 91	1.42	.055 91	1.75	.068 90
1.10	.043 31	1.43	.056 30	1.76	.069 29
1.11	.043 70	1.44	.056 69	1.77	.069 69
1.12	.044 09	1.45	.057 09	1.78	.070 08
1.13	.044 49	1.46	.057 48	1.79	.070 47
1.14	.044 88	1.47	.057 87	1.80	.070 87
1.15	.045 28	1.48	.058 27	1.81	.071 26
1.16	.045 67	1.49	.058 66	1.82	.071 65
1.17	.046 06	1.50	.059 06	1.83	.072 05
1.18	.046 46	1.51	.059 45	1.84	.072 44
1.19	.046 85	1.52	.059 84	1.85	.072 83
1.20	.047 24	1.53	.060 24	1.86	.073 23
1.21	.047 64	1.54	.060 63	1.87	.073 62
1.22	.048 03	1.55	.061 02	1.88	.074 02
1.23	.048 43	1.56	.061 42	1.89	.074 41
1.24	.048 82	1.57	.061 81	1.90	.074 80
1.25	.049 21	1.58	.062 20	1.91	.075 20
1.26	.049 61	1.59	.062 60	1.92	.075 59
1.27	.050 00	1.60	.062 99	1.93	.075 98
1.28	.050 39	1.61	.063 39	1.94	.076 38
1.29	.050 79	1.62	.063 78	1.95	.076 77
1.30	.051 18	1.63	.064 17	1.96	.077 17
1.31	.051 57	1.64	.064 57	1.97	.077 56
1.32	.051 97	1.65	.064 96	1.98	.077 95
1.33	.052 36	1.66	.065 35	1.99	.078 35
				2.00	.078 74

mm/in.

METRIC CONVERSION PLANNER

mm	Inch	mm	Inch	mm	Inch	mm	Inch
2.01	.079 13	2.34	.092 13	2.67	.105 12	3.00	.118 11
2.02	.079 53	2.35	.092 52	2.68	.105 51		
2.03	.079 92	2.36	.092 91	2.69	.105 91		
2.04	.080 32	2.37	.093 31	2.70	.106 30		
2.05	.080 71	2.38	.093 70	2.71	.106 69		
2.06	.081 10	2.39	.094 09	2.72	.107 09		
2.07	.081 50	2.40	.094 49	2.73	.107 48		
2.08	.081 89	2.41	.094 88	2.74	.107 87		
2.09	.082 28	2.42	.095 28	2.75	.108 27		
2.10	.082 68	2.43	.095 67	2.76	.108 66		
2.11	.083 07	2.44	.096 06	2.77	.109 06		
2.12	.083 46	2.45	.096 46	2.78	.109 45		
2.13	.083 86	2.46	.096 85	2.79	.109 84		
2.14	.084 25	2.47	.097 24	2.80	.110 24		
2.15	.084 65	2.48	.097 64	2.81	.110 63		
2.16	.085 04	2.49	.098 03	2.82	.111 02		
2.17	.085 43	2.50	.098 43	2.83	.111 42		
2.18	.085 83	2.51	.098 82	2.84	.111 81		
2.19	.086 22	2.52	.099 21	2.85	.112 20		
2.20	.086 61	2.53	.099 61	2.86	.112 60		
2.21	.087 01	2.54	.100 00	2.87	.112 99		
2.22	.087 40	2.55	.100 39	2.88	.113 39		
2.23	.087 80	2.56	.100 79	2.89	.113 78		
2.24	.088 19	2.57	.101 18	2.90	.114 17		
2.25	.088 58	2.58	.101 57	2.91	.114 57		
2.26	.088 98	2.59	.101 97	2.92	.114 96		
2.27	.089 37	2.60	.102 36	2.93	.115 35		
2.28	.089 76	2.61	.102 76	2.94	.115 75		
2.29	.090 16	2.62	.103 15	2.95	.116 14		
2.30	.090 55	2.63	.103 54	2.96	.116 54		
2.31	.090 94	2.64	.103 94	2.97	.116 93		
2.32	.091 34	2.65	.104 33	2.98	.117 32		
2.33	.091 73	2.66	.104 72	2.99	.117 72		

mm/in.

METRIC CONVERSION PLANNER

mm	Inch	mm	Inch	mm	Inch
3.01	.118 50	3.34	.131 50	3.67	.144 49
3.02	.118 90	3.35	.131 89	3.68	.144 88
3.03	.119 29	3.36	.132 28	3.69	.145 28
3.04	.119 69	3.37	.132 68	3.70	.145 67
3.05	.120 08	3.38	.133 07	3.71	.146 06
3.06	.120 47	3.39	.133 46	3.72	.146 46
3.07	.120 87	3.40	.133 86	3.73	.146 85
3.08	.121 26	3.41	.134 25	3.74	.147 24
3.09	.121 65	3.42	.134 65	3.75	.147 64
3.10	.122 05	3.43	.135 04	3.76	.148 03
3.11	.122 44	3.44	.135 43	3.77	.148 43
3.12	.122 83	3.45	.135 83	3.78	.148 82
3.13	.123 23	3.46	.136 22	3.79	.149 21
3.14	.123 62	3.47	.136 61	3.80	.149 61
3.15	.124 02	3.48	.137 01	3.81	.150 00
3.16	.124 41	3.49	.137 40	3.82	.150 39
3.17	.124 80	3.50	.137 80	3.83	.150 79
3.18	.125 20	3.51	.138 19	3.84	.151 18
3.19	.125 59	3.52	.138 58	3.85	.151 57
3.20	.125 98	3.53	.138 98	3.86	.151 97
3.21	.126 38	3.54	.139 37	3.87	.152 36
3.22	.126 77	3.55	.139 76	3.88	.152 76
3.23	.127 17	3.56	.140 16	3.89	.153 15
3.24	.127 56	3.57	.140 55	3.90	.153 54
3.25	.127 95	3.58	.140 94	3.91	.153 94
3.26	.128 35	3.59	.141 34	3.92	.154 33
3.27	.128 74	3.60	.141 73	3.93	.154 72
3.28	.129 13	3.61	.142 13	3.94	.155 12
3.29	.129 53	3.62	.142 52	3.95	.155 51
3.30	.129 92	3.63	.142 91	3.96	.155 91
3.31	.130 31	3.64	.143 31	3.97	.156 30
3.32	.130 71	3.65	.143 70	3.98	.156 69
3.33	.131 10	3.66	.144 09	3.99	.157 09
				4.00	.157 48

mm/in.

METRIC CONVERSION PLANNER

mm	Inch	mm	Inch	mm	Inch
4.01	.157 87	4.34	.170 87	4.67	.183 86
4.02	.158 27	4.35	.171 26	4.68	.184 25
4.03	.158 66	4.36	.171 65	4.69	.184 65
4.04	.159 06	4.37	.172 05	4.70	.185 04
4.05	.159 45	4.38	.172 44	4.71	.185 43
4.06	.159 84	4.39	.172 83	4.72	.185 83
4.07	.160 24	4.40	.173 23	4.73	.186 22
4.08	.160 63	4.41	.173 62	4.74	.186 61
4.09	.161 02	4.42	.174 02	4.75	.187 01
4.10	.161 42	4.43	.174 41	4.76	.187 40
4.11	.161 81	4.44	.174 80	4.77	.187 80
4.12	.162 20	4.45	.175 20	4.78	.188 19
4.13	.162 60	4.46	.175 59	4.79	.188 58
4.14	.162 99	4.47	.175 98	4.80	.188 98
4.15	.163 39	4.48	.176 38	4.81	.189 37
4.16	.163 78	4.49	.176 77	4.82	.189 76
4.17	.164 17	4.50	.177 17	4.83	.190 16
4.18	.164 57	4.51	.177 56	4.84	.190 55
4.19	.164 96	4.52	.177 95	4.85	.190 95
4.20	.165 35	4.53	.178 35	4.86	.191 34
4.21	.165 75	4.54	.178 74	4.87	.191 73
4.22	.166 14	4.55	.179 13	4.88	.192 13
4.23	.166 54	4.56	.179 53	4.89	.192 52
4.24	.166 93	4.57	.179 92	4.90	.192 91
4.25	.167 32	4.58	.180 32	4.91	.193 31
4.26	.167 72	4.59	.180 71	4.92	.193 70
4.27	.168 11	4.60	.181 10	4.93	.194 09
4.28	.168 50	4.61	.181 50	4.94	.194 49
4.29	.168 90	4.62	.181 89	4.95	.194 88
4.30	.169 29	4.63	.182 28	4.96	.195 28
4.31	.169 69	4.64	.182 68	4.97	.195 67
4.32	.170 08	4.65	.183 07	4.98	.196 06
4.33	.170 47	4.66	.183 46	4.99	.196 46
				5.00	.196 85

mm/in.

METRIC CONVERSION PLANNER

mm	Inch	mm	Inch	mm	Inch
5.01	.197 24	5.34	.210 24	5.67	.223 23
5.02	.197 64	5.35	.210 63	5.68	.223 62
5.03	.198 03	5.36	.211 02	5.69	.224 02
5.04	.198 43	5.37	.211 42	5.70	.224 41
5.05	.198 82	5.38	.211 81	5.71	.224 80
5.06	.199 21	5.39	.212 20	5.72	.225 20
5.07	.199 61	5.40	.212 60	5.73	.225 59
5.08	.200 00	5.41	.212 99	5.74	.225 98
5.09	.200 39	5.42	.213 39	5.75	.226 38
5.10	.200 79	5.43	.213 78	5.76	.226 77
5.11	.201 18	5.44	.214 17	5.77	.227 17
5.12	.201 57	5.45	.214 57	5.78	.227 56
5.13	.201 97	5.46	.214 96	5.79	.227 95
5.14	.202 36	5.47	.215 35	5.80	.228 35
5.15	.202 76	5.48	.215 75	5.81	.228 74
5.16	.203 15	5.49	.216 14	5.82	.229 13
5.17	.203 54	5.50	.216 54	5.83	.229 53
5.18	.203 94	5.51	.216 93	5.84	.229 92
5.19	.204 33	5.52	.217 32	5.85	.230 32
5.20	.204 72	5.53	.217 72	5.86	.230 71
5.21	.205 12	5.54	.218 11	5.87	.231 10
5.22	.205 51	5.55	.218 50	5.88	.231 50
5.23	.205 91	5.56	.218 90	5.89	.231 89
5.24	.206 30	5.57	.219 29	5.90	.232 28
5.25	.206 69	5.58	.219 69	5.91	.232 68
5.26	.207 09	5.59	.220 08	5.92	.233 07
5.27	.207 48	5.60	.220 47	5.93	.233 46
5.28	.207 87	5.61	.220 87	5.94	.233 86
5.29	.208 27	5.62	.221 26	5.95	.234 25
5.30	.208 66	5.63	.221 65	5.96	.234 65
5.31	.209 06	5.64	.222 05	5.97	.235 04
5.32	.209 45	5.65	.222 44	5.98	.235 43
5.33	.209 84	5.66	.222 83	5.99	.235 83
				6.00	.236 22

mm/in.

METRIC CONVERSION PLANNER

mm	Inch	mm	Inch	mm	Inch
6.01	.236 61	6.34	.249 61	6.67	.262 60
6.02	.237 01	6.35	.250 00	6.68	.262 99
6.03	.237 40	6.36	.250 39	6.69	.263 39
6.04	.237 80	6.37	.250 79	6.70	.263 78
6.05	.238 19	6.38	.251 18	6.71	.264 17
6.06	.238 58	6.39	.251 57	6.72	.264 57
6.07	.238 98	6.40	.251 97	6.73	.264 96
6.08	.239 37	6.41	.252 36	6.74	.265 35
6.09	.239 76	6.42	.252 76	6.75	.265 75
6.10	.240 16	6.43	.253 15	6.76	.266 14
6.11	.240 55	6.44	.253 54	6.77	.266 54
6.12	.240 95	6.45	.253 94	6.78	.266 93
6.13	.241 34	6.46	.254 33	6.79	.267 32
6.14	.241 73	6.47	.254 72	6.80	.267 72
6.15	.242 13	6.48	.255 12	6.81	.268 11
6.16	.242 52	6.49	.255 51	6.82	.268 50
6.17	.242 91	6.50	.255 91	6.83	.268 90
6.18	.243 31	6.51	.256 30	6.84	.269 29
6.19	.243 70	6.52	.256 69	6.85	.269 69
6.20	.244 09	6.53	.257 09	6.86	.270 08
6.21	.244 49	6.54	.257 48	6.87	.270 47
6.22	.244 88	6.55	.257 87	6.88	.270 87
6.23	.245 28	6.56	.258 27	6.89	.271 26
6.24	.245 67	6.57	.258 66	6.90	.271 65
6.25	.246 06	6.58	.259 06	6.91	.272 05
6.26	.246 46	6.59	.259 45	6.92	.272 44
6.27	.246 85	6.60	.259 84	6.93	.272 83
6.28	.247 24	6.61	.260 24	6.94	.273 23
6.29	.247 64	6.62	.260 63	6.95	.273 62
6.30	.248 03	6.63	.261 02	6.96	.274 02
6.31	.248 43	6.64	.261 41	6.97	.274 41
6.32	.248 82	6.65	.261 81	6.98	.274 80
6.33	.249 21	6.66	.262 20	6.99	.275 20
				7.00	.275 59

mm/in.

METRIC CONVERSION PLANNER

mm	Inch	mm	Inch	mm	Inch
7.01	.275 98	7.34	.288 98	7.67	.301 97
7.02	.276 38	7.35	.289 37	7.68	.302 36
7.03	.276 77	7.36	.289 76	7.69	.302 76
7.04	.277 17	7.37	.290 16	7.70	.303 15
7.05	.277 56	7.38	.290 55	7.71	.303 54
7.06	.277 95	7.39	.290 95	7.72	.303 94
7.07	.278 35	7.40	.291 34	7.73	.304 33
7.08	.278 74	7.41	.291 73	7.74	.304 72
7.09	.279 13	7.42	.292 13	7.75	.305 12
7.10	.279 53	7.43	.292 52	7.76	.305 51
7.11	.279 92	7.44	.292 91	7.77	.305 91
7.12	.280 32	7.45	.293 31	7.78	.306 30
7.13	.280 71	7.46	.293 70	7.79	.306 69
7.14	.281 10	7.47	.294 09	7.80	.307 09
7.15	.281 50	7.48	.294 49	7.81	.307 48
7.16	.281 89	7.49	.294 88	7.82	.307 87
7.17	.282 28	7.50	.295 28	7.83	.308 27
7.18	.282 68	7.51	.295 67	7.84	.308 66
7.19	.283 07	7.52	.296 06	7.85	.309 06
7.20	.283 46	7.53	.296 46	7.86	.309 45
7.21	.283 86	7.54	.296 85	7.87	.309 84
7.22	.284 25	7.55	.297 24	7.88	.310 24
7.23	.284 65	7.56	.297 64	7.89	.310 63
7.24	.285 04	7.57	.298 03	7.90	.311 02
7.25	.285 43	7.58	.298 43	7.91	.311 42
7.26	.285 83	7.59	.298 82	7.92	.311 81
7.27	.286 22	7.60	.299 21	7.93	.312 20
7.28	.286 61	7.61	.299 61	7.94	.312 60
7.29	.287 01	7.62	.300 00	7.95	.312 99
7.30	.287 40	7.63	.300 39	7.96	.313 39
7.31	.287 80	7.64	.300 79	7.97	.313 78
7.32	.288 19	7.65	.301 18	7.98	.314 17
7.33	.288 58	7.66	.301 58	7.99	.314 57
				8.00	.314 96

mm/in.

METRIC CONVERSION PLANNER

mm	Inch	mm	Inch
8.01	.315 35	8.34	.328 35
8.02	.315 75	8.35	.328 74
8.03	.316 14	8.36	.329 13
8.04	.316 54	8.37	.329 53
8.05	.316 93	8.38	.329 92
8.06	.317 32	8.39	.330 32
8.07	.317 72	8.40	.330 71
8.08	.318 11	8.41	.331 10
8.09	.318 50	8.42	.331 50
8.10	.318 90	8.43	.331 89
8.11	.319 29	8.44	.332 28
8.12	.319 69	8.45	.332 68
8.13	.320 08	8.46	.333 07
8.14	.320 47	8.47	.333 46
8.15	.320 87	8.48	.333 86
8.16	.321 26	8.49	.334 25
8.17	.321 65	8.50	.334 65
8.18	.322 05	8.51	.335 04
8.19	.322 44	8.52	.335 43
8.20	.322 83	8.53	.335 83
8.21	.323 23	8.54	.336 22
8.22	.323 62	8.55	.336 61
8.23	.324 02	8.56	.337 01
8.24	.324 41	8.57	.337 40
8.25	.324 80	8.58	.337 80
8.26	.325 20	8.59	.338 19
8.27	.325 59	8.60	.338 58
8.28	.325 98	8.61	.338 98
8.29	.326 38	8.62	.339 37
8.30	.326 77	8.63	.339 76
8.31	.327 17	8.64	.340 16
8.32	.327 56	8.65	.340 55
8.33	.327 95	8.66	.340 95

mm	Inch
8.67	.341 34
8.68	.341 73
8.69	.342 13
8.70	.342 52
.871	.342 91
8.72	.343 31
8.73	.343 70
8.74	.344 09
8.75	.344 49
8.76	.344 88
8.77	.345 28
8.78	.345 67
8.79	.346 06
8.80	.346 46
8.81	.346 85
8.82	.347 24
8.83	.347 64
8.84	.348 03
8.85	.348 43
8.86	.348 82
8.87	.349 21
8.88	.349 61
8.89	.350 00
8.90	.350 39
8.91	.350 79
8.92	.351 18
8.93	.351 58
8.94	.351 97
8.95	.352 36
8.96	.352 76
8.97	.353 15
8.98	.353 54
8.99	.353 94
9.00	.354 33

mm/in.

METRIC CONVERSION PLANNER

mm	Inch	mm	Inch	mm	Inch
9.01	.354 72	9.34	.367 72	9.67	.380 71
9.02	.355 12	9.35	.368 11	9.68	.381 10
9.03	.355 51	9.36	.368 50	9.69	.381 50
9.04	.355 91	9.37	.368 90	9.70	.381 89
9.05	.356 30	9.38	.369 29	9.71	.382 28
9.06	.356 69	9.39	.369 69	9.72	.382 68
9.07	.357 09	9.40	.370 08	9.73	.383 07
9.08	.357 48	9.41	.370 47	9.74	.383 46
9.09	.357 87	9.42	.370 87	9.75	.383 86
9.10	.358 27	9.43	.371 26	9.76	.384 25
9.11	.358 66	9.44	.371 65	9.77	.384 65
9.12	.359 06	9.45	.372 05	9.78	.385 04
9.13	.359 45	9.46	.372 44	9.79	.385 43
9.14	.359 84	9.47	.372 83	9.80	.385 83
9.15	.360 24	9.48	.373 23	9.81	.386 22
9.16	.360 63	9.49	.373 62	9.82	.386 61
9.17	.361 02	9.50	.374 02	9.83	.387 01
9.18	.361 42	9.51	.374 41	9.84	.387 40
9.19	.361 81	.952	.374 80	9.85	.387 80
9.20	.362 20	9.53	.375 20	9.86	.388 19
9.21	.362 60	9.54	.375 59	9.87	.388 58
9.22	.362 99	9.55	.375 98	9.88	.388 98
9.23	.363 39	9.56	.376 38	9.89	.389 37
9.24	.363 78	9.57	.376 77	9.90	.389 76
9.25	.364 17	9.58	.377 17	9.91	.390 16
9.26	.364 57	9.59	.377 56	9.92	.390 55
9.27	.364 96	9.60	.377 95	9.93	.390 95
9.28	.365 35	9.61	.378 35	9.94	.391 34
9.29	.365 75	9.62	.378 74	9.95	.391 73
9.30	.366 14	9.63	.379 13	9.96	.392 13
9.31	.366 54	9.64	.379 53	9.97	.392 52
9.32	.366 93	9.65	.379 92	9.98	.392 91
9.33	.367 32	9.66	.380 32	9.99	.393 31
				10.00	.393 70

mm/in.

METRIC CONVERSION PLANNER

mm	Inch	mm	Inch	mm	Inch
11	.433 07	44	1.732 28	77	3.031 50
12	.472 44	45	1.771 65	78	3.070 87
13	.511 81	46	1.811 02	79	3.110 24
14	.551 18	47	1.850 39	80	3.149 61
15	.590 55	48	1.889 76	81	3.188 98
16	.629 92	49	1.929 13	82	3.228 35
17	.669 29	50	1.968 50	83	3.267 72
18	.708 66	51	2.007 87	84	3.307 09
19	.748 03	52	2.047 24	85	3.346 46
20	.787 40	53	2.086 61	86	3.385 83
21	.826 77	54	2.125 98	87	3.425 20
22	.866 14	55	2.165 35	88	3.464 57
23	.905 51	56	2.204 72	89	3.503 94
24	.944 88	57	2.244 09	90	3.543 31
25	.984 25	58	2.283 46	91	3.582 68
26	1.023 62	59	2.322 83	92	3.622 05
27	1.062 99	60	2.362 20	93	3.661 42
28	1.102 36	61	2.401 57	94	3.700 79
29	1.141 73	62	2.440 94	95	3.740 16
30	1.181 10	63	2.480 31	96	3.779 53
31	1.220 47	64	2.519 69	97	3.818 90
32	1.259 84	65	2.559 06	98	3.858 27
33	1.299 21	66	2.598 43	99	3.897 64
34	1.338 58	67	2.637 80	100	3.937 01
35	1.377 95	68	2.677 17	200	7.874 0
36	1.417 32	69	2.716 54	300	11.811 0
37	1.456 69	70	2.755 91	400	15.748 0
38	1.496 06	71	2.795 28	500	19.685 1
39	1.535 43	72	2.834 65	600	23.622 1
40	1.574 80	73	2.874 02	700	27.559 1
41	1.614 17	74	2.913 39	800	31.496 1
42	1.653 54	75	2.952 76	900	35.433 1
43	1.692 91	76	2.992 13	1000	39.370 1

=1 metre

THE SI SYSTEM OF METRIC CONVERSIONS
approximate

ENGLISH TO METRIC	METRIC TO ENGLISH
inches (ins.) X 25.4 = millimetres (mm)	mm X 0.04 = ins.
feet (ft.) X 0.3 = metres (m)	m X 3.3 = ft.
yards (yds.) X 0.9 = metres (m)	m X 1.1 = yds.
miles (mi.) X 1.6 = kilometres (km)	km X 0.6 = mi.
sq. inch (in²) X 6.5 = sq. centimetres (cm²)	cm² X 0.16 = in²
sq. feet (ft²) X 0.09 = sq. metres (m²)	m² X 11.00 = ft²
sq. yard (yd²) X 0.8 = sq. metres (m²)	m² X 1.2 = yd²
acre (a) X 0.4 = hectares (ha)	ha X 2.5 = a
cu. in. (in³) X 16.0 = cu. centimetres (cm³)	cm³ X 0.06 = in³
cu. ft. (ft³) X 0.03 = cu. metres (m³)	m³ X 35.0 = ft³
cu. yd. (yd³) X 0.8 = cu. metres (m³)	m³ X 1.3 = yd³
(liq) quart (qt) X 0.9 = litre (l)	l X 1.05 = qt
gallon (gal) X 0.004 = cu. metre (m³)	m³ X264.2 = gal
(avdp) ounce (oz) X 28.3 = grams (g)	g X 0.035 = oz
(avdp) pound (lb) X 0.45 = kilogram (kg)	kg X 2.20 = lb
horsepower (h.p.) X 0.75 = kilowatt (kW)	kw X 1.34 = h.p.
ft. per sec. (ft/s) X 0.304 = met. per sec. (m/s)	m/s X 3.280 = ft/s
ounce-force (ozf) X 0.278 = newtons (N)	N X 3.597 = ozf
pounds-force (lbf) X 4.448 = newtons (N)	N X 0.224 = lbf
foot-pounds (ft. lb) X 1.355 = newton-metres (N.m)	N-m X 0.737 = ft. lb.
foot-pounds (ft. lb) X 1.355 = joules (J)	J X 0.737 = ft. lb.
in.-pounds (in. lb.) X 0.112 = newton-metres (N m)	N-m X 8.850 = in. lb
lb. per foot (lb/ft) X 14.593 = new. per metre (N/m)	N-m X 0.068 = lb/ft
cycles per sec. (cps) X 1.0 = hertz (Hz)	Hz X 1.0 = cps
Brit. Therm Unit (Btu) X 1 055.06 = joules (J)	J X 0.000 94 = Btu
Degrees Fah. (°F) X 5/9 after sub. 32 = deg. Celsius (°C)	°C X 9/5 then add 32 = °F

NOTES

NOTES

DIMENSIONING & TOLERANCING

(INTERPRETATION OF ANSI Y14.5M—1982)

FORWARD

THIS INDEX APPLIES ONLY TO THE SECTION 10M (1982) THAT IT PRECEDES. REFER TO THE MAIN INDEX IN THE FRONT OF THE MANUAL FOR THE INDEX TO SECTION 10 (1973).

CONTENTS

10	DIMENSIONS AND TOLERANCES
10.1	Scope
10.2	Applicable Documents
10.3	Definitions
10.4	General Requirements
10.5	Method of Dimensioning
10.6	General Rules and Interpretations of Dimensions and Tolerances
10.7	Geometric Tolerances
10.8	Datums Targets
10.9	Drawing Application and Interpretation
10.10	Positional Tolerance

INDEX

- A -
ALLOWANCE	10.3.1
ANGLE, RIGHT	10.10.5
ANGULAR DIMENSIONS	10.4.1.2
ANGULARITY	10.9.8
ARCS	10.5.15
AXIS	10.3.2

- B -
BASIC	
dimensions	10.3.5.1
	10.7.4
size	10.3.10
BILATERAL	
tolerance	10.3.13.1

- C -
CENTER LINE, SYMMETRY	10.5.5
CENTER PLANE	10.3.3
CHAMFERS	10.5.16
CIRCULARITY	10.9.3
CLEARANCE FIT	10.3.8.1
CONCENTRICITY	10.9.10
COORDINATE DIMENSIONING	10.5.9
COUNTERBORE	10.5.12
COUNTERSINK	10.5.14
CYLINDRICITY	10.9.4

- D -
DATUM	10.3.4
feature	10.3.7
implied	10.6.2
multiple	10.7.6
plane	10.37
reference	10.7.3
rules	10.6.2
selection	10.6.2.3
surface	10.8.1
symbol	10.7.3
targets	10.3.4.1
	10.8
method	10.8.2
true position	10.10.4
DECIMAL SYSTEM	10.4.1
DESIGN	
size	10.3.6
DIAMETRAL DIMENSIONS	10.5.4
DIMENSION	10.3.5
and tolerance	10.0
angular	10.4.1.2
basic	10.3.5.1
	10.7.4
diametral	10.5.4
limit	10.3.9
locating	10.3.5.2

DIMENSION (continued)	
maximum	10.3.5.3
minimum	10.3.5.4
profile	10.9.5.1
reference	10.3.5.5
size	10.3.5.6
staggered	10.5.3
true	10.5.24
typical	10.4.3
DIMENSIONAL TOLERANCES, DRAWING	10.4.2
DIMENSIONING	
center line	10.4.3
chain	10.4.3
chamfers	10.5.16
coordinate	10.5.9
counterbore	10.5.12
countersink	10.5.14
from datums	10.6.2.2
hidden lines	10.4.3
holes	10.5.21
keyways	10.5.22
knurls	10.5.23
method of	10.5
practices	10.4
repetitive	10.5.19
rounded ends	10.5.8
selective	10.5.20
spherical radii	10.5.11
selection of	10.6.2.3
spotface	10.5.13
square shapes	10.5.10
staggered	10.5.3
tapers	10.5.22
true position	10.10
tubes	10.5.7

- E -
ENDS, ROUNDED, DIMENSIONING	10.5.8

- F -
FASTENER	
fixed	10.10.3
floating	10.10.3
FEATURE	10.3.7
size, regardless of	10.3.16
FIM	10.3.18
FIT	10.3.8
clearance	10.3.8.1
interference	10.3.8.2
line	10.3.8.4
transition	10.3.8.3
FLATNESS	10.9.1
tolerances	10.7.9.1
FORM	

INDEX

FORM (continued)
tolerances — 10.7.10
FORMULAE TRUE POSITION TOLERANCE — 10.10.3
FREE-STATE VARIATION — 10.9.3.2
FULL INDICATOR MOVEMENT — 10.3.18

- G -
GEOMETRIC SYMBOLS — 10.7.2
GEOMETRIC TOLERANCES — 10.7

- H -
HOLE
depth — 10.5.21.1
dimensioning — 10.5.21
locating — 10.5.21.2
thru — 10.5.21.1

- I -
INTERFERENCE FIT — 10.3.8.2

- K -
KEYWAYS — 10.5.22
KNURLS — 10.5.23

- L -
LEADERLINES — 10.8.4
LEAST MATERIAL CONDITION — 10.3.15
LIMIT DIMENSIONS — 10.3.9
LIMITS — 10.3.9.1
— 10.3.9.2
LINES, LEADER — 10.8.4
LMC — 10.3.15
LOCATING
datum targets — 10.8.3
dimensions — 10.3.5.2
holes — 10.5.21.2

- M -
MAXIMUM
dimensions — 10.3.5.3
material condition — 10.3.14
— 10.10.2
MEAN SIZE — 10.3.11
MINIMUM DIMENSIONS — 10.3.5.4
MMC — 10.3.14
— 10.10.2
MULTIPLE DATUM — 10.7.6

- N -
NOMINAL SIZE — 10.3.12

- O -
ORIGIN — 10.5.18

- P -
PARALLELISM — 10.9.6
PERPENDICULARITY — 10.9.7
PLANE
center — 10.3.3
POINT, DATUM TARGET — 10.8
PRACTICES, DIMENSIONING — 10.4
— 10.5
PROFILE, TOLERANCE DEFINITION OF — 10.9.5
PROFILE TOLERANCE — 10.9.5.1
application — 10.9.5.2

- R -
RADII — 10.5.17
RADIUS, SPHERICAL DIMENSIONING — 10.5.11
REFERENCE
datum — 10.7.3
dimension — 10.3.5.5
REGARDLESS OF FEATURE SIZE — 10.3.16
RFS — 10.3.16
ROUNDED ENDS, DIMENSIONING — 10.5.8
ROUNDNESS (SEE CIRCULARITY) — 10.9.3
RULES, TRUE POSITION DIMENSIONING — 10.10
RUNOUT — 10.9.9
circular — 10.9.9.5
total — 10.9.9.6

- S -
SIZE
basic — 10.3.10
design — 10.3.6
dimension — 10.3.5.6
limits of — 10.6.1
mean — 10.3.11
nominal — 10.3.12
stock — 10.5.6
tolerance — 10.3.13
— 10.6.1
SPHERICAL RADII, DIMENSIONING — 10.5.11
SPOTFACE — 10.5.13
STAGGERED DIMENSIONS — 10.5.3
STANDARD
tolerances — 10.4.2
STOCK SIZE — 10.5.6
STRAIGHTNESS — 10.9.2
SURFACE
datum — 10.6.2
SYMBOL — 10.7.2
angularity — 10.7.2

INDEX

SYMBOL (continued)	
basic	10.7.4
circularity	10.7.2
combined	10.7.7
composite	10.7.8
concentricity	10.7.2
cylindricity	10.7.2
datum, identifying	10.7.3
datum target	10.8.2
drawing application and interpretation	10.9
feature control	10.7.5
flatness	10.7.2
former	10.7.11
parallelism	10.7.2
perpendicularity	10.7.2
placement	10.7.4
straightness	10.7.2
symmetry	10.7.2
tolerancing, geometric	10.7.2
true position	10.7.2
SYMMETRICAL OUTLINE	10.5.5
SYMMETRY	10.9.11

- T -

TAPERS	10.5.22
diameter	10.5.22
flat	10.5.22.2
TOLERANCE	10.0
	10.3.13
angularity	10.9.8.1
bilateral	10.3.13.1
circularity	10.9.3.1
concentricity	10.9.10.1
cylindricity	10.9.4.1
flatness	10.9.1.1
formula, true position	10.10.3
gears	10.6.3.2
geometric	10.7
parallelism	10.9.6.1
perpendicularity	10.9.7.1
positional	10.10
profile	10.9.5.1
roundness (see circularity)	10.9.3.1
free-state variation	10.9.8.2
runout	10.9.9.1
screw threads	10.6.3.1
size	10.6.1
splines	10.6.3.2
standard	10.4.2
straightness	10.9.2.1
symmetry	10.9.11.1
true position	10.10.1
unilateral	10.3.13.2
zone	10.9
TRANSITION FIT	10.3.8.3

TRUE	
dimensions	10.5.24
position	10.10
datums	10.10.4
rules	10.10.2
symbol	10.7.2
tolerance formula	10.10.3
TUBES, DIMENSIONING OF	10.5.7
TYPICAL	10.4.3

- U -

UNILATERAL TOLERANCE	10.3.13.2
UNIT OF MEASURE	10.4.1
angular	10.4.1.2
inch	10.4.1.1

- V -

VARIATION, FREE-STATE	10.9.3.2
VIRTUAL CONDITION	10.3.17

- Z -

ZONE	
tolerance	
cylindrical	10.10
true position	10.10

Section 10M — DIMENSIONS AND TOLERANCES FOR ENGINEERING DRAWINGS

NOTE: Section 10M is an interpretation of ANSI Y14.5M-1982, 'Dimensioning and Tolerancing for Engineering Drawings', to be released approximately in early 1983. This DRM reflects the changes and new material added to the 1973 ANSI Y14.5.

Both the 1973 and the 1982 interpretations in Section 10 and 10M respectively are presented since the drawings prepared to the 1973 version will be around for many years until they are phased into and updated to the 1982 version. New drawings should use the 1982 version as soon as the official publication date is released, unless otherwise stated in the contract to the contrary.

10.1 SCOPE — This section establishes and illustrates the methods of specifying dimensions and tolerances on drawings.

10.2 APPLICABLE DOCUMENTS

ANSI B4.1 Preferred Limits and Fits for Cylindrical Parts

ANSI Y14.5M Dimensioning and Tolerancing for Engineering Drawings

10.3 DEFINITIONS

10.3.1 ALLOWANCE — The prescribed difference between the maximum material condition of mating parts.

10.3.2 AXIS — The centering of a cylinder, cone, or other surface of revolution as established by the contacting points of the actual surface of the line formed at the intersection of two planes.

10.3.3 CENTER PLANE — The middle or median plane of a feature.

10.3.4 DATUM — Points, lines, planes, cylinders, assumed to be exact for purposes of computation or reference, as established from actual features, and from which the location or form of other features of a part may be established. When datums are specified on the drawing, they exist not in the part itself but in the much more precisely made manufacturing or inspection equipment.

10.3.4.1 DATUM TARGET — A specified point, line, or area on a part which is identified on a drawing as a datum.

10.3.5 DIMENSION — A numerical value expressed in appropriate units of measure and indicated on a drawing along with lines, symbols and notes, to define a geometrical characteristic of an object.

10.3.5.1 BASIC DIMENSION — A numerical value used to describe the theoretically exact size, shape, or location of a feature or datum target. It is the basis from which permissible variations are established by tolerances on other dimensions, in notes or by feature control symbols. Basic dimensions are shown on the drawing enclosed in a rectangle. $\boxed{.XXX}$ (Older methods were 'BASIC' or 'BSC')

10.3.5.2 LOCATING DIMENSION — One which specifies a position or distance of one feature of an object with respect to another or to a datum.

10.3.5.3 MAXIMUM DIMENSION — A dimension that controls the maximum limit; the minimum limit being controlled by other elements of design.

10.3.5.4 MINIMUM DIMENSION — A dimension that controls the minimum limit; the maximum limit being controlled by other elements of design.

10.3.5.5 REFERENCE DIMENSION

(a) A dimension that has been specified elsewhere on the same drawing or from another drawing or document. These reference callouts state the nominal dimensions, normally without tolerance, or both limits of a limit dimension. The preferred method for indicating reference dimensions on drawings is to enclose the dimensions with parentheses. e.g. (.250).

(b) A dimension that is an accumulation of other dimensions. These reference callouts state either the mean or nominal dimension and the tolerance may be stated when necessary. A reference dimension is not used for manufacturing or inspection purposes.

10.3.5.6 SIZE DIMENSION — The specified value of a diameter, length, width, or other geometric characteristic.

10.3.6 DESIGN SIZE — The ideal or preferred size from which variation is allowed by applying tolerances.

10.3.7 FEATURE — A feature is any component portion of a part that can be used as basis for a datum. An individual feature may be:

(a) A plane surface (in which case there is no consideration of feature size).

(b) A single cylindrical or spherical surface, or two plane parallel surfaces (all of which are associated with a size dimension).

Complex features are composed of two or more individual features as defined above.

10.3.8 FIT — A general term used to signify the range of interference or clearance which results from the application of tolerances in the design of mating parts. Fits for cylindrical parts should be established using the preferred limits specified in ANSI B4.1. Fit dimensions shall be specified on the drawing and not called out by the fit designation such as RC1, LC5, etc.

10.3.8.1 CLEARANCE FIT — One having limits so designed that a clearance always results when the mating parts are assembled.

10.3.8.2 INTERFERENCE FIT — One having limits so designed that an interference always results when mating parts are assembled.

10.3.8.3 TRANSITION FIT — One having limits so designed that either a clearance or an interference may result when mating parts are assembled.

10.3.8.4 LINE FIT — One having limits so designed that surface contact or clearance may result when mating parts are assembled.

10.3.9 LIMITS — Maximum and minimum values prescribed for specific dimensions. (See Figure 1.)

10.3.9.1 All limits or tolerances are considered to be absolute. Dimensional limits, regardless of the number of decimal places, are to be used as if they were continued with zeros.

10.3.9.2 For purposes of determining conformance with limits, the measured value is to be compared directly with the specified value and any deviation, however small, outside of the specified limiting values signifies non-conformance with the limits.

Example:

1.25 means 1.220--0
1.280--0

1.250 means 1.240--0
1.260--0

1.250
1.252 means 1.2500--0
1.2520--0

10.3.10 BASIC SIZE — The exact theoretical size from which the limits of size are derived by the application of allowances and tolerances.

10.3.11 MEAN SIZE — The size midway between the limits of size.

10.3.12 NOMINAL SIZE — The designation used for the purpose of general identification. For example, a rod may be referred to as .250 diameter although the actual dimension on the drawing is .249 diameter. In this case, the .250 diameter is the nominal size.

10.3.13 TOLERANCE — The total permissible variation from design size, form, or location.

10.3.13.1 BILATERAL TOLERANCE — A tolerance in which variation is permitted in both directions from the specified dimension. (See Figure 1.)

10.3.13.2 UNILATERAL TOLERANCE — A tolerance in which variation is permitted in only one direction from the specified dimension. (See Figure 1.)

Bilateral tolerance |← 1.878 +.002/-.001 →|

Unilateral tolerance |← 1.880 +.000/-.003 →|

Limit dimension |← 1.887 / 1.880 →|

The Application of Tolerances
Figure 1

10.3.14   MAXIMUM MATERIAL CONDITION (MMC or Ⓜ) — The condition where the feature of size contains the maximum amount of material. e.g., minimum hole diameter and maximum shaft diameter.

10.3.15   LEAST MATERIAL CONDITION (LMC or Ⓛ) — The condition where the feature of size contains the least (minimum) amount of material. e.g., maximum hole diameter diameter and minimum shaft diameter.

10.3.16   REGARDLESS OF FEATURE SIZE (RFS or Ⓢ) — The condition where tolerance of position or form must be met irrespective of where the feature lies within its size tolerance.

10.3.17   VIRTUAL CONDITION — The boundary generated by the collective effects of the MMC limit of a feature and any applicable form or positional tolerance.

10.3.18   FULL INDICATOR MOVEMENT (FIM) — The total movement of the indicator when applied to a surface in an appropriate manner. The terms Full Indicator Reading (FIR) and Total Indicator Reading (TIR) were formerly used.

10.4   GENERAL REQUIREMENTS

10.4.1   UNITS OF MEASURE — Unless otherwise specified herein or in other DRM sections, all dimensions and tolerances shall be expressed in inches and decimal parts of an inch or in angular units.

10.4.1.1   RULES APPLICABLE TO INCH UNITS — When specifying decimal dimensions and tolerances the following rules shall apply:

   (a)   Zeros shall not be used before the decimal point for values less than one inch. (This is not the case for Metric Dimensioning)

   (b)   All decimals shall have a minimum of two digits following the decimal point.

   (c)   A dimension and its tolerance, or both limits of a limit dimension, shall have an equal number of digits following the decimal point.

   (d)   Both tolerances shall be specified when using unilateral tolerances. On existing drawings where only one tolerance is shown, the unspecified tolerance shall be interpreted to be zero.

   (e)   Unilateral and bilateral tolerances used with dimension lines shall show the tolerances following the dimension. Unilateral or unequal bilateral tolerances shall be shown with the plus tolerance above the minus tolerance. (See Figure 1.)

   (f)   When unilateral or unequal bilateral tolerances are specified in general notes, the tolerances may be shown on the same line with the plus tolerance preceding the minus tolerance, e.g. 1.50 + .03 - .00, 1.500 + .003 - .001.

(g) When unilateral tolerances are used, it is preferred that the dimensions specify the maximum position or MMC size with the tolerances applied to the minimum position or LMC size tolerance.

Example:

Position	2.500	+ .000 / - .005
Shaft	1.000	+ .000 / - .002
Hole	.998	+ .002 / - .000

(h) Normally when the tolerance of a dimension is equal to the standard title block tolerance, limit dimensions or dimensions with unilateral or unequal bilateral tolerances are not used.

(i) Limit dimensions shall be shown on drawings as follows:

(1) Limit dimensions used with dimension lines shall show the maximum value above the minimum value. e.g. 1.887
1.880

(2) When limit dimensions are shown in a single line, the limit dimension that otherwise would be above precedes the other with a dash midway between the digit height, e.g. ⌀.990 - 1.000 THRU

10.4.1.2 RULES APPLICABLE TO ANGULAR UNITS — When specifying angular dimensions and tolerances the following rules shall apply:

(a) Angular dimensions and tolerances shall be expressed in degrees (°) and when necessary, in minutes (') and seconds (") or in decimal parts of a degree.

(b) An angular dimension and its tolerance or both limits of a limit dimension shall be held to the same units of measure,

e.g. 30° ±5°   (cont.)
30° 0' 30" ±0° 0' 15"   30.8° ±.5°
30° 30' 0"   30° 30.0' ±0° .5'
30° 30' 45"

An exception to this rule is when the standard title block tolerance applies and is specified in degrees only and the field of the drawing uses degrees and parts of a degree.

(c) The requirements of 10.4.1.1 (d) through (i) also apply to angular dimensions.

10.4.1.3 Tolerances used in geometric characteristic symbols shall be held to a minimum of three decimal places regardless of the number of decimal places in the position or size dimension.

10.4.2 STANDARD TOLERANCES — Dimensions shown without tolerance are controlled by the standard tolerances in the title block, except stock materials, dimensions on welding symbols, undimensioned angles between lines drawn at 90°, dimensions labeled REF, MAX, MIN, BASIC, and similar dimensions that are otherwise controlled. Some standard tolerances are .XX = ±.03 and .XXX = ±.010.

10.4.2.1 The format tolerances shall not be altered except where a larger tolerance is required for the majority of the dimensions. The revised tolerances cannot be less than the format tolerances and must always be progressively smaller as the number of decimal places increases, e.g., .XXX cannot have a larger tolerance than .XX. When a third tolerance is required, a four place decimal may be added. This is accomplished by adding the following delta note with the applicable correction or addition:

⚠ TOLERANCE ON DECIMALS

.XX ± .-- (Add applicable two place tolerance)

.XXX ± .--- (Add applicable three place tolerance)

.XXXX ± .---- (Add applicable four place tolerance)

The format tolerance(s) will be lined out and the delta note number added in the block, for the revised format tolerance(s).

10.4.3 FUNDAMENTAL RULES OF DIMENSIONING — Dimensioning shall conform to the following rules:

(a) Dimension, extension and leader lines shall not cross each other unless absolutely necessary. When it is unavoidable, a dimension line is never broken except for insertion of the dimension. An extension or leader line shall not run through a dimension nor shall they be broken except where they pass through or adjacent to arrowheads.

(b) Dimensions are shown in the view that most clearly represents the form of the feature.

(c) Sufficient dimensions shall be shown to clearly define size, shape and position of each feature.

(d) A feature shall not be located by more than one toleranced dimension in any one direction.

(e) A dimension shall be enclosed in parentheses '( )' when it is (1) repeated on the same drawing, (2) specified on a subordinate document, (3) an accumulation of other dimensions or (4) shown for information purposes.

(f) Unless clarity is improved, dimensions are shown outside the outline of the part.

(g) Dimensions are selected and arranged to minimize the accumulation of tolerances between related features.

(h) Each dimension shall be expressed clearly so that it can be interpreted in only one way.

(i) Only the end product dimensions and data are shown on drawings unless essential to the definition of engineering requirements. When non-mandatory in-process manufacturing information is shown on the drawing, it shall be marked with a note similar to "NON-MANDATORY, MANUFACTURING DATA".

(j) Chain dimensions should be avoided unless basic dimensions are used.

(k) Center lines, object lines or extension lines should not be used as dimension lines.

(l) Dimensioning to hidden lines shall be avoided.

(m) Maximum and minimum limits must be such that parts will assemble and function under all dimensional conditions that are within limits.

(n) The word "TYPICAL" or the abbreviation "TYP" is not used. Indicate the number of places the dimension applies.

10.5 METHOD OF DIMENSIONING

10.5 METHOD OF DIMENSIONING — Dimensional characteristics are specified on the drawing by the use of symbols. When dimensional symbols do not adequately describe the desired condition, a note may be used, either separately or supplementing the symbol. Table 1 shows the approved symbol for each characteristic.

10.5.1 Parts and features should be dimensioned by the method that most clearly shows the design requirements. Bilateral, unilateral, and limit dimensions may all be used on the same drawing to achieve this requirement.

10.5.2 Figure 2 shows how dimensions are to be spaced.

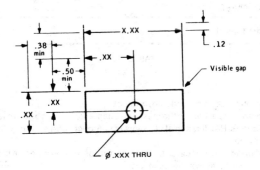

Spacing of Dimensions
Figure 2

10.5.3 Staggered dimensions shall be used to prevent interference with other dimensions. (See Figure 28.)

10.5.4 In all views, diametral dimensions shall be specified as the symbol ⌀.

SYMBOL	DEFINITION	See Note	See Figure
R	RADII	1 & 2	16
SR	SPHERICAL RADII	1	8
SØ	SPHERICAL DIAMETER	1	43
Ø	DIAMETER	1	4
□	SQUARE --- Single dimension applies to square shape.	1	7
∨	COUNTERSINK	1	11
⊔	COUNTERBORE or SPOTFACE	1	9 & 10
⌒	ARC	3	13
↧	DEPTH	1	9
2.000	DIMENSION ORIGIN	4	17
⌀	SYMBOL FOR ALL AROUND --- Used with profile tolerance applications		47a
X	REPETITIVE FEATURES AND DIMENSIONS	5	18
∓	SYMMETRICAL OUTLINES --- May show dimensions only on one side of centerline.		3
─ · ─	CHAINLINE --- Indicates limited length or area of a surface to recieve additional treatment per limits specified on drawing		19
◁	FLAT TAPER		30
▷	BASIC TAPER		28

TABLE 1

NOTES:
1.) Each dimension value is preceded by symbol.
2.) The radial contour of part within the crescent-shaped zone must be a curve without sudden angular deviations. RADII taken at all points of contour of part shall be within the minimum and maximum limits.
3.) Symbol used to indicate a linear dimension is an arc and is measured on the curved surface. The symbol is placed above the dimension.
4.) Origin symbol is used to indicate the end of a toleranced dimension between parallel surfaces that originates a plane established by one of the surfaces. This symbol is not to be used with geometric tolerancing applications.
5.) An "X" may be used to specify repetitive dimensions and features. Along with a numerical value to indicate the number of places or times. The dimension or feature is required. There should be a space between the "X" and the dimension.

10.5.5 SYMMETRICAL OUTLINE — The symmetrical outline symbol, (see Table 1) is used when drawing space is limited and only one-half of the symmetrical shape can be conveniently shown or when quantities of like features are specified for an entire view. See Figure 3. (Previous symbol was ⌐⌡)

10.5.6 STOCK SIZE — When the stock size is specified in the parts list and used as furnished, it shall be indicated on the field of the drawing by the word "STOCK". When two stock dimensions are used and the orientation is not obvious from the picture, the dimensions shall be included in the callout (See Figure 5). The foregoing does not apply for shapes such as I-beams, channels, etc., where it is obvious that the shape is used as furnished. Stock tolerances shall be established by existing commercial standards, such as "MILL" standards and federal specifications.

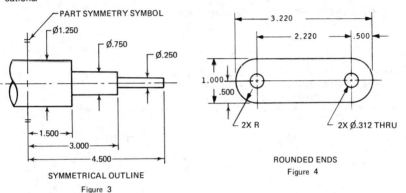

SYMMETRICAL OUTLINE
Figure 3

ROUNDED ENDS
Figure 4

10.5.7 DIMENSIONING OF TUBES — Tubes are dimensioned as illustrated in Section 4. (Tube Drawing)

10.5.8 ROUNDED END PARTS — Parts with rounded ends are dimensioned by giving the overall length and width, and indicating the radius as "R". (See Figure 4.)

10.5.9 COORDINATE DIMENSIONING — Coordinate dimensioning is as shown in Figure 6.

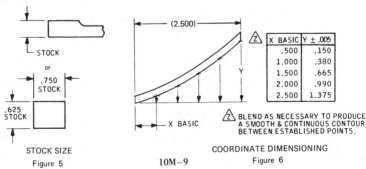

STOCK SIZE
Figure 5

COORDINATE DIMENSIONING
Figure 6

10M−9

10.5.10 SQUARE SHAPES — Square shapes are dimensioned as shown in Figure 7.

10.5.11 DIMENSIONING SPHERICAL RADII — Spherical radii are dimensioned as shown in Figure 8.

10.5.12 COUNTERBORES — Counterbores are dimensioned as shown in Figure 9.

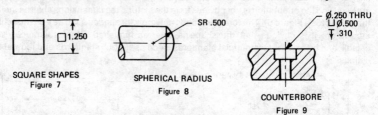

SQUARE SHAPES
Figure 7

SPHERICAL RADIUS
Figure 8

COUNTERBORE
Figure 9

10.5.13 SPOT FACING — Spot facing is the operation of cleaning up the surface around a feature. The diameter and, when required, the maximum depth or minimum remaining thickness shall be specified. Spot faces are called out by note only, except the depth or remaining thickness may be dimensioned on the drawing. (See Figure 10)

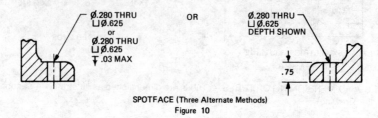

SPOTFACE (Three Alternate Methods)
Figure 10

10.5.14 COUNTERSINKS — Countersinks on a flat surface may be called out by dimensions or notes. (See Figure 11)

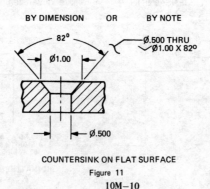

COUNTERSINK ON FLAT SURFACE
Figure 11

10.5.14.1 Countersinks on cylindrical surfaces are directly dimensioned as shown in Figure 12.

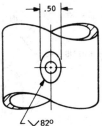

COUNTERSINK ON CURVED SURFACE
Figure 12

10.5.15 ARCS — Arc measurements are dimensioned on the curved surface as shown in Figure 13.

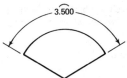

ARC MEASUREMENT
Figure 13

10.5.16 CHAMFERS — Chamfer of fourty-five degree angles are called out by one of the methods shown in Figure 14.

10.5.16.1 Chamfers of other angles are directly dimensioned by one of the methods shown in Figure 15.

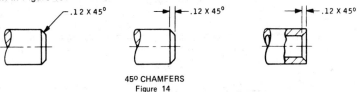

45° CHAMFERS
Figure 14

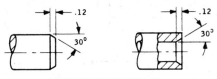

CHAMFERS OTHER THAN 45°
Figure 15

10.5.17 RADII — Radius dimensions are called out as shown in Figure 16.

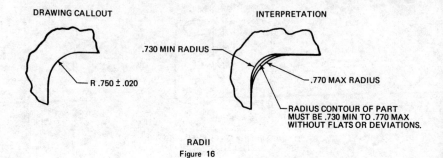

RADII
Figure 16

10.5.18 ORIGIN — A plane which establishes the origin from which dimensions are taken between parallel surfaces. Not to be used with geometric tolerancing applications. See Figure 17.

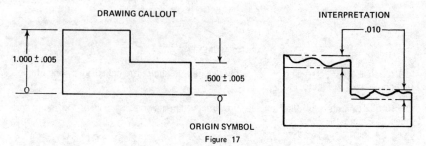

10.5.19 REPETITIVE FEATURES AND DIMENSIONS — Repetitive features by the use of notes or dimensions are shown in Figure 18.

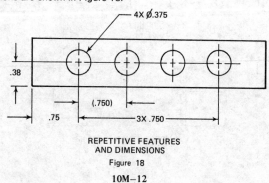

REPETITIVE FEATURES
AND DIMENSIONS
Figure 18

10.5.20 SELECTIVE DIMENSIONING — Areas of limited length or surface that are to receive additional or special treatment are dimensioned as shown in Figure 19.

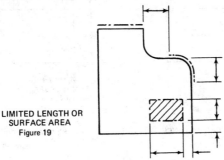

LIMITED LENGTH OR SURFACE AREA
Figure 19

10.5.21 HOLES — Holes are dimensioned in the view where they appear as circles whenever practical.

10.5.21.1 DEPTH OF HOLES — Dimension the depth of holes as shown in Figure 20 Thru holes are either defined by picture or the term "THRU."

10.5.21.2 LOCATING HOLES — Figures 21 through 25 illustrate the positioning of round holes by giving distances, or distances and directions, to the hole centers. These methods can also be used to locate round pins and other features of symmetrical contour. Allowable variations for any of the positioning dimensions may be specified by giving a tolerance with each distance or angle, or by "positional tolerancing dimensioning" explained in 10.10.

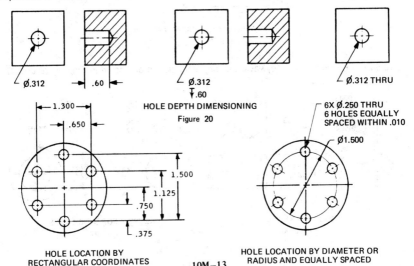

HOLE DEPTH DIMENSIONING
Figure 20

HOLE LOCATION BY RECTANGULAR COORDINATES
Figure 21

HOLE LOCATION BY DIAMETER OR RADIUS AND EQUALLY SPACED
Figure 22

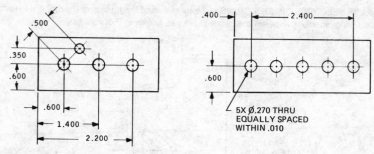

HOLE LOCATION BY LINEAR DISTANCE
Figure 23

HOLE LOCATION EQUALLY SPACED
Figure 24

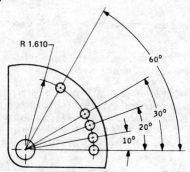

HOLE LOCATION BY RADIUS AND ANGLE
Figure 25

10.5.22 KEYWAYS — Keyways may be dimensioned as shown in Figures 26 and 27.

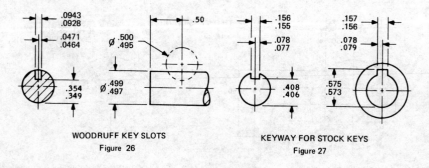

WOODRUFF KEY SLOTS
Figure 26

KEYWAY FOR STOCK KEYS
Figure 27

10.5.22 TAPERS— Basic diameter and basic taper method. A basic taper symbol procedes the basic diameter and taper ratio control frame. The basic diameter is an exact dimension which must be located within specified limits. This diameter method of dimensioning tapers illustrated in Figure 28 controls the size of the tapered section, as well as its axial position in relation to some other surface. As is shown in the interpretation of Figure 28, the tolerance on the location of the BASIC diameter not only controls the axial position of the tapered section, but also sets up a tolerance zone within the form of the taper must fall. The taper may vary from the value given, but must fall within the zone created by the locational tolerances.

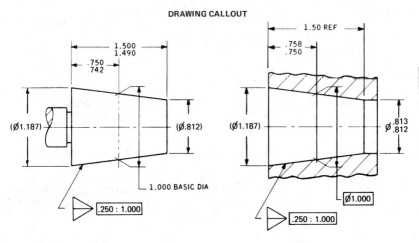

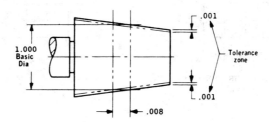

DIMENSIONING TAPERS - BASIC TAPER & BASIC DIAMETER
Figure 28

10.5.22.1 If the accuracy of the taper is non critical, dimension as shown in Figure 29.

10.5.22.2 FLAT TAPERS — Flat tapers are dimensioned as shown in Figure 30.

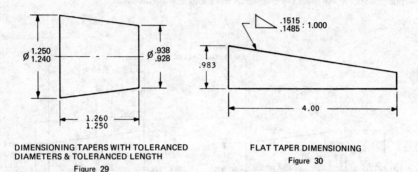

DIMENSIONING TAPERS WITH TOLERANCED
DIAMETERS & TOLERANCED LENGTH
Figure 29

FLAT TAPER DIMENSIONING
Figure 30

10.5.23 KNURLS — Knurls are used to provide a rough surface for gripping, decoration, or for a press fit between mating parts. Knurls for gripping or decorative purposes are called out by type, pitch, and axial length. (See Figure 31.) Knurls for press fits are called out by type, pitch, axial length, diameter before knurling, and should include the minimum diameter after knurling. (See Figure 32.) Types of knurls are diamond, straight and diagonal. Standard pitches are 64, 96, 128 and 164.

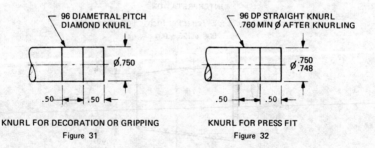

KNURL FOR DECORATION OR GRIPPING
Figure 31

KNURL FOR PRESS FIT
Figure 32

10.5.24 TRUE DIMENSIONS — When dimensioning a surface in a plane not perpendicular to the plane of projection, the term "TRUE" may be used as a matter of convenience to avoid showing an auxiliary view. (See Figure 33.)

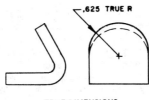

TRUE DIMENSIONS
Figure 33

10.6 GENERAL RULES AND INTERPRETATIONS OF DIMENSIONS AND TOLERANCES

10.6.1 LIMITS OF SIZE AND FORM, RULE 1 — The toleranced dimensions for the size of a feature control the form as well as the size. The basic interpretation of this implied control of form is as follows:

(a) No element of the actual feature (including a datum feature) shall extend beyond the envelope of perfect form at MMC. This envelope is the true form implied by the drawing.

(b) The measured dimensions of the feature at any cross-section shall not be less than the LMC limit of size of an external feature nor greater than the maximum limit of size of an internal feature.

Figure 34 illustrates the extreme variations of form that are permitted by this interpretation.

Note: The above stated interpretation prescribing an envelope of perfect form at MMC applies only to individual features and not to interrelationship of features. such interrelationship should be controlled by form or positional tolerances specified on the drawing.

10.6.1.1 EXCEPTIONS TO RULE 1

(a) The interpretation of Rule 1 does not apply to stock material such as bars, sheets, tubing, etc. These stock materials are controlled by the material specification called out on the drawing or by industry standards for the material when called out commercial grade.

(b) Where it is desired to permit a tolerance of form to exceed the envelope of perfect form at MMC, this may be done by adding to the drawing the suitable form tolerance and a local note or delta note which states "PERFECT FORM NOT REQD AT MMC." The note is applied to the pertinent size dimension(s). When this is done, the form tolerance specified is allowed even through the feature is at its MMC. When this procedure is used, the MMC size of the mating part must be revised (male feature decreased), (female feature increased) by an amount at least equal to the form tolerance.

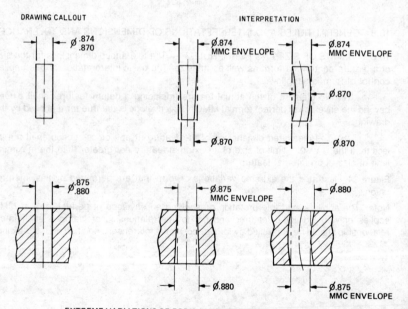

EXTREME VARIATIONS OF FORM ALLOWED BY SIZE TOLERANCES
Figure 34

10.6.2 RULES APPLICABLE TO DATUMS — Datums may be implied or specified on a drawing, but in either case the following rules apply:

10.6.2.1 ACCURACY OF DATUM FEATURES — Datum features are only as accurate as stated in Rule 1, therefore, when their deviation of form or position must be controlled more closely than can be assured by Rule 1, the drawing shall include a note or geometric tolerance symbol stating the required accuracy.

10.6.2.2 DIMENSIONS FROM DATUMS — When features are dimensioned from datums, they are located with respect to the datums, not with respect to one another. (See Figure 35.) When a plane surface is used to establish a datum plane, measurements to other features are taken from the datum plane which contacts the high points of the datum surface and not from the actual datum surface. Thus, the specified form or positional tolerance for a feature does not include any variation which may exist in the datum surface.

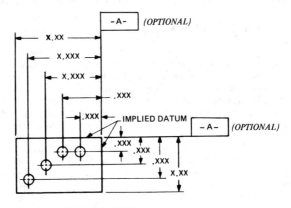

IMPLIED DATUMS
Figure 35

10.6.2.3 SELECTION OF DATUMS — Features that are selected as datums must be easily recognizable or identified with a datum symbol. A datum on an actual part must be accessible during manufacture so that measurements from it can be readily made. Also, corresponding features on mating parts must be used as datums to insure assembly and facilitate tool and fixture design.

10.6.3 RULES APPLICABLE TO THREADS, GEARS, AND SPLINES — The following rules are applicable to all symbols and notes specifying tolerances of form or position involving screw threads, gears, or splines as toleranced features or datums.

10.6.3.1 SCREW THREADS — Where tolerances of form or position are expressed by symbols and notes, each such tolerance applicable to a screw thread and each datum reference to a screw thread shall be understood to apply to the pitch cylinder.. If design requirements necessitate an exception to this general rule, a qualifying notation shall supplement the symbol or note, e.g., MAJOR DIA. In the case of symbol applciation, the qualifying notation shall be shown beneath the feature control symbol where applicable to the feature, and beneath the datum identifying symbol where applicable to the datum.

10.6.3.2 GEARS AND SPLINES — For gears and splines, a qualifying notation must be added to the symbol or note, e.g., MAJOR DIA, MINOR DIA, PITCH CYLINDER.

10.7 GEOMETRIC TOLERANCES

10.7.1 GENERAL — A geometric tolerance is the permissible variation from the specified form of an individual feature of a part. Geometric tolerances control characteristics such as straightness, flatness, roundness, concentricity, perpendicularity, parallelism, angularity, etc. Shapes or forms into which material is fabricated are defined by the use of geometric terms such as a plane, a cylinder, a cone, a square, or a hexagon. The geometric definition assumes a perfect form, but because a perfect form cannot be produced, variation must be controlled if quality is to be maintained.

10.7.2 SYMBOLS — Geometric characteristics are specified on the drawing by the use of symbols. When geometric symbols do not adequately describe the desired condition, a note may be used, either separately or supplementing the symbol, to define the requirement. Figure 36 shows the approved symbol for each characteristic. Figure 37 shows how the symbols are to be drawn with their tolerance, modifiers, and datum references.

10.7.3 DATUM IDENTIFYING SYMBOLS — Each datum (except implied datums) on a drawing is assigned a different identifying reference letter for which letters of the alphabet, except "I", "O", and "Q" are used. Datum assignment begins with the letter "A" and proceeds thorough the alphabet as required. When datum features requiring identification on a drawing are so numerous as to exhaust the single letter series, the double letter series, i.e., "AA" through "AZ", may be used. The datum identifying symbol is formed by a rectangle containing the datum reference letter preceded and followed by a dash. (See Figure 37A).

10.7.4 SYMBOL FOR BASIC — The symbolic means of labeling a basic is by enclosing each such dimension in a rectangular frame. (See Figure 37B.)

10.7.5 SYMBOL PLACEMENT — The feature control symbol shall be associated with the feature(s) being toleranced by one of the methods shown in figure 38.

GEOMETRIC CHARACTERISTIC SYMBOLS

TYPE OF TOLERANCES		SYMBOL	CHARACTERISTIC	FORMER SYMBOL/S AND NOTES
INDIVIDUAL FEATURES	FORM	▱	FLATNESS	∼
		—	STAIGHTNESS	⌒ or —
		○	CIRCULARITY (ROUNDNESS)	
		⌀	CYLINDRICITY	
INDIVIDUAL OR RELATED FEATURES WHERE DATUMS ARE SPECIFIED	PROFILE	⌒	PROFILE OF ANY LINE	SEE NOTE 1.
		⌒	PROFILE OF ANY SURFACE	SEE NOTE 1.
RELATED FEATURES	ORIENTATION	∥	PARALLELISM	∥
		⊥	PERPENDICULARITY (SQUARENESS)	
		∠	ANGULARITY	
	LOCATION	◎ SEE NOTE 3	CONCENTRICITY	⦿ & SEE NOTE 2.
			SYMMETRY	≡
		⌖	POSITIONAL	
	RUNOUT TOLERANCES	↗	CIRCULAR	SEE NOTE 4.
		↗↗	TOTAL	SEE NOTE 5.

NOTES:

1.) PROFILE TOLERANCE CONTROL SIZE AS WELL AS FORM.

2.) WHERE CONCENTRICITY RFS APPLIES, IT IS PREFERRED THAT THE RUNOUT SYMBOL BE USED. WHERE CONCENTRICITY AT MMC APPLIES, IT IS PREFERRED THAT THE POSITIONAL SYMBOL BE USED. FOR AN APPLICATION OF CONCENTRICITY SEE FIGURE 62. THE INNER CIRCLE OF THE SYMBOL IS NOT FILLED SOLID.

3.) WHERE SYMMETRY APPLIES, THE POSITIONAL SYMBOL SHALL BE USED.

4.) THE SYMBOL ↗ WITHOUT THE QUALIFIER "CIRCULAR" FORMERLY DENOTED TOTAL RUNOUT.

5.) "TOTAL" IS NO LONGER SPECIFIED UNDER THE FEATURE CONTROL SYMBOL.

Figure 36

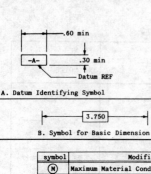
A. Datum Identifying Symbol

B. Symbol for Basic Dimension

symbol	Modifier
Ⓜ	Maximum Material Condition (MMC)
Ⓢ	Regardless of Feature Size (RFS)
Ⓛ	Least Material Condition (LMC)
⌀	Diameter (DIA)
Ⓟ	Projected Tolerance Zone (TOL ZONE PROJ)
(1.250)	Reference (REF)
3.875	Basic

C. Symbols for MMC, RFS and LMC & Others

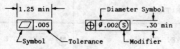
D. Feature Control Symbols

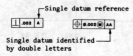
E. Feature Control Symbols Incorporating Datum References

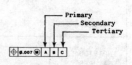
F. Datum Reference Showing Order Of Precedence

G. Multiple Datum Features Establishing Single Datum Reference

H. Combined Feature Control Symbol

J. Composite Feature Control Symbol

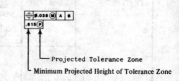
K. Feature Control Symbol With A Projected Tolerance Zone

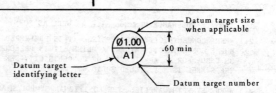

A solid datum target leader indicates nearside 'Datum Target'.
A dashed datum target leader indicates farside 'Datum Target'.

L. Datum Target Symbol

GEOMETRIC TOLERANCE SYMBOLS

Figure 37

10.7.6 MULTIPLE DATUMS — Where reference is made to more than one datum used to control the positional or form tolerance of a feature, the first datum reference letter in the sequence shall be the primary datum, the second letter the secondary datum, etc. The datum reference letters shall be placed in order of importance with respect to establishing the location, form, or position of other features. Thus, the datum reference letters will not necessarily be in alphabetical order. An example of the use of primary, secondary, and tertiary datums would be when it is desired to place a part in three coordinate reference frame so that the dimensions may be given in the three planes. Since an actual part will not have perfect squareness between any of the three datum feature surfaces used for dimensioning, it is necessary for the drawing to show how the datums are applied to the part. This is accomplished as follows: The primary datum is a plane established by any three or more points, (not in a line) that contact the surface identified as the primary datum; the secondary datum plane is established as perpendicular to the primary datum and contacting the part at two or more points on the surface identified as the secondary datum; and the tertiary datum is established as perpendicular to both the primary and secondary datums and contacting the part at one or more points on the surface identified as the tertiary datum. (See Figure 37F.) In some designs the datums obviously are of equal importance. For example, a runout tolerance related to two equally important bearing surfaces. (See Figure 37G.)

10.7.7 COMBINED SYMBOLS — When a feature serves as a datum and is also controlled by a positional or form tolerance, the geometric characteristic symbol and the datum identifying symbol are combined as shown in Figure 37H. In such cases, the length of the box for the datum symbol may be the same as that of the geometric characteristic symbol or .60 inch.

10.7.8 COMPOSITE FEATURE CONTROL SYMBOL — Composite feature control symbol is used where more than one tolerance of a given geometric characteristic applies to the same feature. It is also used to locate a pattern of holes, then refine the locations of the holes within the pattern. The upper portion of the composite feature control symbol specifies the datums related to the pattern (usually three for non-cylindrical datums) and the lower portion indicates the datum which controls the perpendicularity. The tolerance for the interrelated holes in the pattern must be less than the tolerance for the pattern location. A single entry of the geometric characteristic symbol is followed by each tolerance requirement, one above the other, separated by a horizontal line. (See Figure 37J.)

10.7.9 COMBINED FEATURE CONTROL AND PROJECTED TOLERANCE ZONE SYMBOL — Where a positional or perpendicularity tolerance is specified as a projected tolerance zone, a frame containing the projected height followed by the appropriate symbol is placed beneath the feature control symbol. (See Figure 37K.)

10.7.10 RULES APPLICABLE TO USE OF GEOMETRIC FORM TOLERANCE SYMBOLS.

(a) Rule 1 applies even when geometric form tolerances are specified.

(b) Geometric form tolerance control is applied to features only when it is necessary to control form more precisely than the limits established by Rule 1 or its exceptions.

(c) Profile tolerances, in themselves, establish an envelope of perfect form at MMC and are therefore not subject to usage limitations in (b) above.

(d) Runout tolerances are considered form tolerances but they also control position. These tolerance symbols may be used when it is necessary to control the interrelationship between features even when size tolerances adequately control the form of each individual feature.

(e) Form tolerances in feature control symbols are not modified by such terms as DIA, TIR, FIR, or R.

(f) Form tolerances which always require a datum reference are parallelism, perpendicularity, angularity, and runout.

(g) Form tolerances which never utilize a datum reference are flatness, straightness, roundness and cylindricity.

(h) Profile tolerances may or may not utilize datum references depending on the design requirements.

10.7.11 FORMER SYMBOLS — The symbol ⌒ was formerly used for expressing a tolerance on flatness. The symbols ⌒ and — were formerly used for expressing tolerances on straightness and flatness. Whenever these symbols appear on existing drawings, they may be interpreted as specified herein for the specific case shown. The symbol ⌖ A.XXX TIR was formerly used for expressing a tolerance on concentricity. Whenever this symbol appears on existing drawings it shall be interpreted the same as runout on a diameter.

10.7.12 DIAMETER SYMBOL — The symbol used to designate a diameter is as shown in Figure 37C. It precedes the specified tolerance in a feature control symbol. The symbol shall be used elsewhere on a drawing in place of the word DIAMETER or the abbreviation DIA.

10.8 DATUM TARGETS

10.8.1 QUALIFICATION OF DATUM SURFACES — The datum features of certain parts frequently require further qualification before they can be related to the three-plane framework. Examples are surfaces produced by casting, forging, and molding; surfaces adjacent to welds; and thin sheet metal. All of these parts are subject to bowing, warping, and distortion, therefore, it is not recommended that an entire surface be designated as a datum. For example, a cast surface may actually rock or "Teeter-Totter" when placed in contact with a datum plane such as a machine table or surface plate, thereby making accurate and repeatable measurements very difficult. To overcome this problem, the datum target method (formerly referred to as tooling points) should be used.

10.8.2 DATUM TARGET METHOD — The datum target method is a useful technique for relating the above mentioned parts to the three-plane framework. Normally, three datum targets are required to restablish the first or primary plane, two datum targets to prevent rotation of the secondary plane and a single datum target to position the third plane in relation to the first and second plane. Additional datum targets may be indicated when necessary. It is at these points or areas that contact is made with the processing and inspection equipment.

10.8.3 LOCATING DATUM TARGETS — Datum targets are separated as far apart as possible, dimensioned relative to each other, and located on surfaces that will not be machined. On castings and forgings they should be located on one side of the parting line, not too close to a fillet or corner, and not on the parting line or on a gate. If a separate machining drawing is made of the casting or forging, or if a separate machining view is made on the casting or forging drawing, the datum target points shall be shown in the same location as on the casting or forging but shall not be dimensioned. When a separate machining drawing is made, modify the general note to read: " ⊖ Symbol Designates Datum Targets. See Drawing XXXXXX". For an example of locating and dimensioning datum targets see Section 12.

10.8.4 LEADER LINES — A solid leader line is used to indicate that the "Datum Target" is on the nearside. A dashed leader line is used to indicate that the "Datum Target" is on the farside.

10.9 DRAWING APPLICATION AND INTERPRETATION — The following illustrations show the correct method of applying geometric tolerances to drawings and the correct interpretation of the tolerance zone provided by each.

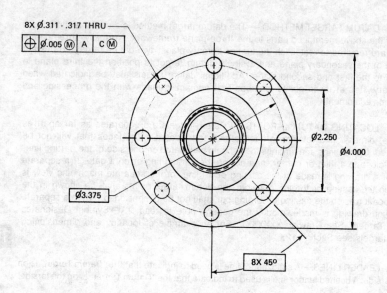

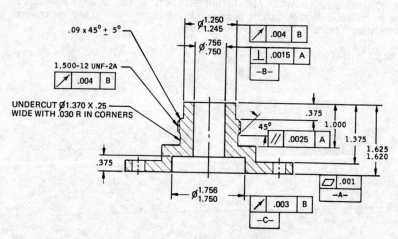

APPLICATION OF GEOMETRIC TOLERANCING SYMBOLS
Figure 38

10.9.1 FLATNESS — Flatness is the condition of a surface having all elements in one plane.

10.9.1.1. FLATNESS TOLERANCE — A flatness tolerance specifies a tolerance zone confined by two parallel plances within which the entire surface must lie. (See Figure 39.) The expression "MUST BE CONCAVE" or "MUST BE CONVEX" may be added if necessary. Due to the fact that flatness does not refer to datums, inspection will have to level the entire surface to the inspection table before checking surface for flatness.

DRAWING CALLOUT

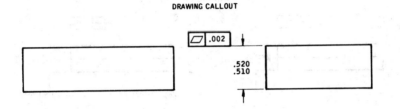

INTERPRETATION

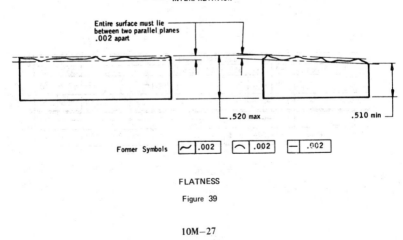

FLATNESS

Figure 39

10.9.2 STRAIGHTNESS — Straightness is a condition where an element of a surface is a straight line.

10.9.2.1 STRAIGHTNESS TOLERANCE — A straightness tolerance specifies a tolerance zone of uniform width along a straight line, within which all points of the considered line must lie. Straightness tolerance symbols are attached to leader lines only and are applied in a view where the surface elements to be controlled are represented as a straight line. (See Figure 40.) This symbol is not used to specify straightness of axis. Existing drawings which specify straightness of an axis are to be interpreted to mean that the axis must lie within a diameter or width zone equal to the specified tolerance.

10.9.2.2 STRAIGHTNESS TOLERANCE EXCEEDING MMC ENVELOPE — When a feature is likely to exceed the envelope of perfect form at MMC and this deviation will not interfere with the fit or function of the feature, it shall be noted as shown in Figure 40.

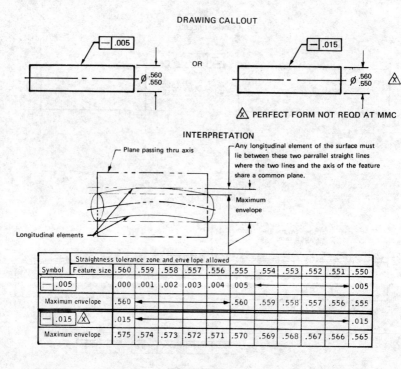

STRAIGHTNESS

Figure 40

10.9.3 CIRCULARITY — Circularity is a condition of a surface of revolution such as a cylinder, cone, or sphere, where all points of the surface intersected by any plane, (1) perpendicular to a common axis (cylinder, cone), or (2) passing through a common center (sphere), are equidistant from the axis.

10.9.3.1 CIRCULARITY TOLERANCE — A Circularity tolerance specifies a tolerance zone bounded by two concentric circles in that plane within which the periphery must lie.

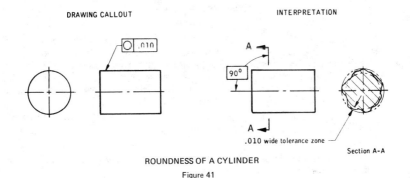

ROUNDNESS OF A CYLINDER
Figure 41

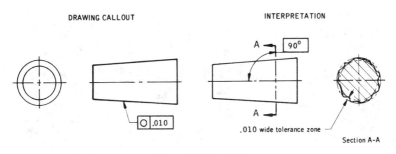

ROUNDNESS OF A CONE
Figure 42

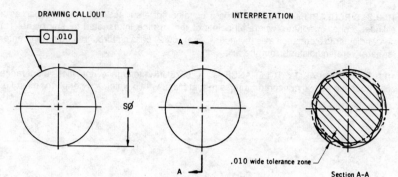

ROUNDNESS OF A SPHERE
Figure 43

10.9.3.2 CIRCULARITY, FREE-STATE VARIATION — Free-state variation can exist in two ways: (1) distortion due to the weight or flexibility of the part, or (2) distortion due to internal stresses set up in fabrication. Parts that are subject to such distortion are referred to as "nonrigid" parts and such distortion is referred to as "free-state variation." The above distortions are accounted for on drawings only when the feature(s) may fall outside the drawing limits and are controlled as follows:

    (a) By adding a note stating "DIMENSIONS APPLY IN RESTRAINED CONDITION" (specify the amount of restraining force allowable to bring feature(s) within drawing limits when necessary).

    (b) State the allowable free-state variation and show average diameter as shown in Figure 44.

Note: The term "average diameter" or "AVG DIA" is the mean of several diameters (not less than four) used to determine conformance of the diameter tolerance only. Average diameter may also be measured with a periphery tape.

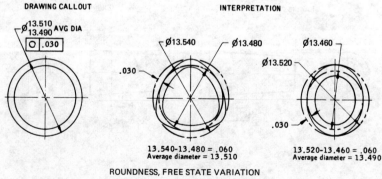

ROUNDNESS, FREE STATE VARIATION
Figure 44

10.9.4 CYLINDRICITY — Cylindricity is a condition of a surface of revolution in which all elements form a cylinder.

10.9.4.1 CYLINDRICITY TOLERANCE — A cylindricity tolerance specifies a tolerance zone confined to the annular space between two concentric cylinders within which the surface must lie. (See Figure 45.)

Note that the cylindricity tolerance controls circularity and straightness, as well as parallelism of the elements and that the specified tolerance is always on a radial basis.

DRAWING CALLOUT

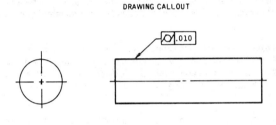

INTERPRETATION

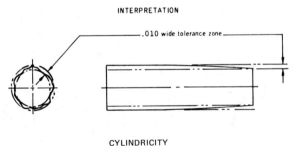

CYLINDRICITY
Figure 45

10.9.5 PROFILE TOLERANCING — Profile tolerancing is a method used where a uniform amount of variation may be permitted along a basic line or surface. The basic line or surface may consist of straight lines or curved lines, the latter being either arcs or irregular curves.

10.9.5.1 PROFILE TOLERANCE — A profile tolerance (either bilateral or unilateral) specifies a tolerance zone, always measured normal to the profile at all points of the profile, within which the specified line or surface must lie.

10.9.5.2 APPLICATION OF PROFILE TOLERANCES — Figures 46 and 47 illustrate methods of dimensioning profiles and comply with the following requirements:

(a) A view or section is drawn which shows the desired basic profile.

(b) The profile is dimensioned by basic dimensions. This dimensioning may be in the form of located radii and angles, or it may consist of coordinate dimensions to points on the profile.

(c) An exaggerated tolerance zone is shown by one phantom line drawn parallel to the profile. The tolerance zone may be shown unilaterally to either side of the profile (See Figure 47). When the profile tolerance is bilateral, that is equally on each side of the basic profile, the tolerance zone is not shown (See Figure 47A).

(d) Line and surface controls may be applied to the same feature when the line elements in one direction need to be controlled more closely than the surface as a whole.

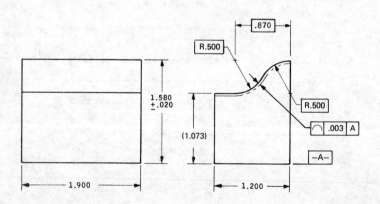

PROFILE OF A LINE
Figure 46

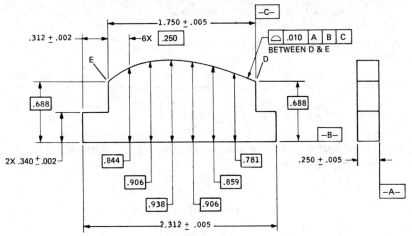

UNILATERAL TOLERANCE ZONE
PROFILE OF A SURFACE
Figure 47

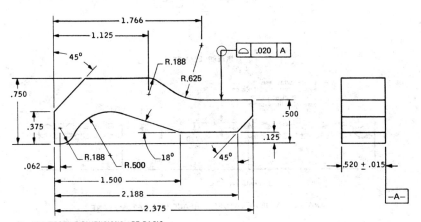

X. ALL PROFILE DIMENSIONS ARE BASIC.

BILATERAL TOLERANCE ZONE
Figure 47a

10.9.6 PARALLELISM — Parallelism is the condition of a surface, axis, or line which is equidistant at all points from a datum plane or axis.

10.9.6.1 PARALLELISM TOLERANCE — A parallelism tolerance specifies one of the following:

(a) A tolerance zone confined by two planes parallel to a datum plane within which the considered feature (surface or axis) must lie. (See Figures 48 and 49.)

(b) A cylindrical tolerance zone parallel to a datum feature axis and within which the axis of a feature must lie. (See Figure 50.)

Note that the parallelism tolerance when applied to a plane surface controls flatness if a flatness tolerance is not specified.

DRAWING CALLOUT

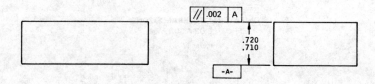

INTERPRETATION

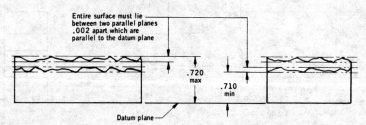

PARALLELISM OF A FEATURE SURFACE TO A DATUM PLANE

Figure 48

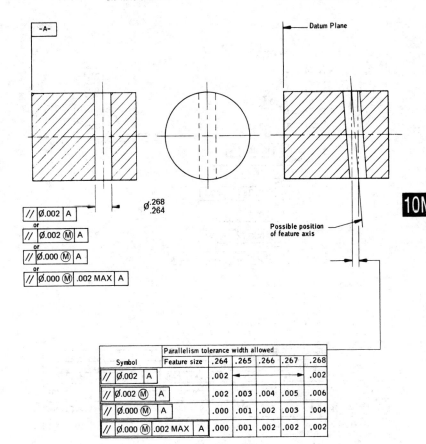

PARALLELISM OF A FEATURE AXIS TO A DATUM PLANE
Figure 49

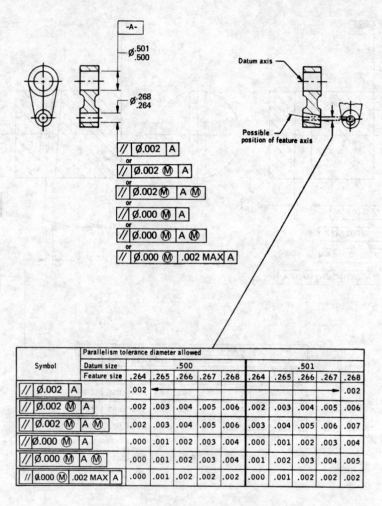

PARALLELISM OF A FEATURE AXIS TO A DATUM AXIS

Figure 50

10.9.7 PERPENDICULARITY — Perpendicularity is the condition of a surface, axis, or line which is at right angles to a datum plane or axis.

10.9.7.1 PERPENDICULARITY TOLERANCE — A perpendicularity tolerance specifies one of the following:

(a) A tolerance zone confined by two parallel planes perpendicular to a datum plane within which the surface of a feature must lie. (See Figures 51).

(b) A tolerance zone confined by two parallel planes perpendicular to a datum plane within which the centerplane of a feature must lie. (See Figure 52.)

(c) A cylindrical tolerance zone perpendicular to a datum plane within which the axis of the feature must lie. (See Figures 53, 54, 55, and 56.)

(d) A tolerance zone confined by two parallel planes perpendicular to a datum axis within which the axis of a feature must lie. (See Figure 57.)

(e) A tolerance zone confined by two parallel straight lines perpendicular to a datum plane or datum axis within which an element of the surface must lie. (See Figure 58.)

Note that the perpendicularity tolerance when applied to a plane surface controls flatness if a flatness tolerance is not specified.

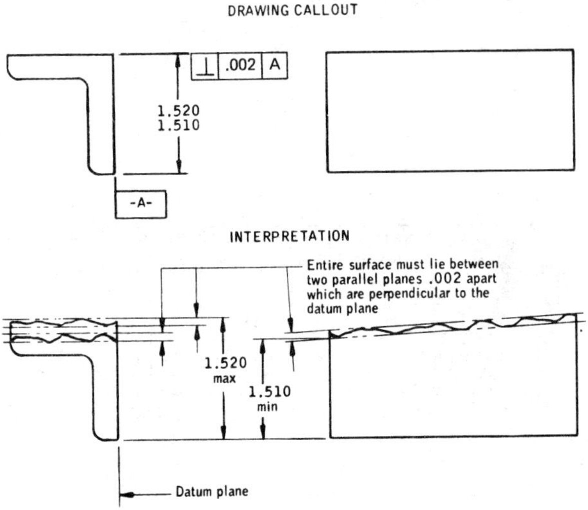

PERPENDICULARITY OF A FEATURE SURFACE TO A DATUM PLANE

Figure 51

10M−37

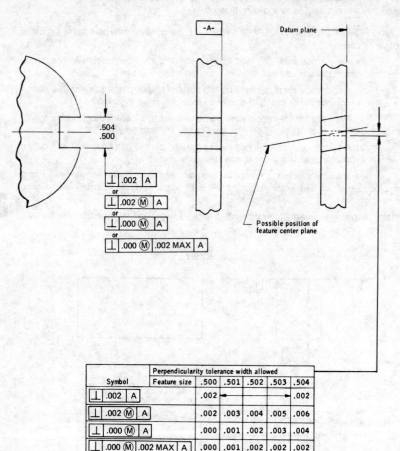

PERPENDICULARITY OF A FEATURE CENTER PLANE TO A DATUM PLANE

Figure 52

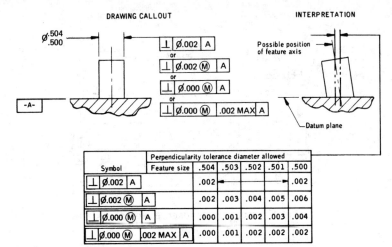

PERPENDICULARITY OF A FEATURE AXIS TO A DATUM PLANE, FIXED PIN
Figure 53

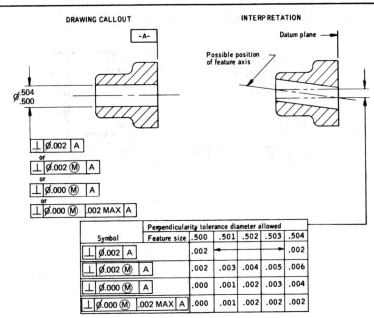

PERPENDICULARITY OF A FEATURE AXIS TO A DATUM PLANE
Figure 54

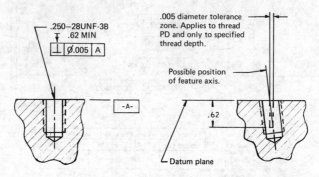

PERPENDICULARITY OF THREADED HOLES AND/OR INSERTS

Figure 55

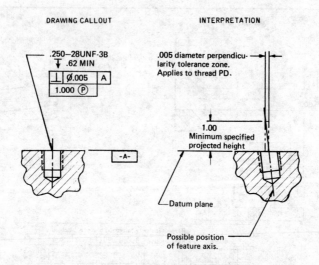

PERPENDICULARITY OF THREADED HOLES AND/OR INSERTS
TOLERANCE ZONE PROJECTED

Figure 56

10M–40

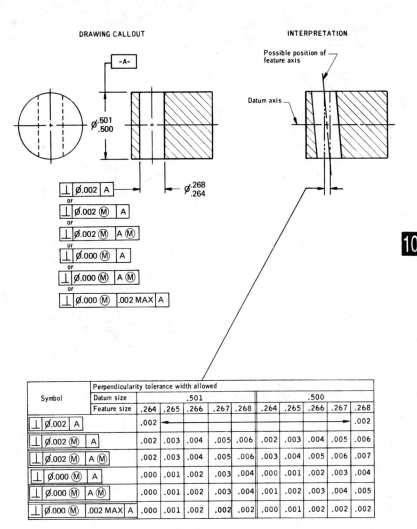

PERPENDICULARITY OF A FEATURE AXIS TO A DATUM AXIS

Figure 57

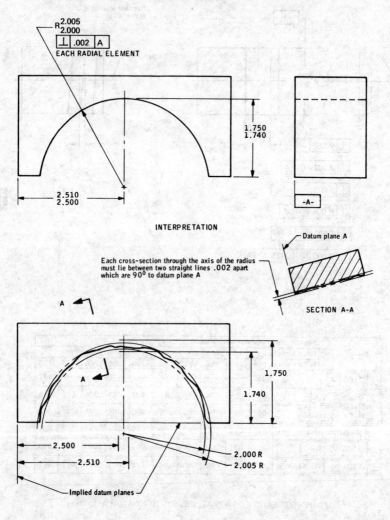

PERPENDICULARITY OF FEATURE ELEMENTS OF A DATUM PLANE
Figure 58

10.9.8 ANGULARITY — Angularity is the condition of a surface or line which is at the specified angle (other than 90°) from a datum plane or axis.

10.9.8.1 ANGULARITY TOLERANCE — An angularity tolerance for a surface specifies a tolerance zone confined by two parallel planes, inclined at the specified angle to a datum plane, within which the toleranced surface must lie. (See Figure 59.) Note that the angularity tolerance when applied to a plane surface controls flatness if a flatness tolerance is not specified.

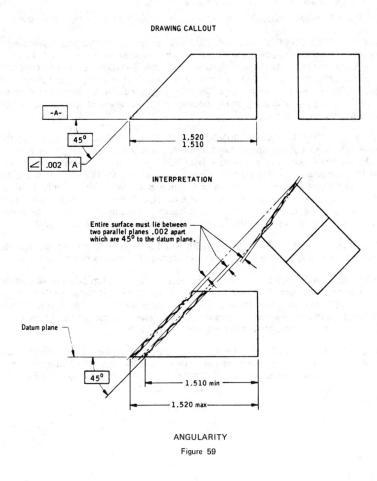

ANGULARITY

Figure 59

10.9.9 RUNOUT — Runout is the condition of perfect form and axial alignment of two or more surfaces of revolution such as cylinders, cones, or contours and may include plane surfaces perpendicular to and generated about a common axis.

10.9.9.1 RUNOUT TOLERANCE — A runout tolerance controls the relationship of two or more features within the allowable errors of concentricity, perpendicularity, and alignment of the features. It also controls variations in roundness, straightness, flatness, angularity, and parallelism of individual surfaces. In essence, runout establishes composite form control of those features of a part having a common axis.

Note: Runout always applies RFS, therefore, the symbol for MMC shall not be used.

10.9.9.2 SELECTION OF RUNOUT DATUMS — To control the relationship of features, it is necessary to establish a datum axis about which the features are to be related. This axis may be established by a diameter of considerable length; two diameters having considerable axial separation; or a diameter and a surface which is at right angles to it. Insofar as possible, surfaces used as datums for establishing axes should be functional and must be accessible during manufacturing and inspection. Pitch diameters of features should be avoided as datums for runout.

10.9.9.3 INTERPRETATION OF RUNOUT TOLERANCES — Figure 60 illustrates the interpretation of runout tolerances. Measurements are taken under a single setup for all runout tolerances related to a common axis. However, features that are functionally related to each other and not to the common axis may be toleranced to reflect this requirement. (See Figure 62). Any two features on a common axis which are individually within their specified runout tolerance are related to each other within the sum of their runout tolerance. Therefore, to ensure 100% interchangeability, the sum of the runout tolerances of two mating diameters shall not exceed the clearance of the diameters at MMC.

10.9.9.4 APPLICATION OF RUNOUT TOLERANCES — Figures 60 through 65 illustrate various methods of specifying datum axes and applying the feature control symbols.

10.9.9.5 CIRCULAR RUNOUT — Circular runout is the maximum permissible surface variation at any *fixed point* during one complete rotation of the part about the datum axis. Control is applied on an individual basis to each circular element rather than total form control of the surface area. (See Figure 60). When circular runout is to be applied at a specific location, it is so stated on the drawing.

10.9.9.6 TOTAL RUNOUT — Total runout is the maximum permissible surface variation at *all* surface elements during one complete rotation of the part about the datum axis. Total runout is indicated by the total runout symbol within the feature control symbol. (See Figure 66)

DRAWING CALLOUT

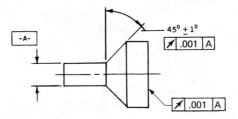

INTERPRETATIONS

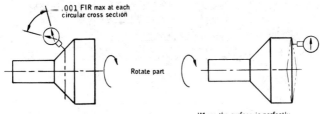

Where the surface is perfectly convex or concave, no circular runout exists, although the surface is not perpendicular to the datum.

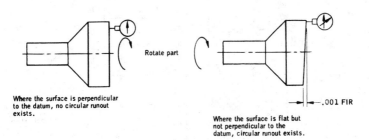

Where the surface is perpendicular to the datum, no circular runout exists.

Where the surface is flat but not perpendicular to the datum, circular runout exists.

CIRCULAR RUNOUT

Figure 60

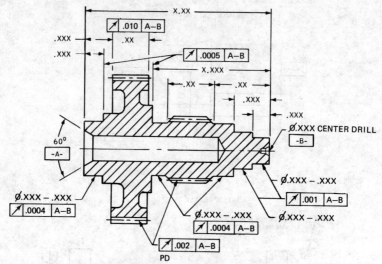

PART MOUNTED ON MACHINING CENTERS
Figure 61

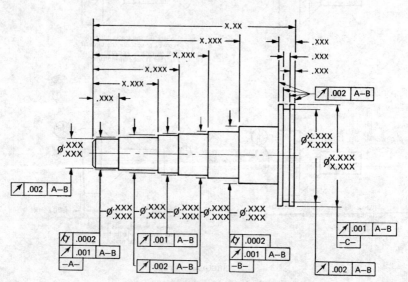

PART MOUNTED ON TWO BEARING SURFACES
Figure 62

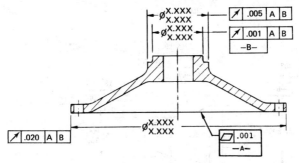

PART MOUNTED ON LARGE FLAT SURFACE WITH NARROW FINISHED DIAMETER
Figure 63

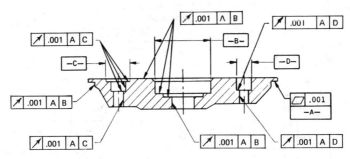

PART MOUNTED ON LARGE FLAT SURFACE WITH MULTIPLE COMMON AXIS
Figure 64

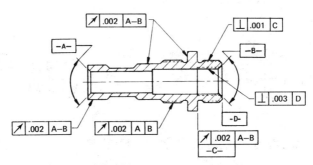

PART MOUNTED ON TAPERED SURFACES
Figure 65

10M−47

DRAWING CALLOUT

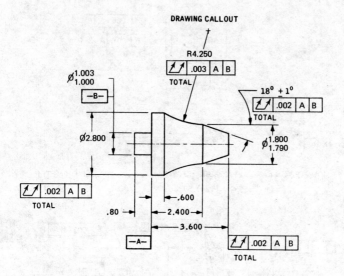

INTERPRETATION

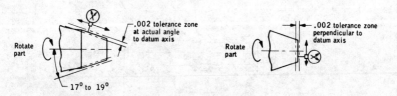

Note: A tolerance may be assigned to the 4.250 R or 18° angle when required, however, basic dimensions are preferred.

INTERPRETATION OF RUNOUT TOLERANCE ZONES

Figure 66

10.9.10 CONCENTRICITY — Concentricity is the condition where the axis of all cross sectional elements of a cylinder, cone or sphere are common to a datum axis. The specified tolerance and datum reference apply only on an RFS basis.

10.9.10.1 CONCENTRICITY TOLERANCE — A concentricity tolerance is the diameter of the cylindrical tolerance zone within which the axis of the feature(s) so toleranced must lie. (See Figure 67). The positional relationship of features on a common axis is not controlled unless concentricity, runout, or true position is specified on the drawing.

Note: Irregularities in the form of the feature to be inspected may make it difficult to actually establish the axis of the feature. For instance, a nominally cylindrical surface may be bowed or out-of-round in addition to being offset from its datum surface; in such cases, finding the axis of the feature may entail a time-consuming analysis of the surface. Therefore, unless there is a definite need for the control of axes (as in the case shown in Figure 67), it is recommended that the control be specified in terms of runout or positional tolerance. (See Figures 60 through 66 and 68.)

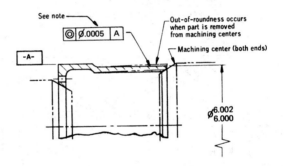

Note: A concentricity tolerance (rather than a runout tolerance, or a true position tolerance) has been applied to the item shown above because of the following supposed conditions:

1. A precise degree of coaxiality is required when part is assembled with mating parts.

2. The toleranced diameter, when removed from supporting tooling (machining centers), is likely to go out-of-round to the full amount permitted by the limits of size. This would preclude verification by either a runout inspection or a true position inspection, and would require a careful "analysis" of the surface; see Paragraph 10.9.10.1

EXAMPLE WHERE CONCENTRICITY TOLERANCE IS REQUIRED
Figure 67

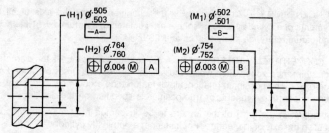

True Position Formula

Where Y = Total positional tolerance for both parts at MMC parts at MMC
H_1 = MMC of female datum
H_2 = MMC of related female diameter
M_1 = MMC of male datum
M_2 = MMC of related male diameter

$Y = (H_1 - M_1) + (H_2 - M_2) = (.503 - .502) + (.760 - .754) = .001 + .006 = .007$
Must be divided between mating diameters, e.g. .004 and .003

A. PARTS WITH TWO MATING DIAMETERS

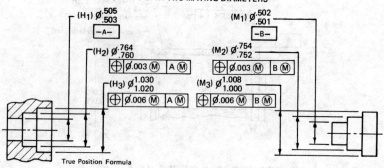

True Position Formula

Where $Y_2, Y_3 \text{---} Y_N$ = Total positional tolerance for mating diameters of both parts at MMC
H_1 = Size tolerance for female datum
$H_2, H_3 \text{---} H_N$ = MMC of related female diameters
M_1 = Size tolerance of male datum
$M_2, M_3 \text{---} M_N$ = MMC of related male diameters

$Y_2 = (H_2 - M_2) - (H_1 + M_1)$ $Y_3 = (H_3 - M_3) - (H_1 + M_1)$ $Y_N = (H_N - M_N) - (H_1 + M_1)$

Example: $Y_2 = (H_2 - M_2) - (H_1 + M_1) - (.760 - .754) - [(.505 - .503) + (.502 - .501)] =$
.006 − (.002 + .001 = .003 Must be divided between mating diameters, e.g. .002 − .001
$Y_3 = (H_3 - M_3) - (H_1 + M_1) - (1.020 - 1.008) - .003$ from First example) −.012 − .003 =
.009 Must be divided between mating diameters, e.g. .005 and .004

B. PARTS WITH THREE OR MORE MATING DIAMETERS WITH MMC DATUM REFERENCES

CONCENTRICITY CONTROLLED BY TRUE POSITION TOLERANCES

Figure 68

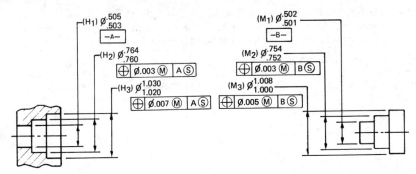

True Position Formula

Where: $Y_2, Y_3 \text{---} Y_N$ = Total true position tolerance for mating diameters of both parts at MMC
H_1 = Female datum
$H_2, H_3 \text{---} H_N$ = MMC of related female diameters
M_1 = Male datum
$M_2, M_3 \text{---} M_N$ = MMC of related male diameters

$$Y_2 = H_2 - M_2 \qquad Y_3 = H_3 - M_3 \qquad Y_N = H_N - M_N$$

Example: $Y_2 = H_2 - M_2 = .760 - .754 = .006$ Must be divided between mating diameters, e.g. .003 and .003
$Y_3 = H_3 - M_3 = 1.020 - 1.008 = .012$ Must be divided between mating diameters, e.g. .007 and .005

C. PARTS WITH THREE OR MORE DIAMETERS WITH RFS DATUM REFERENCES

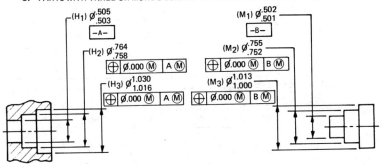

True Position Formula

When using this method, the tolerance is normally preset at .000, therefore, the formula becomes a simple rule as follows:

The clearance between mating diameters at MMC must be equal to or greater than the sum of the size tolerances of both datums.

Example: $(.505 - .503) + (.502 - .501) = .002 + .001 = .003$
$.758 - .755 = .003$
$1.016 - 1.013 = .003$

D. PARTS WITH THREE OR MORE DIAMETERS WITH ZERO TOLERANCE AT MMC

Figure 68 (Continued)

10.9.11 SYMMETRY — Symmetry is a condition wherein a part or a feature has the same contour and size on opposite sides of a central plane, or a condition in which a feature is symetrically disposed about the central plane of a datum feature.

10.9.11.1 SYMMETRY TOLERANCE — Where it is required that a feature be located symmetrically with respect to a datum feature, positional tolerancing s h a l l be used. This permits the tolerance to be expressed on an MMC basis or on an RFS basis. (See Figure 69.)

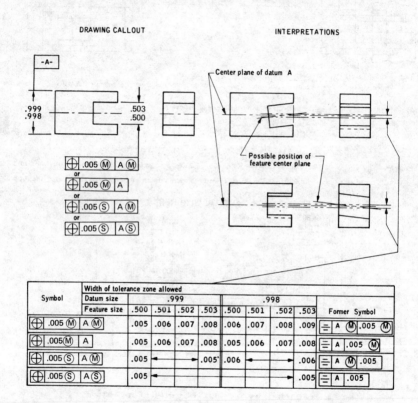

SYMMETRY CONTROLLED BY TRUE POSITION
Figure 69

10.10 POSITIONAL TOLERANCE — Positional tolerance is a term used to describe the perfect or exact locatin of a point, line, or plane of a feature in relationship with a datum or other feature.

10.10.1 POSITIONAL TOLERANCE — A positional tolerance is the total permissible variation in the location of a feature about its true position. For cylindrical features (holes and bosses), the positional tolerance is the diameter (cylinder) of the tolerance zone within which the axis of the feature must lie, the center of the tolerance zone being at the true position. For other features (tabs, slots, etc.), the positional tolerance is the total width of the tolerance zone within which the center plane of the feature must lie, the center plane of the zone being at true position.

10.10.2 APPLICATION OF MMC, LMC AND RFS

(a) Positional tolerancing shall always specify whether RFS, MMC or LMC applies to an individual tolerance, datum reference or both.

(b) The LMC symbol can be used to advantage on drawings of castings, forgings, molded parts, etc., where the positional tolerances are normally calculated and based on the minimum material condition. When the LMC symbol is used, the general note in 10.3.15 must also be included.

10.10.3 FORMULAS FOR POSITIONAL TOLERANCING — The formulas shown below may be used for determining the positional tolerance of round or threaded holes of mating parts. These formulas will result in a "no-interference, no-clearance" fit at maximum material condition of the mating features. They are based on equal positional tolerances for each part; however, the tolerances may be divided unequally when required. For in one part, it is normally more practicable to assign a larger tolerance to the threaded holes in one part, and a smaller tolerance to the corresponding clearance holes in the mating part. The threaded hole or holes for tight-fitting members such as dowels should be specified as "projected tolerance zone XXX" (See Figure 92), otherwise fastener interference may occur. The assembly conditions are commonly referred to as "floating fasteners" (See Figure 70), and "fixed fasteners" (See Figure 71). The "floating fastener formula" is used where two or more mating parts contain clearance holes and the "fixed fastener formula" is used where one part contains threaded holes, or holes for tight-fitting dowels, and the mating part has clearance holes.

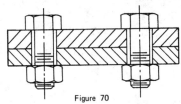

Figure 70
Floating Fastener Formula:

$T = H - F, H = F + Y$

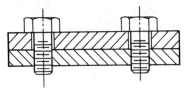

Figure 71
Fixed Fastener Formula:

$T = \dfrac{H - F}{2}, H = F + 2T$

Where: T = Diameter of positional tolerance zone
H = Minimum diameter of clearance hole
F = Maximum diameter of fastener

Note: When using solid dowel pins, the size tolerance of the pin and retaining hole are usually very small ($\pm$ .0005 max), therefore, the "fixed fastener formula" may be used and the tolerances applies at MMC. However, when using rolled spring pins an additional tolerance must be considered since the pin will conform to the actual hole size. To accommodate this additional tolerance, the "fixed fastener formula" may be used with one of the following changes.

LMC Formula:
$$T = \frac{H-F}{2}, H = F + 2T$$

Where: F = Maximum diameter of retaining hole.
T of retaining hole applies at LMC
T of clearance hole applies at MMC

See Figure 94A

MMC Formula:
$$T = \frac{H-F-S}{2}, H = F + 2T + S$$
Where: F = Maximum diameter of retaining hole
S = Size tolerance of retaining hole.

10.10.4 DATUMS FOR POSITIONAL TOLERANCE — Datums, should always be used with positional dimensioning. When two or more circular features could be used as a datum, one feature must be selected and identified with a datum symbol.

10.10.5 RIGHT ANGLE IMPLICATIONS — A 90° basic angle applies whenever centerlines of surfaces or features are shown at right angles and are located or defined by basic dimensions and no angle is specified.

10.10.6 APPLICATION AND INTERPRETATION OF POSITIONAL TOLERANCES — Figure 68 through 94 illustrate various methods of applying positional tolerances and their interpretations. Positional tolerance features such as holes, hole patterns, slots, etc. shall be located by a basic dimension.

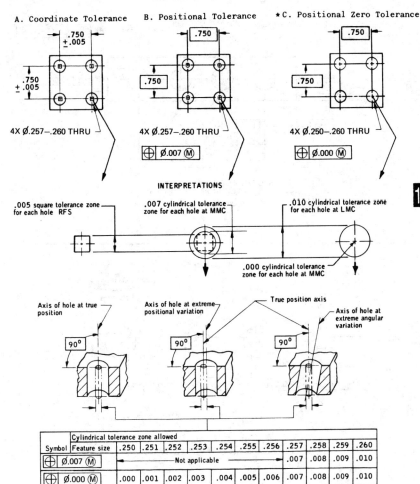

ADVANTAGES OF TRUE POSITION TOLERANCES

Figure 72

DRAWING CALLOUT

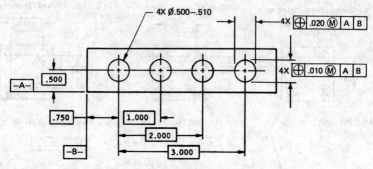

BIDIRECTIONAL DIMENSIONING BY RECTANGULAR COORDINATE METHOD
Figure 73

INTERPRETATION

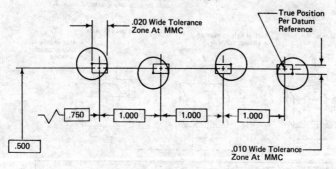

Four rectangular tolerance zones .020 x .010. Axis of each hole must be within or on periphery of rectangular tolerance zone.

10M–56

DRAWING CALLOUT

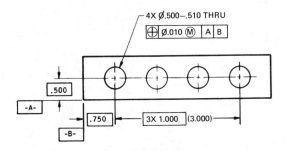

LOCATING ALIGNED HOLES WITH BASIC DIMENSIONS
(DATUMS SPECIFIED)
Figure 74

INTERPRETATION

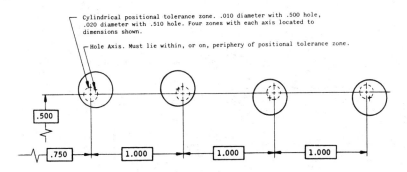

DRAWING CALLOUT

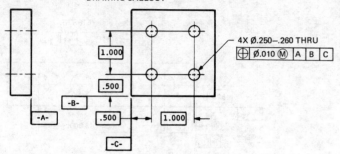

POSITIONAL WITH ALL DIMENSIONS BASIC FROM SPECIFIED DATUMS

Figure 75

INTERPRETATION

Cylindrical positional tolerance zone. .010 diameter with .250 hole, .020 diameter with .260 hole. Four zones with each axis located to dimension shown.

Hole axis. Must lie within, or on, periphery of positional tolerance zone.

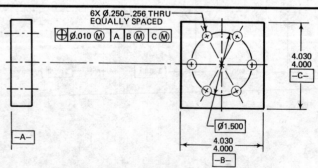

HOLE PATTERN LOCATED SYMMETRICALLY RELATIVE
TO THE CENTER-PLANES OF DATUM FEATURES OF SIZE

Figure 76

DRAWING CALLOUT

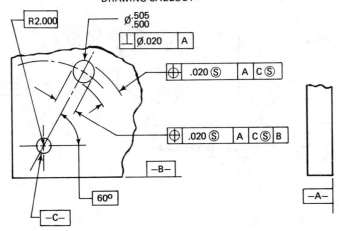

INTERPRETATION

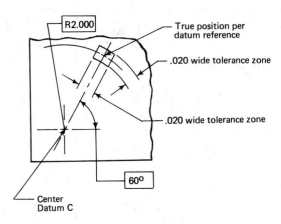

BIDIRECTIONAL DIMENSIONING BY POLAR COORDINATE METHOD
Figure 78

DRAWING CALLOUT

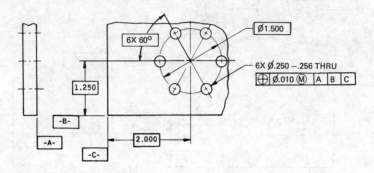

INTERPRETATION

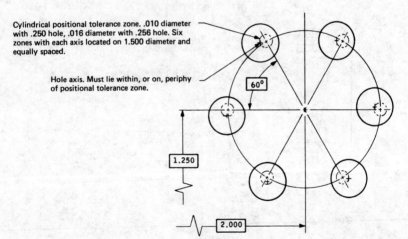

HOLE PATTERN LOCATED BY BASIC DIMENSIONS FROM SPECIFIED DATUMS

Figure 79

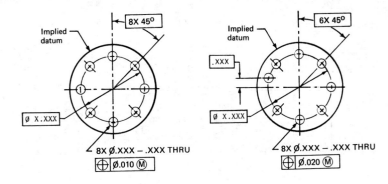
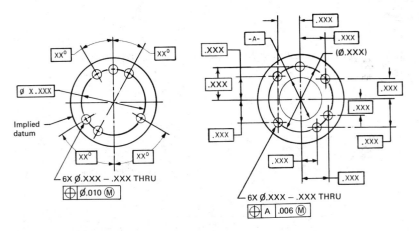

METHODS OF DIMENSIONING HOLE PATTERNS IN RELATION TO A DATUM AXIS

Figure 80

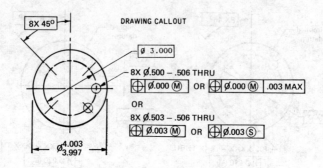

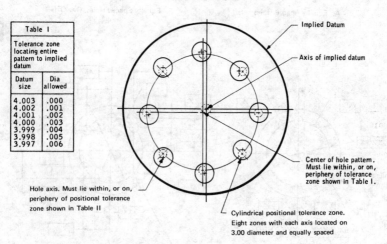

HOLE PATTERN LOCATED TO AN IMPLIED DATUM AXIS

Figure 81

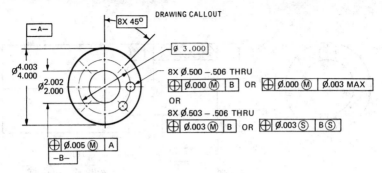

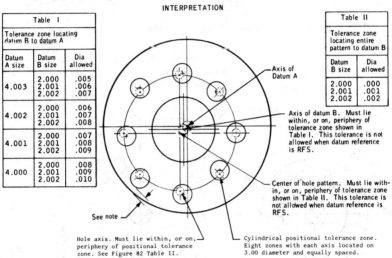

Note: In some designs the remaining wall thickness may be critical. The calculations for determining this are:

(a) Subtract the basic hole pattern diameter from the O.D. at LMC.
(b) From (a), subtract the LMC hole diameter.
(c) Find the sum of the positional tolerances and related size tolerances.
(d) Subtract (c) from (b) and divide by 2.

Example using ⌖ ⌀.003 Ⓜ B

(a) 4.000 datum A
    -3.000 hole pattern dia
     1.000
(b) -.506 LMC hole
      .494
(d) -.018
    2)̄.476 = .238 remaining wall

(c) TP of B to A = .005 + size tol of A (.003) + size tol of B (.002) = .010
    TP of hole pattern to B = .003 + size tol of B (.002) =   .005
    Size tol of hole =   .003
                                                                       .018

HOLE PATTERN LOCATED TO A SPECIFIED DATUM
Figure 82

DRAWING CALLOUT

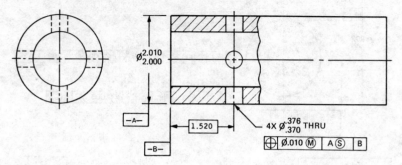

NOTE: 90° Basic position of holes applies. See paragraph 10.10.5

INTERPRETATION

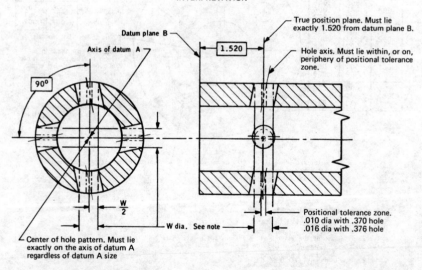

NOTE: W diameter equals the MMC hole minus the positional tolerance, e.g. .370−.010 =.360. The holes may deviate from true position or true direction provided W diameter is not violated and the holes are within limits of size

RADIAL HOLES LOCATED BY BASIC DIMENSION FROM SPECIFIED DATUM
Figure 83

DRAWING CALLOUT

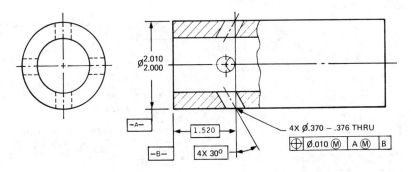

INTERPRETATION

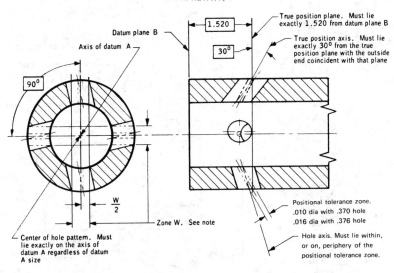

Note: Zone W equals the MMC hole minus the positional tolerance, e.g. .370-.010 = .360. The holes may deviate from true position or true direction provided zone W is not violated and the holes are within limits of size.

ANGULAR RADIAL HOLES LOCATED BY BASIC DIMENSION AT O.D.
Figure 84

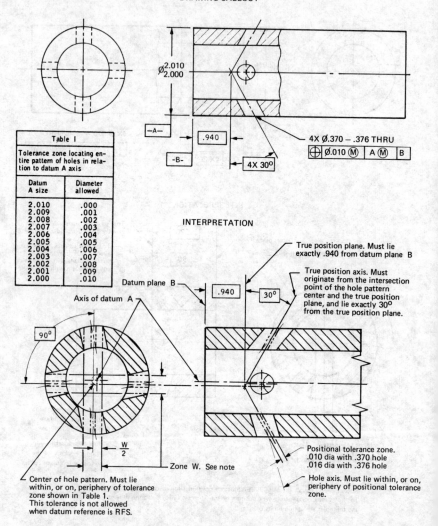

ANGULAR RADIAL HOLES LOCATED BY BASIC DIMENSION AT AXIS OF O.D.
Figure 85

DRAWING CALLOUT

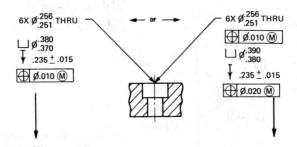

INTERPRETATION

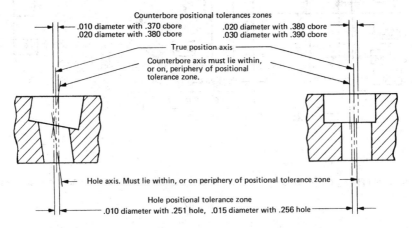

Note: When a mating part must fit a counterbored or spotfaced hole, the calculations shown in Figure 68A may be used to determine the positional tolerance or clearance.

POSITION OF MULTIFEATURES IN A SINGLE CALLOUT

Figure 86

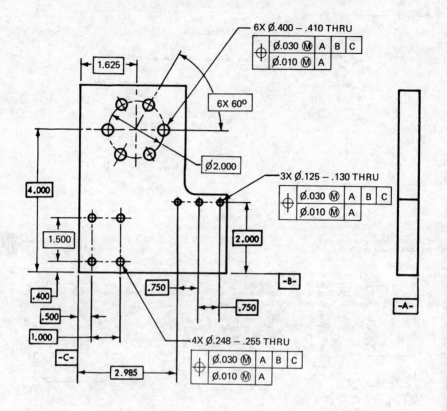

HOLE PATTERNS LOCATED BY COMPOSITE POSITIONAL TOLERANCING, SPECIFIED DATUMS

Figure 87

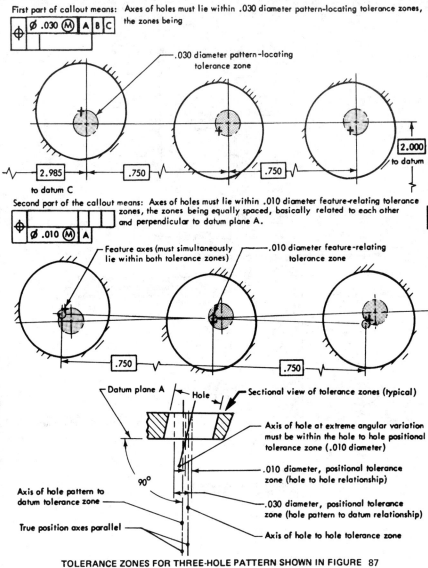

TOLERANCE ZONES FOR THREE-HOLE PATTERN SHOWN IN FIGURE 87

Figure 88

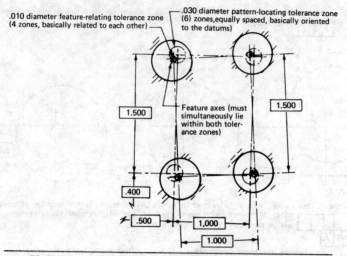

TOLERANCE ZONES FOR FOUR-HOLE PATTERN SHOWN IN FIGURE 87
Figure 89

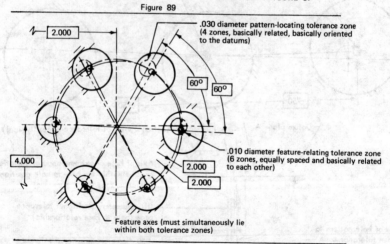

TOLERANCE ZONES FOR SIX-HOLE PATTERN SHOWN IN FIGURE 87
Figure 90

DRAWING CALLOUT

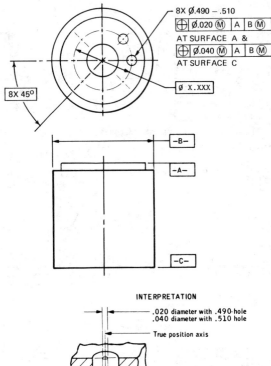

INTERPRETATION

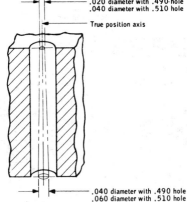

LARGER POSITIONAL TOLERANCE AT ONE END OF HOLE THAN AT THE OTHER END
Figure 91

ASSEMBLY CONDITIONS

A. Spring Pin
B. Bolt
C. Stud

DRAWING CALLOUTS

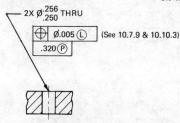

Note: Projected tolerance height should equal maximum pin height above surface.

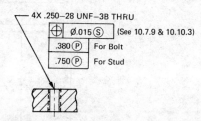

Note: Projected tolerance height for bolt should equal maximum thickness of mating part; height for stud should equal maximum stud height above surface.

INTERPRETATIONS

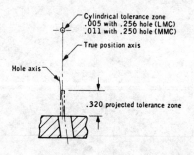

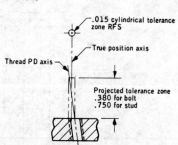

PROJECTED TOLERANCE ZONE AND LMC APPLICATION
Figure 92

ASSEMBLY CONDITION

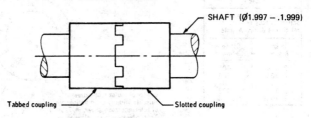

SHAFT (⌀1.997 − .1.999)

Tabbed coupling   Slotted coupling

DRAWING OF TABBED PART

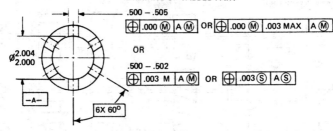

DRAWING OF SLOTTED PART

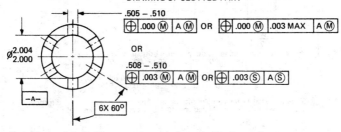

Note: The formula for fixed fastners may be used for determining the tolerance of a pattern of mating tabs or slots.

Example:

$$Y = \frac{H - F}{2} \qquad Y = \frac{.505 - .505}{2} = .000 \quad \text{or} \quad Y = \frac{.508 - .502}{2} = .003$$

Where: Y = permissible positional tolerance
H = minimum slot width
F = maximum tab width

POSITIONAL TOLERANCING OF TABS AND SLOTS

Figure 93

Table I

Tolerance zone locating entire pattern of tabs or slots in relation to datum A axis

Datum A size	Diameter allowed
2.000	.000
2.001	.001
2.002	.002
2.003	.003
2.004	.004

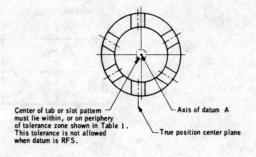

Center of tab or slot pattern must lie within, or on periphery of tolerance zone shown in Table I. This tolerance is not allowed when datum is RFS.

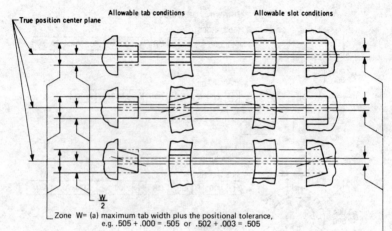

Zone W = (a) maximum tab width plus the positional tolerance,
e.g. .505 + .000 = .505 or .502 + .003 = .505

(b) minimum slot width minus the positional tolerance,
e.g. .505 − .000 = .505 or .508 − .003 = .505

Side surfaces of each tab or slot may deviate from true position or true direction, provided zone W is not violated and the tab or slot is within limits of size.

Table II

Symbol	Tab size	.505	.504	.503	.502	.501	.500
	Slot size	.505	.506	.507	.508	.509	.510
⊕ .000 Ⓜ A Ⓜ		.000	.001	.002	.003	.004	.005
⊕ .000 Ⓜ .003 MAX A Ⓜ		.000	.001	.002	.003	.003	.003
⊕ .003 Ⓜ A Ⓜ		Not applicable			.003	.004	.005
⊕ .003 Ⓢ A Ⓢ		Not applicable			.003	.003	.003

INTERPRETATION OF FIGURE 93
Figure 94

NOTES

NOTES

Section 11 — SURFACE TEXTURE

11.1 SCOPE — This section establishes the method of specifying geometric characteristics of surface irregularities with respect to their height, width, and direction. It provides, through the use of symbols and numerical value classifications, a uniform system for accurately expressing the desired surface requirements (roughness, waviness, contact area, and lay) of solid materials on drawings. Surfaces normally excluded from these requirements are those which are controlled by their manufacturing process and are usually acceptable for most applications, i.e., textile, rubber, optical glass, plastics, felt, sheet metal, tubing, etc.

11.2 APPLICABLE DOCUMENTS

ANSI B46.1	Surface Texture, Surface Roughness, Waviness and Lay
ANSI Y14.36	Surface Texture Symbols

11.3 DEFINITIONS

11.3.1 SURFACE — The surface of an object is the boundary which separates that object from another object, substance, or space.

11.3.2 PROFILE — The profile is the contour of a surface in a plane perpendicular to a surface, unless some other angle is specified.

11.3.3 CENTER LINE (MEAN LINE) — A mean line is a theoretical line parallel to the general surface profile at a mid-point between the high and low measurements of the surface features. (See Figure 1)

11.3.4 SURFACE IRREGULARITIES — Surface irregularities are deviations from the center line, including roughness, waviness, and lay.

11.3.5 FLAWS — Flaws are unintended surface imperfections which occur as a direct result of a manufacturing process including defects singularly or in groups such as cracks, blow holes, checks, ridges, scratches, gouges, nicks, cuts, punctures, dimples, chemical corrosion, etc. (See paragraph 11.4.6; Figure 1)

11.3.6 MICROMETER — A micrometer is one millionth of a meter (.000001 meter).

11.3.7 ROUGHNESS — Roughness is the surface feature of random and repetitively spaced minute or finer deviations from the "center line" having roughness height and width as included within a typical sampling length known as the Roughness-Width Cutoff. (See Figure 1)

11.3.8 ROUGHNESS HEIGHT RATING — *Roughness height rating is a value in micrometers for the (arithmetical) average height between peaks and valleys of the surface roughness of a typical "Roughness-Width Cutoff" length. (See paragraph 11.5.2; Figures 1 and 2).

*Metrication values of ISO terminology and symbolism to U.S. usage was adopted 16 MAR 1978. This chapter will reflect the metric values and retain the inch values for reference only during the changeover period.

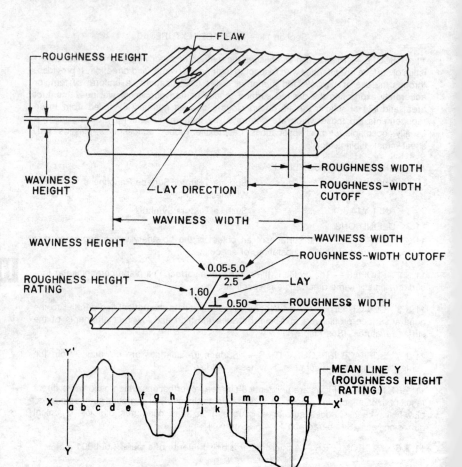

$$Y = \frac{y_a + y_b + y_c + y_d + y_e \ldots + y_n}{n}$$

Surface Texture Terms and Definitions
Figure 1

11.3.9 ROUGHNESS-WIDTH RATING — Roughness-width rating is a value in metric units the intended distance between successive peaks or ridges which constitute the predominant surface features as roughness. (See paragraph 11.5.3; Figures 1 and 2)

11.3.10 ROUGHNESS-WIDTH CUTOFF RATING — Roughness-width cut-off rating is a value in metric units of a sampling length used in measuring the average roughness height. (See paragraph 11.5.4; Figures 1 and 2)

11.3.11 WAVINESS — Waviness is the surface feature of the usually wider-spaced characteristics deviating from the "center line" having waviness height and width is generally wider spacing than the roughness-width cutoff length. Roughness may be considered as superimposed on a "wavy" surface. (See Figure 1). When a percentage of contact area is specified, the waviness height and width designations are not shown. (See paragraph 11.3.4)

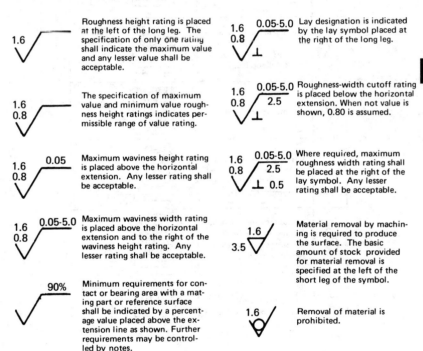

Surface Texture Control Symbols
Figure 2

Lay Symbol	Meaning	Example Showing Direction of Tool Marks
=	Lay approximately parallel to the line representing the surface to which the symbol is applied.	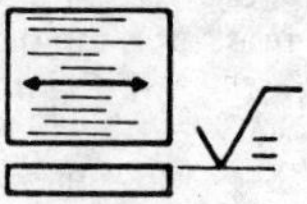
⊥	Lay approximately perpendicular to the line representing the surface to which the symbol is applied.	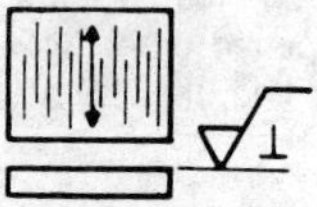
X	Lay angular in both directions to line representing the surface to which the symbol is applied.	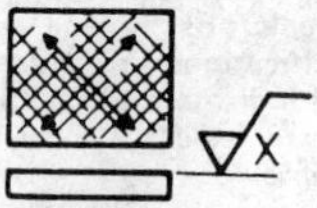
M	Lay multidirectional.	
C	Lay approximately circular relative to the center of the surface to which the symbol is applied.	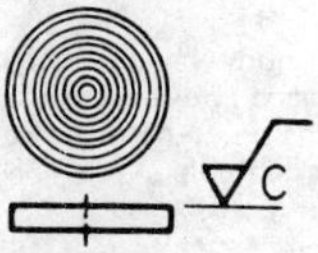
R	Lay approximately radial relative to the center of the surface to which the symbol is applied.	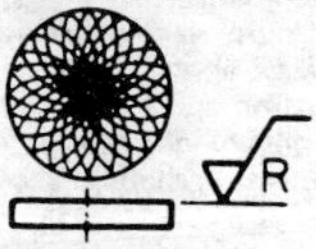
P	Lay particulate, non-directional, or protuberant.	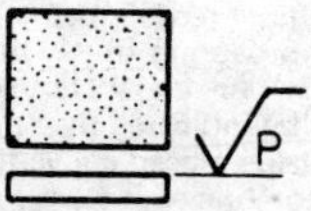

The "P" symbol is not currently shown in ISO Standards. American National Standards Commitee B46 (Surface Texture) has proposed its inclusion in ISO 1302—"Methods of indicating surface texture on drawings."

Figure 3

11.3.12 WAVINESS HEIGHT RATING — Waviness height rating is a value in metric units for the peak-to-valley distance of surface waviness. (See paragraph 11.5.5; Figure 1)

11.3.13 WAVINESS WIDTH RATING — Waviness width rating is a value in metric units for the spacing of successive wave peaks or valleys. (See paragraph 11.5.6; Figure 1)

11.3.14 CONTACT AREA — Contact area is a value in percentages for the surface required to effect contact with its mating component surface. Unless otherwise specified, the contact area shall be distributed over the surface with approximate uniformity. (See paragraph 11.5.7; Figure 1)

11.3.15 LAY — Lay is the direction of the predominant surface pattern determined by the production method used (tool marks or grain). (See Figure 3)

11.4 SURFACE TEXTURE CONTROL APPLIED TO DRAWINGS

11.4.1 GENERAL INSTRUCTIONS — To insure efficient and uniform drafting practices, the application of surface control requirements shall be made in the manner specified by this section.

11.4.2 LIMITATION — Where no surface control is specified, it is to be assumed that the surface produced by the manufacturing operation will be satisfactory. If a surface is to be controlled, the maximum acceptable value rating for the specific surface feature(s) should be indicated. It is to be noted that the surface symbol and value ratings do not control the geometric flatness of the surface. (See Section 10)

11.4.3 GENERAL COVERAGE — The surface control applying to all or most of the surfaces of the part is indicated in a general note similar to the following, as applicable:

SURFACE ROUGHNESS $^{xx}\!\!\sqrt{}$ UNLESS OTHERWISE SPECIFIED.

11.4.3.1 All surfaces of a part that are the result of a shop operation and require control of surface texture are defined by the surface symbol in the general notes and/or on the field of drawing. The general application of the symbol does not affect surfaces specified on previous drawings, "stock" condition surfaces, or the surfaces resulting from a welding process.

11.4.4 INTERPRETATION REFERENCE — When an interpretation of surface symbols is required, the following general note, as applicable, may be included on the drawing.

△ SURFACE TEXTURE PER ANSI B46.1

11.4.5 PLATED OR COATED SURFACES — Surface control designations, unless otherwise specified, apply to the completed surface. Drawings or specifications for plated or coated parts should definitely indicate whether the surface control designations apply before plating (coating), apply after plating (coating), or apply before and after plating or coating.

11.4.6 FLAWS — The effect of flaws is not included in the measurement of roughness height, unless otherwise specified.

11.4.6.1 When flaws are to be considered a characteristic of a surface, a note stating specific requirements should be included on the drawing.

    Example:
        ⚠ SURFACES INDICATED TO BE FREE FROM SCRATCHES, DENTS, STEPS OR FLAWS EXCEEDING XXX **MICROMETERS IN DEPTH**.

11.4.7 SURFACE CONTROL SYMBOLS — The surface control system used to designate the limits of surface features is a check mark with a horizontal top extension as shown in Figure 4. The long leg and extension are to the right as the drawing is read. The top extensions may be lengthened to the right as far as needed. The surface texture symbol is modified when necessary to require or prohibit removal of material. (See 11.4.7.3 thru 11.4.7.5)

11.4.7.1 The lettering should be the same size as that of the drawing dimensions. The line weight should approximate that of the lettering.

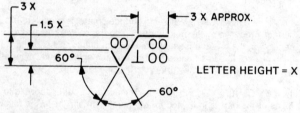

Surface Texture Control Symbol Size
Figure 4

11.4.7.2	✓	Where Roughness Height only is indicated, the horizontal extension is omitted.
11.4.7.3	∇	Where Material Removal By Machining Is Required. The horizontal bar indicates that material removal by machining is required to produce the surface and that material must be provided for that purpose.
11.4.7.4	3.5 ∇	Where Material Removal Allowance. The number indicates the amount of stock to be removed by machining in millimeters (or inches). Tolerances may be added to the basic value shown or in a general note.
11.4.7.5	ⱱ	Where Material Removal Prohibited. The circle in the vee indicates that the surface must be produced by processes such as casting, forging, hot finishing, cold finishing, die casting, powder metallurgy or injection molding without subsequent removal of material.
11.4.7.6	✓	Where any surface characteristics are specified above the horizontal line or to the right of the symbol. Surface may be produced by any method except when the bar or circle is specified. As in 11.4.7.3 and 11.4.7.5

11.4.8 SYMBOL PLACEMENT — Whenever practicable, the apex of the symbol is drawn by touching the surface to be controlled. When this is not practical, the symbol may be placed with the apex touching an extension line from the surface to which it applies. Surface symbols are placed in a horizontal reading position (see Figure 5). Surface control indications for an individual surface are in the one place only and not repeated in another view or section. Where the symbol is used with a dimension, it affects all surfaces defined by the dimension. Areas of transition, such as chamfers and fillets, conform with the roughest adjacent surface area unless otherwise indicated.

11.5 RECOMMENDED VALUE RATINGS

11.5.1 GENERAL CONSIDERATIONS — Usually value ratings selected to qualify surface features will have an associated relationship and will not combine opposite extremes, such as a fine surface roughness height and a wide surface roughness width. Care should be exercised to insert numerical values in the precise position as shown in Figure 4. The following ratings, when indicated with a surface symbol, represent the maximum condition acceptable and are the preferred or most common value used. Other values, not shown, between those indicated may be used as exception to the usual.

11.5.2 ROUGHNESS HEIGHT VALUES — In Micrometers.

ROUGHNESS AVERAGE RATING VALUES
MICROMETERS, μm (MICROINCHES, μin)*

μm	μin	μm	μin	μm	μin	μm	μin
0.012	(.5)	0.25	(10)	1.25	(50)	6.3	(250)
0.025	(1)	0.32	(13)	1.6	(63)	8.0	(320)
0.050	(2)	0.40	(16)	2.0	(80)	10.0	(400)
0.075	(3)	0.50	(20)	2.5	(100)	12.5	(500)
0.100	(4)	0.63	(25)	3.2	(125)	15.0	(600)
0.125	(5)	0.80	(32)	4.0	(160)	20.0	(800)
0.15	(6)	1.00	(40)	5.0	(200)	25.0	(1000)
0.20	(8)						

*Boldface values preferred

11.5.3 ROUGHNESS WIDTH VALUES — In Millimeters.

Minimum	Maximum
0.12	0.50

11.5.4 ROUGHNESS WIDTH CUTOFF — In Millimeters. When no value is specified, the value **0.80** is assumed. Standard values are:

STANDARD ROUGHNESS - WIDTH CUTOFF
VALUES MILLIMETERS, mm (INCHES, in.)*

mm	in.	mm	in.	mm	in.
0.08	(.003)	0.80	(.030)	8.0	(.300)
0.25	(.010)	2.50	(.100)	25.0	(1.000)

*Boldface values preferred

11.5.5 WAVINESS HEIGHT VALUES — In Millimeters.

WAVINESS HEIGHT VALUES
MILLIMETERS, mm   (INCHES, in.)*

mm	in.	mm	in.	mm	in.
0.0005	(.00002)	0.008	(.0003)	**0.12**	(.005)
0.0008	(.00003)	**0.012**	(.0005)	0.20	(.008)
0.0012	(.00005)	0.020	(.0008)	**0.25**	(.010)
0.0020	(.00008)	0.025	(.0010)	0.38	(.015)
0.0025	(.00010)	0.05	(.002)	**0.50**	(.020)
0.005	(.0002)	0.08	(.003)	0.80	(.030)

*Boldface values preferred

11.5.6 WAVINESS WIDTH VALUES — In Millimeters. Waviness height and waviness width values are separated by a dash, no standard values are established.

11.5.7 CONTACT AREA — Percentage (approximate).

90%   75%   50%

11.6 RECOMMENDED RATINGS FOR STANDARD PROCESSES

11.6.1 The ratings listed in Table 1 represent the acceptable surface qualities resulting from standard commercial manufacturing processes.

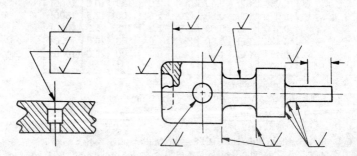

Placement of Surface Control Symbol
Figure 5

Table I
Surface Texture Vs. Process

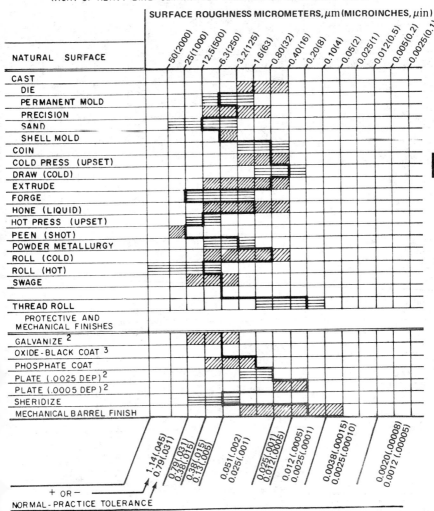

ependent on Previous Finishes, Grit and Grade of Abrasive.
oughness Increases with Thickness of Deposit.
irface on Which Applied Does Not Change.

Table I (Continued)
Surface Texture Vs. Process

SURFACE ROUGHNESS — APPROXIMATE VALUES - WILL VARY WITH MATERIAL AND EQUIPMENT USED - FOR SPECIFIC VALUES CHECK WITH FABRICATION ENGR.

LEFT OF HEAVY LINE : PRACTICAL FINISHES AT COMMERCIAL COSTS
RIGHT OF HEAVY LINE : OBTAINABLE FINISHES AT INCREASED COSTS

SURFACE ROUGHNESS MICROMETERS, μm (MICROINCHES, μin)

MACHINE FINISHES	50(2000)	25(1000)	12.5(500)	6.3(250)	3.2(125)	1.6(63)	0.80(32)	0.40(16)	0.20(8)	0.10(4)	0.05(2)	0.025(1)	0.012(0.5)	0.005(0.2)	0.0025(0.1)
AUTO. SCREW MACHINE				▓	▓	▓	░								
BORE				▓	▓	▓	░	░							
BORE (DIAMOND & PRECISION)							▓	▓	░	░					
BOX TOOL						▓	▓	░							
BROACH					▓	▓	░	░							
BURNISH (ROLLER)							▓	░							
CHIP			▓	░											
COUNTERBORE				▓	▓	░									
COUNTERSINK				▓	░										
CUT-OFF ABRASIVE				▓	▓	░	░								
GAS		▓	░												
PARTING		▓	▓	░	░										
SAND	▓	▓	░												
DRILL				▓	▓	░	░								
DRILL (CENTER)				▓	░										
EXTRUDE				▓	▓	░	░								
FACE				▓	▓	▓	░	░							
FILE					▓	▓	░	░							
GRIND COMMERICAL						▓	▓	░	░						
CYLINDRICAL						▓	▓	░	░	░					
DIAMOND								▓	▓	░	░	░	░		
DISC					▓	▓	░	░							
HAND			▓	░	░										
SNAG		▓	░												
SURFACE					▓	▓	░	░							
GEAR CUTTING MILL					▓	▓	░	░							
HOB				▓	▓	░									
SHAPE				▓	▓	░	░								

+ OR − NORMAL - PRACTICE TOLERANCE:
1.14(.045) / 0.79(.031)
0.79(.031) / 0.38(.015)
0.38(.015) / 0.13(.005)
0.051(.002) / 0.025(.001)
0.025(.001) / 0.012(.0005)
0.012(.0005) / 0.0025(.0001)
0.0038(.00015) / 0.0025(.00010)
0.0020(.00008) / 0.0012(.00005)

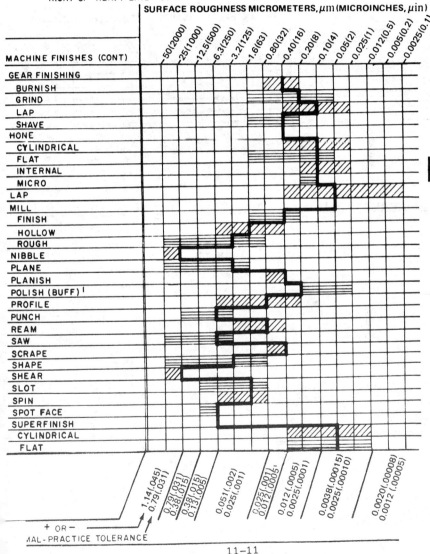

Table I (Continued)
Surface Texture Vs. Process

SURFACE ROUGHNESS — APPROXIMATE VALUES - WILL VARY WITH MATERIAL AND EQUIPMENT USED - FOR SPECIFIC VALUES CHECK WITH FABRICATION ENGINEERING.
LEFT OF HEAVY LINE : PRACTICAL FINISHES AT COMMERCIAL COSTS
RIGHT OF HEAVY LINE : OBTAINABLE FINISHES AT INCREASED COSTS

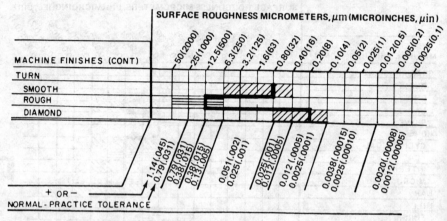

NOTES

Section 12 — CASTING DRAWINGS

12.1 SCOPE — This section establishes the procedures to be followed in the preparation of casting drawings. Data on design practices is incidental to subject presentation and is intended only as a guide.

12.2 APPLICABLE DOCUMENTS — None. (See paragraph 12.9)

12.3 DEFINITIONS

12.3.1 CASTING — A part produced by a process which introduces a molten (plastic, liquid, etc.) material, by gravity or under pressure, into a mold, allowing it to solidify therein.

12.3.2 PARTING LINE — The plane on which the mold is split to facilitate the removal of the pattern or the casting from the mold.

12.3.3 DRAFT — The angle or taper given to a pattern or die to facilitate removal of either the pattern or casting from the mold.

12.3.4 TIE-BAR — A rib or reinforcement, not functionally required, added to casting to prevent warping or distortion of casting.

12.3.5 HOLD-DOWN LUGS — Projections or pads, not functionally required, added to casting to provide clamping surfaces to retain part during machining.

12.3.6 DATUM TARGET (TOOLING POINT) — See Section 10.

12.3.7 GATE — An opening into the mold cavity that is used for filling the mold. When the casting is removed from the mold, a sprue formed by the gate remains attached, but is subsequently removed.

12.3.8 FIN — A thin projection on the CASTING, formed by mold sections not making intimate contact at the parting line. It may also be known as FLASH.

12.3.9 PATTERN — A form that is used to shape the mold. It may be made from wood, metal, plastic, etc., and reflects the size and shape of casting with necessary allowances for shrinkage.

12.3.10 MOLD — A matrix of sand, metal, etc., into which molten material is poured or injected to form the cast piece.

12.3.11 CORE — A loose part of the mold used to form a cavity, opening or hole in the casting.

FOUNDRY PROCESS	PROCESS DESCRIPTION	METALS CAST	PATTERN TYPE	AVERAGE CASTING SIZE and COMPLEXITY	SURFACE FINISH (ROUGHNESS HEIGHT RATING)
SAND CASTING (GREEN SAND OR DRY SAND)	SAND CASTINGS ARE MADE BY POURING MOLTEN METAL INTO AN EXPENDABLE MOLD, THAT WAS PREPARED BY PACKING A MIXTURE OF LOOSE SAND AND CLAY AROUND A PATTERN	ALL FERROUS AND NON-FERROUS CASTING ALLOYS	WOOD OR METAL	OUNCES TO TONS, DEPENDING ON THE ALLOY COMPLEXITY VARIES FROM SIMPLE TO VERY DIFFICULT	ALUMINUM AND MAGNESIUM 250 TO 500 SOMETIMES 125 STEEL 500 TO 1000.
SHELL MOLD CASTING	SHELL MOLD CASTINGS ARE MADE BY POURING MOLTEN METAL INTO AN EXPENDABLE MOLD, THAT WAS PREPARED BY CURING A LAYER OF A REFRACTORY MATERIAL AND THERMOSETTING RESIN MIXTURE ON A HEATED METAL PATTERN	MOST CASTING ALLOYS EXCEPT LOW CARBON STEELS	METAL PATTERNS WITH EJECTOR PINS AND HEATING ELEMENTS	OUNCES TO 100 POUNDS COMPLEX SHAPES CAN BE CAST	125 TO 500
PERMANENT MOLD CASTING	PERMANENT MOLD CASTINGS ARE MADE BY POURING MOLTEN METAL INTO A SEPARABLE REUSABLE MOLD, THAT WAS PREPARED BY MACHINING A CAVITY IN METAL BLOCKS	ALUMINUM, MAGNESIUM, AND COPPER-BASE ALLOYS-CAST IRON	NO PATTERN IRON, STEEL, OR ALUMINUM REUSABLE MOLD.	OUNCES TO SEVERAL POUNDS COMPLEXITY IS LIMITED BUT MAY BE INCREASED BY SAND CORES OR COLLAPSABLE METAL CORES	125 TO 250
PLASTER MOLD CASTING	PLASTER MOLD CASTINGS ARE MADE BY POURING MOLTEN METAL INTO AN EXPENDABLE MOLD, THAT WAS PREPARED BY POURING A PLASTER MIXTURE INTO A PATTERN AND ALLOWING IT TO HARDEN	ALUMINUM, MAGNESIUM, COPPER-BASE, AND OTHER LOW-MELTING ALLOYS	WOOD OR METAL	OUNCES TO SEVERAL POUNDS COMPLEX SHAPES CAN BE CAST	125 TO 250 SOMETIMES 63
DIE CASTING	DIE CASTINGS ARE MADE BY INJECTING MOLTEN METAL, UNDER A HIGH PRESSURE, INTO A SEPARABLE, REUSABLE MOLD, THAT WAS PREPARED BY MACHINING A CAVITY IN METAL DIE BLOCKS	ALUMINUM, MAGNESIUM, COPPER-BASE, AND OTHER LOW-MELTING ALLOYS	NO PATTERN ALLOY STEEL DIE MOLD.	OUNCES TO SEVERAL POUNDS COMPLEXITY LIMITED ONLY BY THE FACT THAT THE CASTING MUST BE WITHDRAWN FROM THE DIE	63 TO 125 SURFACES CAN BE BUFFED OR POLISHED SMOOTH
INVESTMENT CASTING	INVESTMENT CASTINGS ARE MADE BY POURING MOLTEN METAL INTO AN EXPENDABLE MOLD, THAT WAS PREPARED BY SURROUNDING AN EXPENDABLE PATTERN (WAX, PLASTIC, FROZEN MERCURY, ETC) WITH A REFRACTORY MIXTURE AND HEATING THE MOLD TO REMOVE THE PATTERN	MOST CASTING ALLOYS	METAL DIE USED TO MAKE AN EXPENDABLE WAX, PLASTIC OR FROZEN MERCURY PATTERN.	OUNCES TO SEVERAL POUNDS COMPLEX SHAPES CAN BE CAST BY USE OF APPROPRIATE DIES	63 TO 250
CERAMIC MOLD CASTING	CERAMIC MOLD CASTINGS ARE MADE BY POURING MOLTEN METAL INTO AN EXPENDABLE MOLD, THAT WAS PREPARED BY POURING A LAYER OF REFRACTORY OVER A PATTERN AND FUSING THE MOLD AT A HIGH TEMPERATURE	STEELS	WOOD OR METAL	OUNCES TO SEVERAL POUNDS COMPLEXITY LIMITED SINCE MOLDS AND CORES ARE FRAGILE	125 TO 250

General Casting Information
Figure 1

12.4 DESIGN PRACTICE — General information pertaining to casting design and foundry practice can be found in reference books listed in the bibliography. (See 12.11) There is no "rule of thumb" which can apply to all casting designs and processes, see Figure 1; therefore, each design must be considered separately.

12.5 MARKINGS

12.5.1 All cast symbols and markings should be located on a flat, or nearly flat, surface parallel to the parting line whenever possible and preferably on a surface that will not be machined. (See Section 14)

12.5.2 The preferred location for the casting part number is adjacent to the foundry trademark.

12.5.3 When the drawing locates the markings on a surface that will not be machined, the part number shall be preceded by the word CASTING or CSTG as follows:

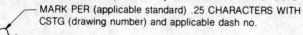

MARK PER (applicable standard) .25 CHARACTERS WITH
CSTG (drawing number) and applicable dash no.

12.5.4 The foundry trademark location is indicated and called out as follows:
MARK PER (applicable standard) WITH FOUNDRY SYMBOL.

12.5.5 All quality control and other markings, not specified above, are located on a raised pad suitable for stamping, and called out as follows:
MARK PER (applicable standard) WITH QUALITY ACCEPTANCE STAMPS AND ASSIGNED SERIAL NO.

12.5.6 When all surfaces of a casting are machined, the method of marking the machined part must be compatible with the designed use of the part. Cast part identification may be specified in addition to other markings.

12.5.7 When the physical size of a casting prevents the application of legible markings, the markings are applied to a tag or to the container.

12.6 MATERIAL REQUIREMENTS

12.6.1 The parts list of the casting drawing shall provide a complete description of the material and its specification. (See Section 9 and Figure 3)

12.6.2 Each casting of the same configuration but of different material shall have separate identification, i.e., different dash numbers, or separate drawings identified with different drawing numbers.

12.6.3 Unless specific design requirements exist, the method of casting (sand, shell-mold, investment, etc.) is not called out. The casting should be designed with a particular process in mind; however, in most cases, more than one process will satisfactorily produce the part.

12.7 DRAWING TITLE. (See Section 6)

12.8 DRAWING NOTES — The following general notes are used on casting drawings as applicable.

12.8.1 MACHINED SURFACES

12.8.1.1 Those surfaces which are to be subsequently machined are indicated on the drawing by a delta, cross-reference/ to the following general note:
⚠ SURFACE INDICATED TO BE SUBSEQUENTLY MACHINED.

12.8.1.2 A part to be machined all over is indicated by the following general note:
THIS CASTING TO BE SUBSEQUENTLY MACHINED ALL OVER.

12.8.2 CAST SURFACES

12.8.2.1 Those surfaces which are to be left as cast or require minimal clean up operations may be covered by a general note as follows:
REMOVE BURRS, GATES, FINS, ETC., FLUSH WITH
CONTOUR +.XX - .XX.

12.8.2.2 Fillets and corner radii which are not dimensioned on the drawing are covered by a general note as follows:
CAST FILLETS TO BE .XX R AND CORNER RADII .XX R
UNLESS OTHERWISE NOTED.

12.8.2.3 Surface roughness acceptable by casting the part may be listed by a general note as follows:
CAST SURFACES XX

12.8.3 ACCEPTANCE REQUIREMENT NOTES

12.8.3.1 HEAT TREATMENT — May be listed by a general note as follows:
AFTER HEAT TREATMENT AND BEFORE RADIOGRAPHIC INSPECTION, FLUORESCENT PENETRANT INSPECT PER (applicable specification). INTERPRET PER (applicable specification or note acceptance standards).

12.8.3.2 RADIOGRAPHIC INSPECTION — May be listed by a general note as follows:
THIS AREA SHALL CONFORM TO (applicable class, level, etc.)
RADIOGRAPHIC QUALITY LEVEL PER (applicable specification).

(If applicable, the casting is zoned into different quality level areas, to denote the critically stressed sections, by using phantom line.)

12.8.3.3 DISCONTINUITIES INSPECTION — May be listed by a general note as follows:
DYE (or FLUORESCENT PENETRANT or MAGNETIC PARTICLE) INSPECT PER (applicable specification and type, class, condition, etc.). INTERPRET PER (applicable specification or note acceptance standards).

12.8.3.4 CLASSIFICATION — May be listed by a general note as follows: CASTING SHALL CONFORM TO (applicable specification, grade, type, class, etc.).

12.8.3.5 CASTING TO BE STRAIGHTENED PRIOR TO FLAW INSPECTION.

12.8.3.6 NO SANDBLASTING OR SHOT PEENING PRIOR TO FLAW INSPECTION.

12.8.3.7 EXCEPT FOR THE DIMENSIONS MARKED "CSTG" THE CONTOUR OF THE CASTING IS NOT RESTRICTED BY THE ENVELOPE SHOWN IN PHANTOM.

12.8.3.8 MACHINED CASTINGS SHALL WITHSTAND XXX $\pm$ XX PSI. HOLD FOR X MINUTES MINIMUM. NO PERMANENT SET IN EXCESS OF 0.2% PERMITTED.

12.9 REFERENCE DATA

12.9.1 The following bibliography will provide general information on casting design and foundry processes.

Aluminum Alloy Casting, Reynolds metals Co., Louisville, Kentucky.

Cast Metals Handbook, American Foundrymen's Society, Des Plaines, Ill.

Magnesium Design, The Dow Chemical Co., Magnesium Department, Midland, Michigan.

Handbook on Designing for Quality Production, Case, H., McGraw-Hill Book Co., Inc., New York, N.Y.

Materials and Processes, Young, James F., Wiley Publishing Co., New York, N.Y.

CASTING DRAWINGS

DESCRIPTION:
    A casting drawing shows the molded condition and requirements for a part made of a specific material. See Figure 2.

USE:

    This type of a drawing is normally used when a part due to material features and physical shape can be fabricated by a molded method to a related size of the desired finished part at a lesser cost than by machining from bulk or stock material.

DRAWING REQUIREMENTS:
1. Separate drawings are normally required for the casting and the item made from the casting.
2. Drawing shows sufficient number of principal, auxiliary, sectional and detail views with adequate dimensions and notes for every feature of the casting.
3. Features of a casting are defined by dimensions to surfaces, intersecting planes, etc.; rather than to centerline of a corner radius or fillet.
4. Use surfaces which will not be subsequently machined for implied or specified datums (see Section 10).
5. Use correlated datums between casting and machine drawings. Indicate bv a general note: CORRELATED DATUMS INDICATED BY -X- -Y- -Z- .
6. Parting lines are generally not shown but when required by design, a local note is used to indicate the location: PARTING LINE (or) PARTING LINE IN THIS AREA ONLY.
7. Datum targets may be used when it is necessary to coordinate measurements for fabrication (see Section 10 and Figures 2 and 3).
8. Draft is normally not shown but the requirements may be indicated by a general note: DRAFT ANGLE X° MAX MAY BE USED WITH DRAWING DIMENSIONS TO INCREASE STOCK UNLESS OTHERWISE NOTED. MATCH DRAFT WHERE NECESSARY.
9. All cast symbols and markings should be located on a flat or nearly flat surface parallel to the parting line whenever possible and preferably on a surface that will not be machined (see Section 14).
10. Each casting of the same configuration but of different material is to have separate identification, i.e., different dash numbers, or separate darawings.
11. Unless specific design requirements exist, the method of casting (sand, shellmold, etc.) is not called out.
12. Castings are preferably drawn full size.

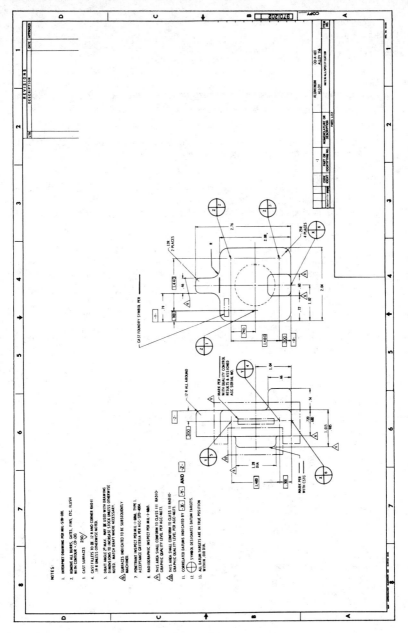

Cast Part Drawing
Figure 2

CASTING MACHINE DRAWING

DEFINITION:
    A drawing that defines the machining and other requirements of a part made from a casting. See Figure 3.

USE:
    The purpose of a casting machine drawing is the same as a detail drawing of an item to be made from raw stock.

DRAWING REQUIREMENTS:
1. Sufficient views, dimensions and notes are given to adequately define the part to be machined from the casting.
2. The casting part number is entered in the material column of the parts list. (See Sections 3 and 9)
3. Markings to identify part after machining.
4. Cast and machine drawing general notes as per 12.8.
5. Designations of datum targets to be carried over from the casting drawing are shown and noted (see Section 10).
6. Correlation of datums is indicated by a general note:

    CORRELATED DATUMS INDICATED BY |-X-| |-Y-| |-Z-|

7. The surface which is to be machined first is given a starting dimension from the datum or datum target.
8. All other machined surface, including bored or drilled holes, are dimensioned from that machined surface.

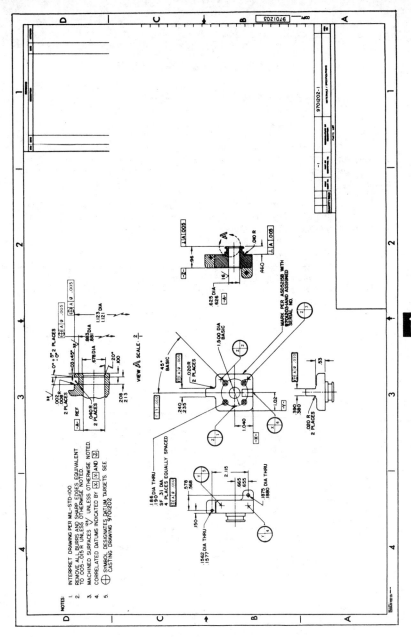

Machining Drawing of a Cast Part
Figure 3

COMBINATION CAST AND MACHINE DRAWING

DESCRIPTION:
    A drawing that shows the cast and machine features of a part on the same drawing. See Figure 4.

USE:
    This type of drawing may be utilized where essential controls of the cast part can adequately be shown in conjunction with the machined part and on the same drawing.

DRAWING REQUIREMENTS:
1. The part is drawn as a "machined" item and the casting outline is shown on the machined detail drawing by phantom lines with limiting dimensions.
2. All cast dimensions are indicated as CSTG.
3. The casting and machined parts are identified with different dash numbers, e.g., -1 and -2.
4. Sufficient views, dimensions and notes are given to adequately define the casting and the part machined from the casting.
5. An alternate method would be to show the casting and the machine part in separate views on the same drawing.

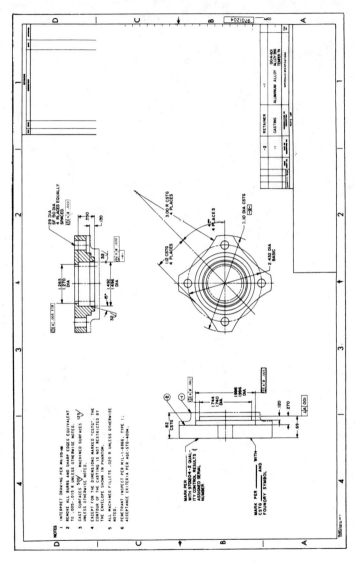

Combination Cast Part and Machining Drawing
Figure 4

NOTES

Section 13 — FORGING DRAWINGS

13.1 SCOPE — This section establishes the procedures to be followed in the preparation of forging drawings. Data on design practices is incidental to subject presentation and is intended only as a guide.

13.2 APPLICABLE DOCUMENTS — None. (See paragraph 13.12)

13.3 DEFINITIONS

13.3.1 FORGING — A part produced by the plastic deformation of a metal, or alloy, into a predetermined shape and size, using a compressive force exerted by a die or dies.

13.3.2 FORGING DIE — A forming tool that determines the size and shape of the forged part.

13.3.3 PARTING LINE — The plane of separation between the die parts that permits removal of the forging. The parting line plane is generally flat, but may be contoured when a complex shape is being forged.

13.3.4 FORGING PLANE — The theoretical plane perpendicular to the direction of die travel and is normally shown only when the parting line and forging planes do not coincide.

13.3.5 DRAFT — The angle or taper added to die cavity surfaces, perpendicular to the direction of die travel and is normally shown only when the parting line and forging planes do not coincide.

13.3.6 MATCH DRAFT — Additional draft allowance permitted for matching surfaces at parting line, when normal draft allowance would result in an offset of surfaces at the parting line.

13.3.7 FLASH — Excess metal on the forging at the parting line, resulting from metal being forced out of the die.

13.3.8 MISMATCH — A defect resulting from die misalignment, producing an offset on the surfaces of the forging at the parting line.

13.3.9 GRAIN FLOW — The directional characteristics of grain structure resulting from the flow of metal into the die contour during the forging operation.

13.3.10 DATUM TARGET (TOOLING POINTS) — See Section 10.

13.3.11 STOCKED FORGING — A forged item of specified material(s), standard shape(s) and nominal size(s), normally kept in stock for ready availability. i. e., ring billet, etc.

13.4 DESIGN PRACTICE

General information pertaining to forging design and shop practices can be found in reference books listed. (See paragraph 13.9) There is no "rule of thumb" that can be applied to all forging designs; therefore, each design must be considered separately and coordinated with the responsible fabrication specialists.

13.5 MARKINGS

13.5.1 All forged symbols and markings should be located on a flat or nearly flat surface parallel to the forging plane whenever possible. Raised markings are preferred over depressed markings. (See Section 14)

13.5.2 The preferred location for the forging part number is a surface that will not be machined and, if possible, adjacent to the forge shop trademark. When located on a surface that will not be machined, the part number shall be preceded by the word FORGING or FORG as follows:

    MARK PER (applicable standard) WITH .25 CHARACTERS
    WITH FORG (drawing number and applicable dash no.)

13.5.3 The Forge shop trademark should be located so that it will not be removed by machining. Its location is indicated and called out as follows:

    MARK PER (applicable standard) WITH FORGE SHOP SYMBOL

13.5.4 All quality control and other applicable markings are located so that they will not be removed in machining. A raised stamping pad on the forging is desirable and, if shown, will be dimensioned. The location is indicated and markings called out as follows:

    MARK PER (applicable standard) WITH QUALITY ACCEPTANCE
    STAMPS AND ASSIGNED SERIAL NO.

13.5.5 Forgings subject to the removal of markings, by the machining of all surfaces, may require forged part identification and forge shop symbol markings, but a record of the above and all other required information markings must also be maintained. When the physical size of a forging prevents the application of legible markings, the required information will be recorded by other acceptable methods, i.e., tag, etc. (See Section 14)

13.6 MATERIAL REQUIREMENTS

13.6.1 The parts list of the forging drawing shall provide a complete description of the material and its specification. (See Section 9 and Figure 1)

13.6.2 Each forging of the same configuration but of different material is to have separate identification, i.e., different dash numbers, or separate drawings identified with different drawing numbers.

13.6.3 Unless specific design requirements exist, the method of forging (Drop, Impact, Upset, Roll, etc.) is not called out. The forging should be designed with a particular process in mind; however, in most cases, more than one process will satisfactorily produce the part.

13.7 DRAWING TITLE (see Section 6)

13.8 DRAWING NOTES — The following general notes are used on forging drawings as applicable.

13.8.1 MACHINE SURFACES

13.8.1.1 Those surfaces which are to be subsequently machined are indicated on the drawing by a delta, cross-referenced to the following general note:
    ⚠ SURFACES INDICATED TO BE SUBSEQUENTLY MACHINED.

13.8.1.2 A part to be machined all over is indicated by the following general note:
    THIS FORGING TO BE SUBSEQUENTLY MACHINED ALL OVER.

13.8.2 FORGED SURFACES

13.8.2.1 Those surfaces which are to be left as forged or require minimal clean up operations may be covered by a general note as follows:
    TRIM FLASH FLUSH WITH CONTOUR +.XX - .XX.

13.8.2.2 Fillets and corner radii which are not dimensioned on the drawing are covered by a general note as follows:
    FORGED FILLETS TO BE .XX R AND CORNER RADII .XX R
    UNLESS OTHERWISE NOTED.

13.8.2.3 Surface roughness acceptable by forging the part may be listed by a general note as follows:
    FORGED SURFACES   XX∨

13.8.3 ACCEPTANCE REQUIREMENTS

13.8.3.1 HEAT TREATMENT — May be listed by a general note as follows:
    AFTER HEAT TREATMENT AND BEFORE RADIOGRAPHIC INSPECTION, FLUORESCENT PENETRANT INSPECT PER (applicable specification).
    INTERPRET PER (applicable specificcation).

13.8.3.2 RADIOGRAPHIC INSPECTION — May be listed by a general note as follows:
    THIS AREA SHALL CONFORM TO (applicable class, level, etc.)
    RADIOGRAPHIC QUALITY LEVEL PER (applicable specification).

(If applicable, the forging is zoned into different quality level areas to denote the critically stressed sections by using phantom lines.) (See Figure 1)

13.8.3.3 DISCONTINUITIES INSPECTION — May be listed by a general note as follows:
    DYE (or FLUORESCENT PENETRANT or MAGNETIC PARTICLE) INSPECT PER (applicable specification and type, class, condition, etc.). INTERPRET PER (applicable specification or note acceptance standards).

13.8.3.4 CLASSIFICATION — May be listed by a general note as follows: FORGINGS SHALL CONFORM TO (applicable specification, grade, type, class, etc.).

13.8.3.5 FORGING TO BE STRAIGHTENED PRIOR TO FLAW INSPECTION.

13.8.3.6 NO SANDBLASTING OR SHOT PEENING PRIOR TO FLAW INSPECTION.

13.8.3.7 EXCEPT FOR THE DIMENSIONS MARKED "FORG", THE CONTOUR OF THE FORGING IS NOT RESTRICTED BY THE ENVELOPE SHOWN IN PHANTOM.

13.8.3.8 MACHINED FORGING SHALL WITHSTAND XXX±XX PSI. HOLD FOR X MINUTES MINIMUM. NO PERMANENT SET IN EXCESS OF X.X% PERMITTED.

13.9 STOCKED FORGING

13.9.1 A stocked forging is called out in the parts list by the forged material and applicable specification. List size and form only when required by design. General notes are used as applicable and should indicate application as to the forged or machined part. See Figure 1

Figure 1

13.10 REFERENCE DATA

13.10.1 The following bibliography will provide general information on forging design and forging processes.

Magnesium Design Notes, Dow Chemical Company, Midland, Michigan.

Stainless Steel Fabrication, Allegheny Ludlum Steel Corporation, Pittsburgh, Pennsylvania.

Design for Forging, Aluminum Corporation of America, Pittsburgh, Pennsylvania.

Forging Handbook, W. Navjaks and D. C. Fabel, The American Society for Metals.

Standard Practices and Tolerances for Impression Forgings, The Drop Forging Association.

Tool Engineers Handbook, A.S.T.E. Handbook Committee, McGraw-Hill Book Co., Inc., New York, N.Y.

Metals Handbook, Metals Handbook Committee, American Society for Metals.

Plastic Working Presses, E. V. Crane, John Wiley and Sons, Inc.

Handbook on Designing for Quality Production, H. Case, McGraw-Hill Book Co., Inc., New York, N.Y.

Materials and Processes, J. F. Young, John Wiley and Sons, Inc.

Forging and Forming Metals, S. S. Rusinoff, 1952, American Technical Society, Chicago, Illinois.

FORGING DRAWING

DESCRIPTION:
A forging drawing shows the molded condition and requirements for a part made of a specific material. See Figure 2.

USE:
This type of a drawing is normally used when a part due to material features and physical shape can be fabricated by a forged method to a related size of the desired finished part at a lesser cost than by machining from bulk or stock material.

DRAWING REQUIREMENTS:
1. Separate drawings are normally required for the forging and the item made from the forging.
2. Drawing shows sufficient number of principal, auxiliary, sectional and detail views with adequate dimensions and notes for every feature of the forging.
3. Features of a forging are defined by dimensions to surfaces, intersecting planes, etc.; rather than to centerline of a corner radius or fillet.
4. Use surfaces which will not be subsequently machined for implied or specified datums (see Section 10).
5. Use correlated datums between forging and machine drawings. Indicate by a general note: CORRELATED DATUMS INDICATED BY -X- -Y- -Z- .
6. Parting lines are generally not shown but when required by design, a local note is used to indicate the location: PARTING LINE (or) PARTING LINE IN THIS AREA ONLY.
7. Datum target may be used when it is necessary to coordinate measurements for fabrication (see Figures 2 and 3 and Section 10).
8. Draft is normally not shown but the requirements may be indicated by a general note: DRAFT ANGLE X° MAX MAY BE USED WITH DRAWING DIMENSIONS TO INCREASE STOCK UNLESS OTHERWISE NOTED. MATCH DRAFT WHERE NECESSARY.
9. All forged symbols and markings should be located on a flat or nearly flat surface parallel to the parting line whenever possible and preferably on a surface that will not be machined (see Section 14).
10. Each forging of the same configuration but of different material is to have separate identification, i.e., different dash numbers, or separate drawings.
11. Unless specific design requirements exist, the method of forging (drop, impact, upset, rolled) is not called out.
12. Forgings are preferably drawn full size.

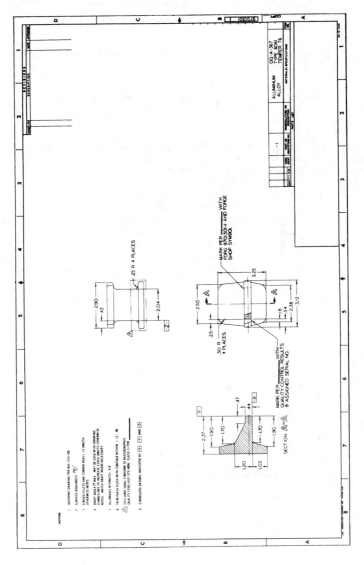

Forged Part Drawing
Figure 2

FORGING MACHINE DRAWINGS

DEFINITION:
    A drawing that defines the machining and other requirements of a part made from a forging. See Figure 3.

USE:
    The purpose of a forging machine drawing is the same as a detail drawing of an item to be made from raw stock.

DRAWING REQUIREMENTS:
1. Sufficient views, dimensions and notes are given to adequately define the part to be machined from the forging.
2. The forging part number is entered in the material column of the parts list. (See Sections 3 and 9)
3. Markings to identify part after machining.
4. Forge and machine drawing general notes as per paragraph 13.8.
5. Designations of datum targets to be carried over from the forging drawing are shown and noted (see Section 10).
6. Correlation of datums is indicated by a general note:

    CORRELATED DATUMS INDICATED BY   -X- -Y- -Z-

7. The surface which is to be machined first is given a starting dimension from the datum or datum target.
8. All other machined surface, including bored or drilled holes, are dimensioned from the machined surface.

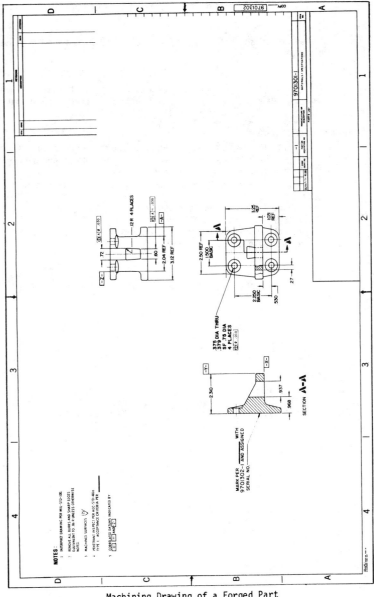

Machining Drawing of a Forged Part
Figure 3

COMBINATION FORGED AND MACHINE DRAWING

DESCRIPTION:
    A drawing that shows the forged and machine features of a part on the same drawing. See Figure 4.

USE:
    This type of drawing may be utilized where essential controls of the forged part can adequately be shown in conjunction with the machined part and on the same drawing.

DRAWING REQUIREMENTS:
1. The part is drawn as a "machined" item and the forging outline is shown on the machined detail drawing by phantom lines with limiting dimensions.
2. All forging dimensions are indicated as FORG.
3. The forging and machined parts are identified with different dash numbers, e.g., -1 and -2.
4. Sufficient views, dimensions and notes are given to adequately define the forging and the part machined from the forging.
5. An alternate method would be to show the forging and the machine part in separate views on the same drawing.

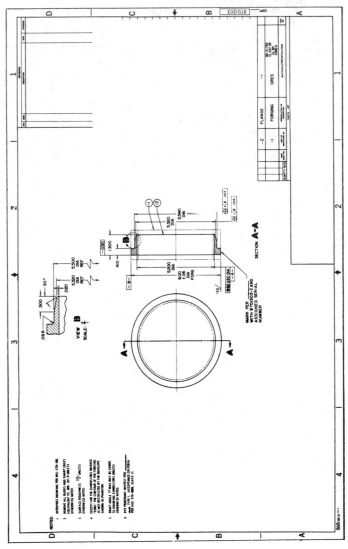

Combination Forged Part and Machining Drawing
Figure 4

NOTES

Section 14 — IDENTIFICATION MARKING

14.1 SCOPE — This section establishes the engineering drawing requirements for specifying the marking of part identification, serialization and other information on hardware.

14.2 APPLICABLE DOCUMENTS

DOD-STD-100	Drawings, Engineering and Associated Lists
MIL-STD-130	Identification Marking of U.S. Military Property
H4-1	Federal Supply Code for Manufacturers (Name to Code)
H4-2	Federal Supply Code for Manufacturers (Code to Name)
H6	Federal Item Identification Guide for Supply Cataloging
AS478	Identification Marking Methods

14.3 DEFINITIONS

14.3.1 CONFIGURATION ELEMENT IDENTIFIER — The number assigned to identify a configuration element, an item subject to configuration management such as a Contract or Configuration End Item (CEI).

14.3.2 END ITEM — A part, assembly, unit, set or system that is the finished product or the prime level of assembly.

14.3.3 MANUFACTURER — An individual, firm, company. or corporation engaged in the fabrication of finished or semi-finished products.

14.3.4 PART — One piece, or two or more pieces joined together which are not normally subject to disassembly without destruction of designed use. (Examples: Gear, screw, electron tube.)

14.3.5 PART OR IDENTIFYING NUMBER — A marking applied to an item, or its package, for the purpose of identifying the item as a unique part or assembly.

14.3.6 PERMANENT MARKING — A method of identification which will remain legible during the normal service life of an item.

14.3.7 PROCUREMENT IDENTIFICATION NUMBER — The government procuring activity's contract or purchase order number. When an order shows both a contract number and a purchase order number, the Procurement Identification number shall be as specified by the procuring activity.

14.3.8 SERIAL NUMBER — An alpha and/or numeric code assigned to an item to differentiate that item from any other item of the same part or identification number.

14.3.9 SET — A unit or units and necessary assemblies and parts connected or associated together to perform an operational function. Also used to denote a collection of parts, such as "tool set."

14.3.10 SPECIAL CHARACTERISTICS — The pertinent rating, operating characteristics, and other information necessary to describe the item.

14.3.11 STOCK NUMBER — The National Stock Number, without prefix or suffix, as specified by the procuring activity.

14.3.12 SYSTEM — A combination of parts, assemblies and sets joined together to perform an operational function.

14.3.13 TEMPORARY MARKING — A method of in-process identification which can be removed without defacing or damaging the item.

14.3.14 UNIT — An assembly or any combination of parts, subassemblies and assemblies mounted together, normally capable of indpendent operation in a variety of situations.

14.3.15 US — The abbreviation used to denote Government ownership.

14.4 GENERAL REQUIREMENTS

14.4.1 Each drawing of a part, assembly, unit, set or other item of supply shall specify marking, using an acceptable method that is not detrimental to the hardware or will not adversely affect its life, utility or function.

14.4.2 Materials, e.g., powders, liquids, etc., which by their physical nature cannot be marked, will be identified by marking the container.

14.4.3 Part identification for components of subassemblies and assemblies which are not subject to disassembly or repair will be specified by temporary markings.

14.4.4 Drawings of parts which do not have a suitable surface or sufficient space for marking will specify tagging or marking the container.

14.4.5 Government and Industry standard parts or vendor items are not reidentified, except as specified in the requirements for altered or selected parts.

14.4.6 Supplier items controlled by specification or source control drawings are identified as specified in the drawing type requirements of Section 4. Marking requirements are not specified by Company standard, but will be controlled by reference to a Government standard, e.g., MIL-STD-130 or Industry standard, e.g., AS478.

14.5 MARKINGS

14.5.1 The height of permanent marking characters is established by the applicable standard and is not specified on the drawing, except when necessary to satisfy design or contractual requirements.

14.5.2 Whenever practicable, the permanent marking of a part or assembly is located to be visible after installation.

14.5.3 Item numbers, used for reference purposes on the drawing, shall not be used as part identification numbers.

14.6 DRAWING APPLICATION

14.6.1 When design conditions permit, one or more optional methods of marking should be specified.

14.6.2 The marking process name is not specified on the drawing (electroetch, etc.).

14.6.3 Hardware cannot be identified with subsequent part numbers. That is, a detail part cannot be marked with the assembly part number.

14.6.4 Parts are marked with one part number only.

14.6.4.1 In-process identification for parts is not normally specified on the engineering drawing; when necessary, it is specified by a temporary marking process that will not affect the surface to which the marking is applied and can be readily removed.

14.6.4.2 Parts containing more than one part number will distinguish between the numbers. For example, an assembly containing a machined casting could have three part numbers: a casting part number, a machined part number and an assembly part number. The casting part number would have the prefix CSTG and the assembly part number would have the prefix "ASSY" to distinguish between the numbers.

14.6.5 Permanent part identification markings are normally specified on the field of drawing, but may also be specified in a general note with a delta referencing the location. The same note is used for either the field of drawing or in the general notes. The area within which the markings must be confined is indicated by a phantom block, but is not dimensioned except when design conditions require such control.

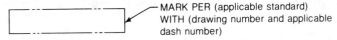

MARK PER (applicable standard)
WITH (drawing number and applicable dash number)

14.6.6 When hardware is required to be permanently marked with a serial number, the requirement is specified by the addition of "AND ASSIGNED SERIAL NUMBER" to the marking note.

> MARK PER (applicable standard) WITH (drawing number) AND APPLICABLE DASH NUMBER AND ASSIGNED SERIAL NUMBER.

14.6.6.1 When an inseparable assembly contains a critical or major component that requires carry over serialization, the notation "AND SERIAL NUMBER OF (applicable part number)" is added to the marking note.

> MARK PER (applicable standard) WITH (drawing number and applicable dash number) AND SERIAL NUMBER OF (applicable part number).

14.6.7 Markings for additional requirements, such as quality control results, foundry trademark, forge shop symbol, etc., are shown in Sections 12 and 13.

14.6.8 Matched sets or parts require markings to identify each component separately and as a part of the matched set. The complete matched set is assigned an assembly number and the detail part identification is correlated to the assembly part identification and serial number. The matched set is identified on the next assembly by the matched set assembly number. The part identification for each of the components comprising the matched set is specified by the following note:

> MARK PER (applicable standard) WITH (applicable part number) PT OF ASSY (applicable assembly number) AND ASSIGNED SERIAL NUMBER.

14.6.9 Altered and/or selected parts are reidentified with a company part number and the original identifying number is obliterated or removed. The requirement for the obliteration or removal is specified in the marking note, but the specific method or process to be used is left to the discretion of the fabrication group.

> REMOVE ORIGINAL PART NUMBER WITHOUT DAMAGE TO HARDWARE.
> MARK PER (applicable standard) WITH (applicable drawing number and applicable dash number).

14.6.10 Temporary markings are normally specified in a general note, but may be specified on the field of drawing when necessary for clarity.

> TEMPORARY MARKING PER (applicable specification) WITH (applicable drawing number and applicable dash number).

14.7 IDENTIFICATION PLATES, DECALCOMANIAS, LABELS AND OTHER MARKING DEVICES

14.7.1 The use of identification or information plates, decalcomanias, labels and other marking devices on assemblies, units, sets, systems or other end items is established by company policy and/or contractual requirements. It is the designer's responsibility to determine and specify the requirements for his project.

14.7.2 The identification plate or other marking device used is specified in the parts list and on the field of drawing. Attaching hardware or other requirements are also specified when necessary.

14.7.3 When markings are to be applied to the marking device, a note or detail and note will specify the markings and the method used. (See Figure 1)

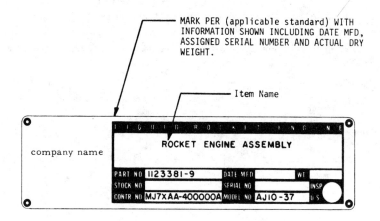

Identification Plate Example
Figure 1

14.7.3.1 The marking method selected should produce markings that are as permanent as the life expectancy of the device to which they are affixed.

14.7.3.2 When conditions will not permit marking-the-device-in-place, the marking note will indicate that the markings are to be affixed prior to assembly.

14.7.3.3 The marking device is not reidentified when markings are affixed at assembly.

14-5

14.8 APPLICATION OF MIL-STD-130 — When a contract invokes MIL-STD-130, the following additional instructions apply to the marking of items.

14.8.1 Marking information on parts when the design activity is also the manufacturer.

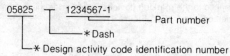

```
05825 ── 1234567-1
            └──────── Part number
            └── *Dash
     └── * Design activity code identification number
```

* Not required on parts in an assembly which is not normally subject to disassembly or repair.

14.8.2 Marking information on parts when the manufacturer is other than the design activity.

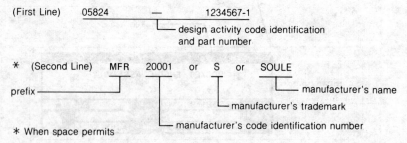

```
(First Line)   05824   —   1234567-1
                           └── design activity code identification
                               and part number

*  (Second Line)  MFR   20001   or   S   or   SOULE
prefix ──────────────┘                          └── manufacturer's name
                                        └── manufacturer's trademark
                              └── manufacturer's code identification number
```

* When space permits

14.8.3 Marking information subassemblies, minor assemblies.

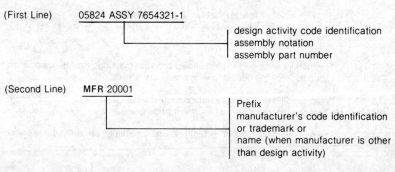

```
(First Line)   05824 ASSY 7654321-1
                     └── design activity code identification
                         assembly notation
                         assembly part number

(Second Line)   MFR 20001
                    └── Prefix
                        manufacturer's code identification
                        or trademark or
                        name (when manufacturer is other
                        than design activity)
```

14.8.4 Inseparable assemblies and separable assemblies which are not normally subject to disassembly or repair will be marked as parts (See 14.8.1 and 14.8.2).

14.8.5 Marking information on units, groups or sets and major assemblies or assemblies supplied with identification plates. (See Figure 2)

X   required
--   not required
*   when specified by procurement activity

Type of Information	DRM Para. or Sect.	Units	Groups or Sets	Major Assemblies or Assemblies Supplied With Name Plates
Nomenclature (item name)	Section 6	X	X	X
Type designation	Section 6	X	X	--
Design activity's code ident and part no.	Section 3, 4 & 5	X	*	X
Special characteristics	14.3.10	X	X	*
Stock number	14.3.11	*	*	*
Manufacturer's code ident number (if other than design activity)	14.3.3	X	X	X
Procurement ident contract or purchase order no.	14.3.7	X	X	*
Serial number	14.3.8	X	X	*
Configuration element identifier	14.3.1	*	*	*
Contractor's name or code ident no.	if the manufacturer is not the contractor	--	X	--
Unit of ____	nomenclature of set	*	--	--
Part of ____	nomenclature of set (group only)	--	*	--
Consists of ____	nomenclature of units of a set or group	--	*	--
US	14.3.15	X	X	X

Specific Requirements per MIL-STD-130
Figure 2

NOTES

Section 15 — WELDING SYMBOLOGY

15.1 SCOPE — This section delineates the application of the approved welding symbols for use on drawings.

15.2 APPLICABLE DOCUMENTS

AWS A1.1   Metric Practice Guide for the Welding Industry
AWS A2.4   Symbols for Welding and Nondestructive Testing
AWS A3.0   Welding Terms and Definitions

15.3 DEFINITIONS — Not applicable.
15.4 WELDING SYMBOLS

15.4.1 Welding symbols provide the means of placing complete welding information on drawings. (See Figure 1, AWS A3.0 and AWS A2.4)

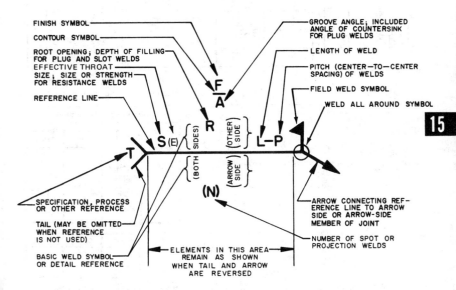

Standard Location of Elements of a Welding Symbol
Figure 1

15-1

15.4.2 A weld symbol is an ideograph used to indicate the desired type of weld. (See Figures 2, 3, and 4)

Basic weld symbols

Groove							
Square	Scarf*	V	Bevel	U	J	Flare-V	Flare-bevel
![]	![]	![]	![]	![]	![]	![]	![]

Fillet	Plug or slot	Spot or projection	Seam	Back or backing	Surfacing	Flange	
						Edge	Corner

*Used for brazed joints only

Figure 2

Basic Weld Symbols Shall Be Drawn on the Reference Line Shown Dotted.

Type of weld					
Plug or spot	Arc spot or arc seam	Resistance spot	Projection	Resistance seam	Flash or upset	Field

OBSOLETE SYMBOLS; USE PREFFERED SYMBOL WITH PROCESS REFERENCE IN THE TAIL

Figure 3

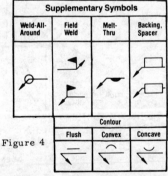

Figure 4

15.4.2.1 Length, spacing and/or size of welds shall be specified in decimal inch or metric units tolerance applied in title block is not applicable to weld callouts, and the decimal or metric unit weld callouts will in no way affect or alter present fabrication and/or inspection procedures.

Example:

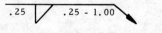

15.4.2.2 Location significance or arrow shall connect the welding symbol reference line to one side of the joint. The term "Arrow Side", "Other Side" and "Both Sides", are used to locate the weld with respect to the joint.

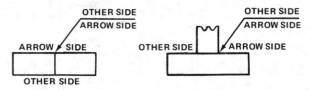

15.4.3 Finish symbols indicate the method of finishing and not the degree of finish, i.e., C = chip, M = machine, G = grind, R = roll, H = hammer.

15.4.4 STANDARD SIZE OF WELD SYMBOL FOR DRAWING APPLICATION— Symbols should be consistent in size on any one drawing, and should be reasonably close to the sizes shown. Dimensions are for the draftsman's information in making weld symbols and are in agreement with standard templates. Units used in this section are in U.S. inch system. However dimensions on symbols may be shown in decimal or metric units whichever is applicable. See AWS A2.3 and A4.4.7 in Appendix A4 for guidance in the use of metric units for the welding industry.

15.4.4.1 BASIC WELD SYMBOLS

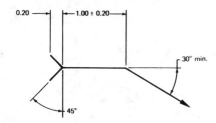

15.4.4.1.1 GROOVE WELDS

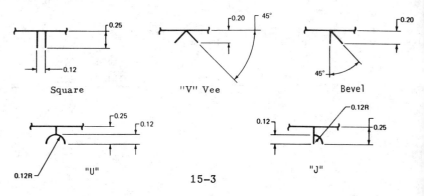

15-3

15.4.4.1.1 GROOVE WELDS (Continued)

Flare "V" Flare Bevel

15.4.4.1.2 MISCELLANEOUS WELDS

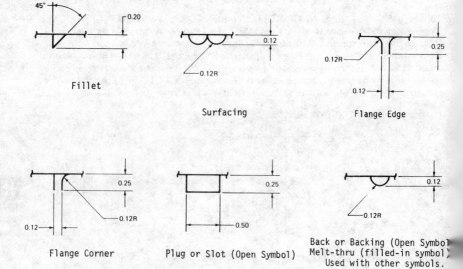

Fillet

Surfacing

Flange Edge

Flange Corner

Plug or Slot (Open Symbol)

Back or Backing (Open Symbol)
Melt-thru (filled-in symbol)
Used with other symbols.

15.4.4.2 SPOT, SEAM AND PROJECTION WELD SYMBOLS

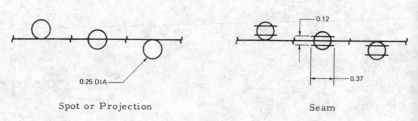

Spot or Projection

Seam

15.4.4.3 SUPPLEMENTARY SYMBOLS

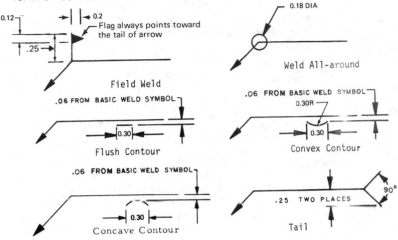

15.4.5 AMERICAN WELDING SOCIETY SYMBOLS AUTHORIZED FOR USE ON DRAWINGS

15.4.5.1 MELT-THRU WELDS

15.4.5.1.1 The melt-thru symbol is used where at least 100% joint penetration of the weld through the material is required in welds made from one side only. A melt-thru weld is shown by placing the melt-thru symbol on the side of the reference line opposite the groove or flange weld symbol as follows:

15.4.5.1.2 The dimensions of a melt-thru weld are not shown on the welding symbol. If it is desirable to specify dimensions, they are shown in a separate detail of the welded joint.

15.4.5.1.3 Melt-thru welds that are to be welded approximately flush without mechanical finishing are shown by adding the flush-contour symbol to the melt-thru symbol:

15.4.5.1.4 If mechanical finishing is desired, the standard finish symbol is added:

15.4.5.2 FLARE-V AND FLARE-BEVEL GROOVE WELDS

15.4.5.2.1 Flare groove welds extend only to the tangent points as indicated below by dimension lines. The extension beyond the point of tangency is treated as an edge or lap joint:

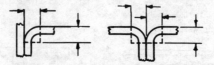

15.4.5.3 FLANGE WELDS

15.4.5.3.1 Flange weld symbols have no both-sides significance and are intended to be used for light gage metal joints involving flaring or flanging of the edges to be joined.

Edge Flange Symbol   Corner Flange Symbol

15.4.5.3.2 The radius and height above the point of tangency is indicated by showing these dimensions separated by a plus mark to the left of the symbol.

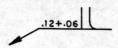

15.4.5.3.3 The radius and height reads from left to right along the reference line. the size of flange welds is shown by a dimension placed outward of the flange dimension as follows:

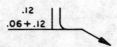

15.4.5.3.4 If one or more pieces are inserted between two outer pieces, the same symbol as for the two outer pieces is used regardless of the number of pieces inserted.

15–6

15.5 APPLICATION FOR FUSION WELDS

15.5.1 MELT-THRU WELD SYMBOL

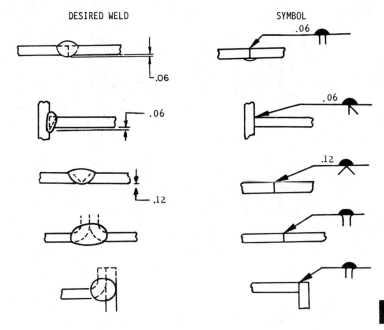

15.5.2 FLARE-V AND FLARE BEVEL GROOVE WELD SYMBOLS

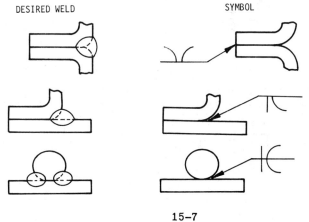

15.5.3 FLANGE WELD SYMBOLS

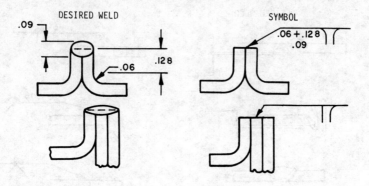

15.5.4 STANDARD EDGE PREPARATIONS

15.5.4.1 Thickness range from .005 to .125 inclusive.

15.5.4.2 Thickness range from .060 to .375, when it is desirable or necessary to weld from one side.

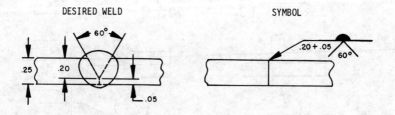

15.5.4.3 Thickness range from .060 to .375 where both sides of joint are accessible for welding. It should be noted that for the heavier gages, an equal double "V" preparation is preferred to minimize distortion of base material by dictating effective throat (E).

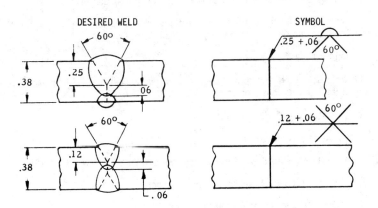

15.5.5 WELDED ASSEMBLIES — When a weld assembly is shown on a next assembly, the welding is normally not shown again; the unit is now a solid part and should be treated as such.

15.6 SUGGESTED WELD ROD CALLOUTS

15.6.1 Information relative to welding process (see Figure 5) and weld rod callouts should be incorporated in the notes and parts list (where applicable).

15.6.2 Call weld rod material out in parts list. (See DRM Section 9)

15.6.3 When a specification, process or other reference is used with a welding symbol, the reference may be placed in the tail.

MIL-W-6858 CL A ⟩⎯⎯⎯↗

15.6.4 When use of a definite process is required, the process may be indicated by one or more of the letter designations shown in Figure 6.

GMAW-AU⟩⎯⎯↗ ↖⎯⎯⟨GTAW

15.7 TYPICAL DRAWING NOTES

15.7.1 ALL FILLET WELDS (weld size dimension) UNLESS OTHERWISE NOTED.

15.7.2 FUSION WELD PER (applicable specification).

15.7.3 RESISTANCE WELD PER (applicable specification).

NOTES

MASTER CHART OF WELDING AND ALLIED PROCESSES

gas metal arc welding	GMAW
— pulsed arc	GMAW-P
— short circuiting arc	GMAW-S
gas tungsten arc welding	GTAW
— pulsed arc	GTAW-P
plasma arc welding	PAW
shielded metal arc welding	SMAW
stud arc welding	SW
submerged arc welding	SAW
— series	SAW-S

arc brazing	AB
block brazing	BB
diffusion brazing	DFB
dip brazing	DB
flow brazing	FLB
furnace brazing	FB
induction brazing	IB
infrared brazing	IRB
resistance brazing	RB
torch brazing	TB
twin carbon arc brazing	TCAB

electron beam welding	EBW
electroslag welding	ESW
flow welding	FLOW
induction welding	IW
laser beam welding	LBW
thermit welding	TW

air acetylene welding	AAW
oxyacetylene welding	OAW
oxyhydrogen welding	OHW
pressure gas welding	PGW

air carbon arc cutting	AAC
carbon arc cutting	CAC
gas metal arc cutting	GMAC
gas tungsten arc cutting	GTAC
metal arc cutting	MAC
plasma arc cutting	PAC
shielded metal arc cutting	SMAC

- ARC WELDING (AW)
- BRAZING (B)
- OTHER WELDING
- OXYFUEL GAS WELDING (OFW)
- ADHESIVE BONDING (ABD)
- ARC CUTTING (AC)
- WELDING PROCESSES
- SOLID STATE WELDING (SSW)
- SOLDERING (S)
- RESISTANCE WELDING (RW)
- ALLIED PROCESSES
- THERMAL SPRAYING* (THSP)
- OXYGEN CUTTING (OC)
- THERMAL CUTTING (TC)
- OTHER CUTTING

electron beam cutting	EBC
laser beam cutting	LBC

*Sometimes a welding process

atomic hydrogen welding	AHW
bare metal arc welding	BMAW
carbon arc welding	CAW
— gas	CAW-G
— shielded	CAW-S
— twin	CAW-T
— electrogas	EGW

cold welding	CW
diffusion welding	DFW
explosion welding	EXW
forge welding	FOW
friction welding	FRW
hot pressure welding	HPW
roll welding	ROW
ultrasonic welding	USW

dip soldering	DS
furnace soldering	FS
induction soldering	IS
infrared soldering	IRS
iron soldering	INS
resistance soldering	RS
torch soldering	TS
wave soldering	WS

flash welding	FW
high frequency resistance welding	HFRW
percussion welding	PEW
projection welding	RPW
resistance seam welding	RSEW
resistance spot welding	RSW
upset welding	UW

electric arc spraying	EASP
flame spraying	FLSP
plasma spraying	PSP

chemical flux cutting	FOC
metal powder cutting	POC
oxyfuel gas cutting	OFC
— oxyacetylene cutting	OFC-A
— oxyhydrogen cutting	OFC-H
— oxynatural gas cutting	OFC-N
— oxypropane cutting	OFC-P
oxygen arc cutting	AOC
oxygen lance cutting	LOC

Welding Processes
Figure 5

15-11

Designation of Welding and Allied Processes by Letters

Welding and Allied Processes	Letter Designation	Welding and Allied Processes	Letter Designation
ADHESIVE BONDING	ABD	**RESISTANCE WELDING**	RW
ARC WELDING	AW	Flash Welding	FW
Atomic Hydrogen Welding	AHW	High Frequency Resistance Welding	HFRW
Bare Metal Arc Welding	BMAW	Percussion Welding	PEW
Carbon Arc Welding	CAW	Projection Welding	RPW
Electrogas	EGW	Resistance Seam Welding	RSEW
Flux Cored Arc Welding	FCAW	Resistance Spot Welding	RSW
Gas Metal Arc Welding	GMAW	Upset Welding	UW
Gas Tungsten Arc Welding	GTAW	**SOLDERING**	S
Plasma Arc Welding	PAW	Dip Soldering	DS
Shielded Metal Arc Welding	SMAW	Furnace Soldering	FS
Stud Arc Welding	SW	Induction Soldering	IS
Submerged Arc Welding	SAW	Infrared Soldering	IRS
		Iron Soldering	INS
		Resistance Soldering	RS
ARC WELDING PROCESS VARIATIONS		Torch Soldering	TS
		Wave Soldering	WS
Gas Carbon Arc Welding	CAW-G	**SOLID STATE WELDING**	SSW
Gas Metal Arc Welding – Pulsed Arc	GMAW-P	Cold Welding	CW
		Diffusion Welding	DFW
Gas Metal Arc Welding – Short Circuiting Arc	GMAW-S	Explosion Welding	EXW
		Forge Welding	FOW
Gas Tungsten Arc Welding – Pulsed Arc	GTAW-P	Friction Welding	FRW
		Hot Pressure Welding	HPW
Series Submerged Arc Welding	SAW-S	Roll Welding	ROW
Shielded Carbon Arc Welding	CAW-S	Ultrasonic Welding	USW
Twin Carbon Arc Welding	CAW-T	**THERMAL CUTTING**	TC
BRAZING	B	ARC CUTTING	AC
Arc Brazing	AB	Air Carbon Arc Cutting	AAC
Block Brazing	BB	Carbon Arc Cutting	CAC
Diffusion Brazing	DFB	Gas Metal Arc Cutting	GMAC
Dip Brazing	DB	Gas Tungsten Arc Cutting	GTAC
Flow Brazing	FLB	Metal Arc Cutting	MAC
Furnace Brazing	FB	Plasma Arc Cutting	PAC
Induction Brazing	IB	Shielded Metal Arc Cutting	SMAC
Infrared Brazing	IRB	ELECTRON BEAM CUTTING	EBC
Resistance Brazing	RB	LASER BEAM CUTTING	LBC
Torch Brazing	TB	OXYGEN CUTTING	OC
Twin Carbon Arc Brazing	TCAB	Chemical Flux Cutting	FOC
OTHER WELDING PROCESSES		Metal Powder Cutting	POC
Electron Beam Welding	EBW	Oxyfuel Gas Cutting	OFC
Electroslag Welding	ESW	oxyacetylene cutting	OFC-A
Flow Welding	FLOW	oxyhydrogen cutting	OFC-H
Induction Welding	IW	oxynatural gas cutting	OFC-N
Laser Beam Welding	LBW	oxypropane cutting	OFC-P
Thermit Welding	TW	Oxygen Arc Cutting	AOC
OXYFUEL GAS WELDING	OFW	Oxygen Lance Cutting	LOC
Air Acetylene Welding	AAW	**THERMAL SPRAYING**	THSP
Oxyacetylene Welding	OAW	Electric Arc Spraying	EASP
Oxyhydrogen Welding	OHW	Flame Spraying	FLSP
Pressure Gas Welding	PGW	Plasma Spraying	PSP

Suffixes for Optional Use in Applying Welding and Allied Processes

Automatic .... AU
Machine .... ME
Manual .... MA
Semiautomatic .... SA

Figures 6

Alphabetical Listing of Welding Designations and Allied Processes by Letters

Letter Designation	Welding and Allied Processes	Letter Designation	Welding and Allied Processes
AAC	Air Carbon Arc Cutting	GTAW	Gas Tungsten Arc Welding
AAW	Air Acetylene Welding	GTAW-P	Gas Tungsten Arc Welding– Pulsed Arc
ABD	Adhesive Bonding		
AB	Arc Brazing	HFRW	High Frequency Resistance Welding
AC	Arc Cutting	HPW	Hot Pressure Welding
AHW	Atomic Hydrogen Welding	IB	Induction Brazing
AOC	Oxygen Arc Cutting	INS	Iron Soldering
AW	Arc Welding	IRB	Infrared Brazing
B	Brazing	IRS	Infrared Soldering
BB	Block Brazing	IS	Induction Soldering
BMAW	Bare Metal Arc Welding	IW	Induction Welding
CAC	Carbon Arc Cutting	LBC	Laser Beam Cutting
CAW	Carbon Arc Welding	LBW	Laser Beam Welding
CAW-G	Gas Carbon Arc Welding	LOC	Oxygen Lance Cutting
CAW-S	Shielded Carbon Arc Welding	MAC	Metal Arc Cutting
CAW-T	Twin Carbon Arc Welding	OAW	Oxyacetylene Welding
CW	Cold Welding	OC	Oxygen Cutting
DB	Dip Brazing	OFC	Oxyfuel Gas Cutting
DFB	Diffusion Brazing	OFC-A	Oxyacetylene Cutting
DFW	Diffusion Welding	OFC-H	Oxyhydrogen Cutting
DS	Dip Soldering	OFC-N	Oxynatural Gas Cutting
EASP	Electric Arc Spraying	OFC-P	Oxypropane Cutting
EBC	Electron Beam Cutting	OFW	Oxyfuel Gas Welding
EBW	Electron Beam Welding	OHW	Oxyhydrogen Welding
EGW	Electrogas Welding	PAC	Plasma Arc Cutting
		PAW	Plasma Arc Welding
ESW	Electroslag Welding	PEW	Percussion Welding
EXW	Explosion Welding	PGW	Pressure Gas Welding
FB	Furnace Brazing	POC	Metal Powder Cutting
FCAW	Flux Cored Arc Welding	PSP	Plasma Spraying
FLB	Flow Brazing	RB	Resistance Brazing
FLOW	Flow Welding	RPW	Projection Welding
FLSP	Flame Spraying	RS	Resistance Soldering
FOC	Chemical Flux Cutting	RSEW	Resistance Seam Welding
FOW	Forge Welding	RSW	Resistance Spot Welding
FRW	Friction Welding	ROW	Roll Welding
FS	Furnace Soldering	RW	Resistance Welding
FW	Flash Welding	S	Soldering
GMAC	Gas Metal Arc Cutting	SAW	Submerged Arc Welding
GMAW	Gas Metal Arc Welding	SAW-S	Series Submerged Arc Welding
		SMAC	Shielded Metal Arc Cutting
	Electrogas	SMAW	Shielded Metal Arc Welding
GMAW-P	Gas Metal Arc Welding– Pulsed Arc	SSW	Solid State Welding
		SW	Stud Arc Welding
GMAW-S	Gas Metal Arc Welding– Short Circuiting Arc	TB	Torch Brazing
GTAC	Gas Tungsten Arc Cutting		

FIGURE 7

NOTES

Section 16 — THREAD REPRESENTATION

16.1 SCOPE — This section establishes the method of representing and specifying threads on Engineering Drawings.

16.2 APPLICABLE DOCUMENTS

MIL-STD-9	Screw Threads Conventions and Methods of Specifying
DOD-STD-100	Engineering Drawing Practices
MIL-P-7105	Pipe Threads, Taper, Aeronautical National Form, Symbol ANPT
MIL-S-7742	Screw Threads, Standard, Optimum Selected Series
MIL-B-7838	Bolt, Internal Wrenching, 160 KSI FTU
MIL-S-8879	Screw Threads, Controlled Radius Root with Increased Minor Diameter
FED-STD-H28	Screw-Thread Standards for Federal Services
ANSI B1.13M	Metric Screw Threads - M Profile
ANSI B1.21M	Metric Screw Threads - MJ Profile
ANSI Y14.6 & Y14.6aM	Screw Thread Representation
SAE Handbook AS 1338	Aerospace Metric 60° Screw Thread Profile and Tolerance Classes

16.3 DEFINITIONS

16.3.1 NOMINAL SIZE — The nominal size is the designation used for general identification of threads.

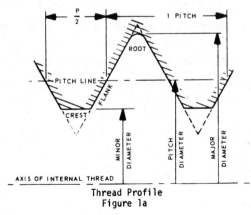

Thread Profile
Figure 1a

16-1

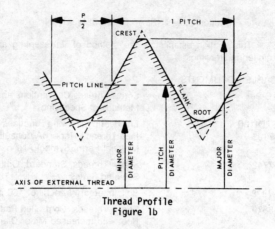

**Thread Profile
Figure 1b**

16.3.2 TERMS RELATING TO DIMENSIONS OF SCREW THREADS — (See Figures 1a and 1b.)

16.3.2.1 PITCH — The pitch of a thread is the distance, measured parallel to its axis, between corresponding points on adjacent thread forms in the same axial plane and on the same side of the axis.

16.3.2.2 MAJOR DIAMETER — The major diameter is the diameter of the coaxial cylinder that would bound the crest of an external thread or the root of an internal thread.

16.3.2.3 MINOR DIAMETER — The minor diameter is the diameter of the coaxial cylinder that would bound the root of an external thread or the crest of an internal thread.

16.3.2.4 PITCH DIAMETER (SIMPLE EFFECTIVE DIAMETER) — The pitch diameter is the diameter of the coaxial cylinder, the surface of which would pass through the thread profiles at such points as to make the width of the groove equal to one-half of the basic pitch.

16.3.3 UNIFIED FORM THREADS — Unified Form Threads are those which have been agreed upon by the standards bodies of Canada, the United States and the United Kingdom. They are mechanically interchangeable with American National threads of the same diameter and pitch and are designated as UN, UNC, UNF, UNEF, UNS, or UNM. (For limits of sizes see **FED-STD-H28/2 Table 2.1**

16.3.4 AMERICAN NATIONAL FORM THREADS — The principal difference between the Unified and the American National Form threads is in the application of allowance, differences in amount of pitch diameter tolerance applied to internal and external threads, and the variation of tolerance with size. These threads are designated: N, NC, NF, NEF, or NS.

16.3.5 STANDARD SERIES THREADS — Standard Series Threads are threads of Unified Form, having diameter and pitch combinations listed in Table II.

16.3.5.1 Controlled radius root threads are used to avoid stress conditions at root and are unified screw threads. Classes 3A and 3B, altered to include mandatory continuous radius at the root of the external thread, and with the minor diameter of both the external and internal threads increased over the unified thread values to accommodate the root radius. UNJ external threads will not assemble with UN internal threads. (See Table V and MIL-S-8879)

16.3.6 AMERICAN STANDARD TAPER PIPE THREADS — The American Standard taper pipe thread is a form of screw thread used on pipe and pipe fittings and it is characterized by a fine pitch and a taper of 1 in 16 (.75 inch per foot) on the diameter. This thread is designated NPT. (See **FED-STD-H28/7 Table 7.2**)

16.3.7 AERONAUTICAL NATIONAL FORM TAPER PIPE THREAD — Aeronautical National form pipe threads are used for Air Force pipe thread requirements. This thread is designated ANPT. (See MIl-P-7105)

16.3.8 ACME SCREW THREADS — An Acme screw thread is a thread form designed for a high stress in a traversing motion and power transmission. (**FED-STD-H28/12 Table 12.1**)

16.3.9 BUTTRESS SCREW THREADS — The Buttress form of thread is designed for applications involving exceptionally high stresses, in one direction only, along the thread axis. Standards for Buttress threads are presented in **FED-STD-H28/14 Table 14.1** These threads are designated (←—— N BUTT or ←——(N BUTT. The arrow and single parenthesis indicates whether the screw is to push or pull. The former indicates that the screw will push and the latter indicates the screw will pull.

16.3.10 DRYSEAL AMERICAN STANDARD PIPE THREADS

16.3.10.1 THE DRYSEAL AMERICAN STANDARD PIPE THREAD — A pipe thread in which metal to metal contact, at the crest and root prior to or coincident with flank contact, is ensured by dimensional controls, producing a leakproof and pressure-tight connection without the use of sealants. All external dryseal pipe threads are tapered and the internal threads may be either straight or tapered. They are separated into the four types described below and can be interchanged within the limitations shown in **FED-STD-H28/8 Table 8.4**

16.3.10.2 TYPE 1 — NPTF DRYSEAL AMERICAN STANDARD TAPER PIPE THREAD — Both internal and external threads tapered. Generally considered superior to NPT and ANPT for strength and sealing.

16.3.10.3 TYPE 2 — PTF-SHORT DRYSEAL SAE SHORT TAPER PIPE THREAD — Same as NPTF except that there is one thread less at large end on internal threads and one thread less on small end of external threads.

16.3.10.4 TYPE 3 — NPSF DRYSEAL AMERICAN STANDARD FUEL INTERNAL STRAIGHT PIPE THREAD — These are straight internal threads, intended to mate with tapered external threads and are generally used in soft and ductile material, which will deform at assembly.

16.3.10.5 TYPE 4 — NPSI DRYSEAL AMERICAN STANDARD INTERMEDIATE INTERNAL STRAIGHT PIPE THREAD — These are straight internal threads, intended to mate with tapered external threads and are generally used in hard material of heavy sections, where there is a minimum thread expansion at assembly.

16.3.11 INCOMPLETE THREADS — Sometimes referred to as runout, may be defined as the imperfect portion of thread, extending from the fully formed thread portion to the completely unthreaded shank or hole.

16.3.12 ROLLED THREADS — These are threads made between suitable hardened-steel dies by displacement of metal to conform with the die contours. No material is removed from the original blank. These threads are for high production. They are superior to that produced by machining. The tensile and shear strength as well as fatigue resistance is increased to a marked degree. In addition, they possess smooth, hard, burnished surfaces and the process produces accurate threads and forms on soft, tough and stringy materials which are impossible to machine without tearing. Internal threads are not normally made by this process. MIL-B-7838 is a typical specification for a bolt having the thread form conforming to MIL-S-7742 with the exception that the thread roots are controlled and produced by a single rolling process after heat treatment. MIL-S-8879 specifies rolled external threads with a high root radius for added strength. See Table V.

16.4 DRAWING APPLICATION

16.4.1 THREAD REPRESENTATION

16.4.1.1 The MIL-STD-9 and ANSI Y14.6 simplified method of thread representation shall be used except where design requirements make detailed representation desirable. (See Figure 2)

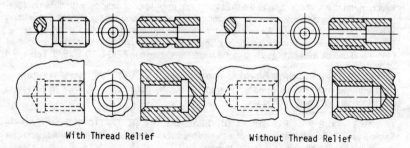

With Thread Relief   Without Thread Relief

Thread Representation
Figure 2

16.4.1.2 TAPERED PIPE THREADS — Shall be drawn at an included angle of approximately 4 degrees. (.125 inches taper on the diameter per inch of length.) (See Figure 3)

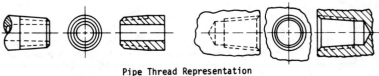

Pipe Thread Representation
Figure 3

16.4.1.3 MATING STRAIGHT THREADS — An assembly of a male and female straight thread in cross section is shown as in Figure 4.

16.4.1.4 THREAD INSERTS — Thread inserts are shown by the convention illustrated in Figure 5.

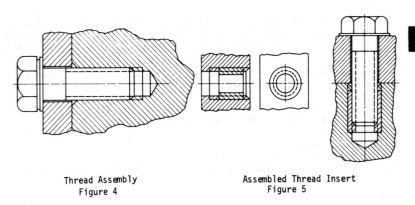

Thread Assembly
Figure 4

Assembled Thread Insert
Figure 5

16-5

16.4.2 THREAD DESIGNATION

16.4.2.1 BASIC DESIGNATION — The necessary information required in the designation of standard series threads consists of the following groups, in sequence, each group being separated by a dash: nominal size, number of threads per inch with thread series symbol and thread class.

Example:
```
.250 - 20 UNC - 2A
  │    │   │    └──── Thread Class
  │    │   └───────── Thread Form and Series
  │    └───────────── No. of Threads per Inch (Pitch)
  └────────────────── Nominal Size
```

16.4.2.1.1 NOMINAL SIZE — The nominal size for threads is designated by the decimal equivalent. The decimal in this callout does not denote a tolerance nor is any tolerance applicable.

Example:
.250-28 UNF-3A (external)
.250-28 UNF-3B (internal)

16.4.2.1.2 NUMBER OF THREADS — The number of threads, preceded by a dash (-), designates the number of single threads per inch (pitch equal to lead). Multiple threads are specified by pitch and lead or by the number of "starts" affixed to the thread designation.

Example:
1. 125-12 UNF-3B-3 START (use for Unified and American national thread forms only)
2. 250-.4P-.8L-ACME-4C (use for all thread forms other than Unified and American National)

16.4.2.1.3 THREAD CLASS — Classes of threads are distinguished from each other by the amount of tolerance, or the amount of tolerance and allowance. For threads computed to Unified formulation, Classes 1A, 2A and 3A apply to external threads and Classes 1B, 2B and 3B apply to internal threads. For the American National series, the class of thread is designated, for external and internal threads, by a number only, preceded by a dash, e.g., -2, -3.

16.4.2.1.4 HAND DESIGNATION — Screw threads are interpreted to mean right hand, unless LH, for left hand, is included in the callout.

Example:
.250-20 UNC-2A-LH

16.4.2.1.5 CONTROLLED RADIUS ROOT THREADS — Are designated as shown below:
.250-28 UNJF-3A   or   .250-28 UNJF-3B

16.4.2.2 SPECIAL THREADS — This term is generally applied to threads of unified form when either the pitch or nominal diameter or both do not conform to sizes in the standard or unified series. Formulas or dimensions and tolerances and general information on these threads may be found in **FED-STD-H28**

16.4.2.2.1 The following is the method of designating unified special threads (unified formula formulations) on the field of drawing:

Examples:

External Thread

(A) .250-24 UNS-3A

MAJOR DIA .2500 / .2428

PD .2229 / .2201

(B) .495-20 UNS-3A

MAJOR DIA .4950 / .4869

PD .4625 / .4593

Internal Thread

1.200-10 UNS-2B

MINOR DIA 1.092 / 1.113

PD 1.1350 / 1.1432

16.4.2.2.2 In addition to the more common screw threads, FED–STD–H28 also includes data on the lesser known International Metric, Butress, Acme and Stub Acme and others. Additional data on still other threads such as the British Standards, Whitworth, Trapezoidal Metric, Dardelot Self–locking etc. may be found in the SAE Handbook.

16.4.2.2.3 The drawing callout for the lesser known threads included in FED–STD–H28 will include the 'Identification Designation' and other information necessary to establish the specific requirements. Any other screw thread, not included in FED–STD–H28, will require either a complete drawing delineation or an appropriate reference to an acceptable established standard.

Example:
THREAD 18MM-1.5

PITCH DIA .6645 / .6695

SPARK PLUG THREAD PER SAE HANDBOOK

16.4.3 COATED THREADS

16.4.3.1 Coatings, as defined herein, is interpreted to mean an additive protective metallic coating. Methods of specifying coated threads necessarily vary with the intended applications as described in **FED-STD-H28** Examples of designations for coated threads follow:

16.4.3.2 Class 2A—Coated
Example:
(A) (For general purpose application).
.250-20 UNC-2A
*MAJOR DIA .2500 MAX ⎤
*PITCH DIA .2175 MAX ⎦ ⎯ AFTER COATING

(B) (For critical applications where uncoated thread and coating buildup must be closely controlled).
.250-20 UNC-2A
*MAJOR DIA .2500 MAX ⎤
*PITCH DIA .2175 MAX ⎦ ⎯ AFTER COATING
**MAJOR DIA .2489
             .2408
**PITCH DIA .2164
             .2127 ⎯ BEFORE COATING

16.4.3.3 Class 3A—Coated
Example:
(A) (For general purpose application).
.250-28 UNF-3A
*MAJOR DIA .2500 MAX ⎤
*PITCH DIA .2268 MAX ⎦ ⎯ AFTER COATING

(B) (For critical applications where uncoated thread and coating buildup must be controlled.)
.250-28 UNF-3A
*MAJOR DIA .2500 MAX ⎤
*PITCH DIA .2268 MAX ⎦ ⎯ BEFORE COATING
***MAJOR DIA .2488 SPL
              .2427
**PITCH DIA   .2256 SPL
              .2236 ⎯ BEFORE PLATING

NOTE: *Major and pitch dia limits are those selected from table for Class 3A.
       **Major and pitch dia limits are those selected from table for Class 2A.
      ***Calculated by reducing the amount of the Class 2A allowance whenever this is adequate.

16.4.4 THREAD INSERTS — An example of a drawing callout for a threaded hole to receive a thread insert from Military Standards (example: MS122083) with a .375-16 UNC-3B threaded hole, and a one diameter length of insert with tang removed, is as follows:

Example:

$\frac{.390}{.398}$ DIA X .75 DEEP

CSK 120° ± 5° X $\frac{.440}{.480}$ DIA

THD .440 MIN FOR .375-16 UNC-3B HELICAL COIL INSERT MS122083
INSTALL PER MS33537 TANG REMOVED
2 PLACES

16.4.5 THREAD CALLOUT

16.4.5.1 Screw threads are called out on drawings as shown in Figure 6.

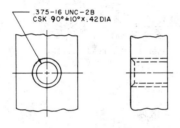

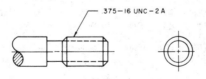

Thread Call-out
Figure 6

16.4.5.2 Aeronautical and national form tapered and dryseal pipe threads are called out on the drawing as shown in Figure 7.

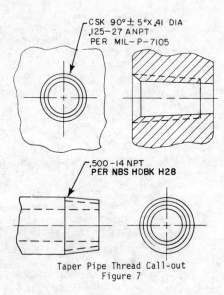

Taper Pipe Thread Call-out
Figure 7

16.4.6 THREAD DIMENSIONING

16.4.6.1 GENERAL APPLICATION

16.4.6.1.1 For Standard series threads, identified in paragraph 16.3.5, the dimensional limits are obtained from Table II or by reference to NBS H28, except those dimensions pertaining to length of engagement, chamfers, reliefs, and depth.

16.4.6.1.2 Controlled radius root threads are prepared to the dimensional limits of Table V as specified in MIL-S-8879.

16.4.6.1.3 Drawing representations of pipe threads, identified in paragraphs 16.3.6, 16.3.7 and 16.3.10, are prepared as shown in Table III. An appropriate callout, as shown in Figure 7, establishes the dimensional values, used in design computations or to establish fabrication and acceptance requirements, as specified in **FED-STD-H28** or MIL-P-7105, as applicable.

16.4.6.2 FULL FORM THREADS — The length of fully formed threads is dimensioned. When only one dimension is used to specify the length of threads, it is interpreted to mean the length of fully formed threads, excluding runout. Where a chamfer not exceeding 2 pitch in length exists at the entering end of the thread, it is included in the length of fully formed threads. Figure 8 shows methods of delineating with no limit on runout.

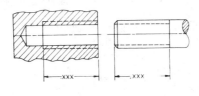

Thread Length
Figure 8

16.4.6.3 INCOMPLETELY FORMED THREADS — The length of incompletely formed threads is not dimensioned. This length may vary according to the method of manufacture. Whenever the number of incompletely formed threads allowed on the entering end is more restrictive than permitted by **FED-STD-H28** dimensions imposing such restrictions shall be specified as follows,
.XX MAX TO FULL FORM THREAD.

16.4.6.4 CONTROLLED LENGTH AND RUN-OUT OF FULL FORM THREADS — Figure 9 illustrates an alternate method of drawing when both the length of full form threads and runout must be controlled.

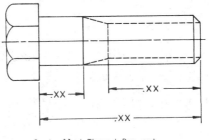

Controlled Thread Run-out
Figure 9

16-11

16.4.6.5 BLIND HOLES — If a blind hole is unavoidable, the allowance for tool chamfer, partial threads and tap clearance at the bottom of the hole should not be less than that suggested in Table I.

16.4.6.6 CHAMFERS — Chamfers are specified on the drawing as shown in Figure 10. Whenever practicable, the chamfer angle shall be 45° ± 5°. The chamfer specified must be a minimum of one-half of the value of relief constant B, Table I, rounded-off to two decimal places.

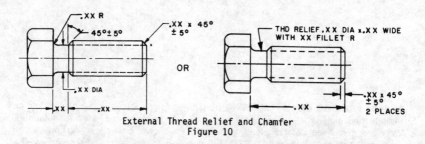

External Thread Relief and Chamfer
Figure 10

16.4.6.7 RELIEFS — External and internal thread reliefs, when required, may be dimensioned as shown in Figures 10 and 11. For relief data, see Table I.

16.4.6.8 COUNTERSINKS — Countersinks are designated by an angle and a diameter as shown in Figure 11. It is recommended that the minimum countersink diameter be the nominal major diameter plus relief constant A of Table I.

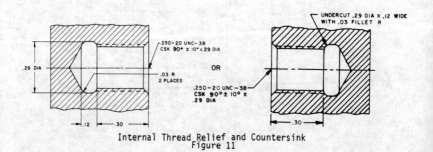

Internal Thread Relief and Countersink
Figure 11

16.4.6.9 TAP DRILL SIZE AND DEPTH — Except when required by specific design considerations, neither the size nor depth of the tap drill is included in the thread callout. Where the depth of the tap drill must be controlled, it may be dimensioned or called out in the note form as illustrated in Figure 12. When the size of the tap drill is specified on the drawing, the internal minor diameter limits shown in Table II shall be used and held to the same number of decimal places as shown in the table. If any other limits are used the thread must be designated "MOD", for example:

$\frac{.3100}{.3240}$ DIA X .97 DEEP MAX

CSK 90° ± 10° X .42 DIA

.375-16 UNC-2B MOD X .62 DEEP MIN

MINOR DIA $\frac{.3100}{.3240}$ MOD

16.4.6.10 COMBINED DIMENSIONING — Complete dimensioning (hole size or tap) drill, depth, thread relief (countersink, counterbore, etc.), thread designation, thread depth, etc. can be accomplished by a note to a single pictorial representation. An example of combined dimensioning is shown in Figure 12.

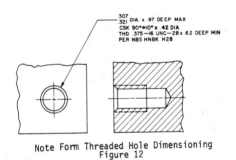

Note Form Threaded Hole Dimensioning
Figure 12

16.4.7 DRAWING NOTES — Drawings reference the documents used for interpreting thread dimensions and designations. This reference is in the form of either a local note on the field of drawing, or a General Note.

Example:

THREADS PER **FED-STD-H28** (or MIL-P-7105, MIL-S-8879, MIL-S-7742, etc., as applicable)

16-13

Table I
Suggested Screw Thread Reliefs

THREADS PER INCH	LENGTH OF RELIEF OR ALLOWANCE BEYOND FULL FORM THREADS				RELIEF CONSTANT	
	PITCH INTERVAL				INTERNAL	EXTERNAL
	1P	3P	5P	6P	A	B
	INCHES	INCHES	INCHES	INCHES	INCHES	INCHES
80	.012	.038	.062	.075	.0052	.0196
72	.014	.042	.069	.083	.0056	.0216
64	.016	.047	.078	.094	.0058	.0241
56	.018	.054	.089	.107	.0066	.0272
48	.021	.062	.104	.125	.0074	.0315
44	.023	.068	.114	.136	.0078	.0341
40	.025	.075	.125	.150	.0084	.0373
36	.028	.083	.139	.167	.0090	.0411
32	.031	.094	.156	.188	.0099	.0460
28	.036	.107	.179	.214	.0114	.0526
24	.042	.125	.208	.250	.0126	.0607
20	.050	.150	.250	.300	.0144	.0722
18	.056	.167	.278	.333	.0162	.0814
16	.062	.188	.312	.375	.0180	.0902
14	.071	.214	.357	.429	.0201	.1026
13	.077	.231	.385	.462	.0215	.1103
12	.083	.250	.417	.500	.0232	.1195
11	.091	.273	.455	.545	.0249	.1299
10	.100	.300	.500	.600	.0272	.1427
9	.111	.333	.556	.667	.0300	.1583
8	.125	.375	.625	.750	.0332	.1776
7	.143	.429	.714	.857	.0376	.2026
6	.167	.500	.833	1.000	.0442	.2367
5	.200	.600	1.000	1.200	.0521	.2830
4.5	.222	.667	1.111	1.333	.0575	.3141
4	.250	.750	1.250	1.500	.0641	.3528

MAXIMUM EQUALS NOMINAL MAJOR DIA MINUS CONSTANT B

MINIMUM EQUALS NOMINAL MAJOR DIA PLUS CONSTANT A

1P MIN — PLUG END TAP FOLLOWED BY BOTTOMING END TAP
3P MIN — INTERMEDIATE END BOTTOMING TAP
5P MIN — .250 AND LARGER } PLUG END TAP
6P MIN — .216 AND SMALLER

TABLE II
MIL-S-7742 Selected Series
Standard Series Limits of Size – Unified Screw Threads
For Reference Only (NBS Handbook H28)

Equivalent Size Numbers Based on Length of Engagement Equal to 1 to 1.5 Diameters

NOMINAL SIZE & THREADS PER INCH	SERIES DESIGNA-TION	CLASS	ALLOW-ANCE	EXTERNAL								INTERNAL						MAJOR DIAM-ETER MIN
				MAJOR DIAMETER LIMITS			*PITCH DIAMETER LIMITS			MINOR DIAM-ETER	CLASS	MINOR DIAM-ETER LIMITS		*PITCH DIAMETER LIMITS			
				MAX	MIN	MIN	MAX	MIN	TOL			MIN	MAX	MIN	MAX	TOL	
			INCHES	INCHES	INCHES	INCHES	INCHES	INCHES	INCHES	INCHES		INCHES	INCHES	INCHES	INCHES	INCHES	INCHES
.060-80 (#0-80)	UNF	2A 3A	0.0005 .0000	0.0595 .0600	0.0563 .0568	-- --	0.0514 .0519	0.0496 .0506	0.0018 .0013	0.0442 .0447	2B 3B	0.0465 .0465	0.0514 .0514	0.0519 .0519	0.0542 .0536	0.0023 .0017	0.0600 .0600
.073-64 (#1-64)	UNC	2A 3A	.0006 .0000	.0724 .0730	.0686 .0692	-- --	.0623 .0629	.0603 .0614	.0020 .0015	.0532 .0538	2B 3B	.0561 .0561	.0623 .0623	.0629 .0629	.0655 .0648	.0026 .0019	.0730 .0730
.086-56 (#2-56)	UNC	2A 3A	.0006 .0000	.0854 .0860	.0813 .0819	-- --	.0738 .0744	.0717 .0728	.0021 .0016	.0635 .0641	2B 3B	.0667 .0667	.0737 .0737	.0744 .0744	.0772 .0765	.0028 .0021	.0860 .0860
.099-48 (#3-48)	UNC	2A 3A	.0007 .0000	.0983 .0990	.0938 .0945	-- --	.0848 .0855	.0825 .0838	.0023 .0017	.0727 .0734	2B 3B	.0764 .0764	.0845 .0845	.0855 .0855	.0885 .0877	.0030 .0022	.0990 .0990
.112-40 (#4-40)	UNC	2A 3A	.0008 .0000	.1112 .1120	.1061 .1069	-- --	.0950 .0958	.0925 .0939	.0025 .0019	.0805 .0813	2B 3B	.0849 .0849	.0939 .0939	.0958 .0958	.0991 .0982	.0033 .0024	.1120 .1120
.138-32 (#6-32)	UNC	2A 3A	.0008 .0000	.1372 .1380	.1312 .1320	-- --	.1169 .1177	.1141 .1156	.0028 .0021	.0989 .0997	2B 3B	.104 .1040	.114 .1140	.1177 .1177	.1214 .1204	.0037 .0027	.1380 .1380
.164-32 (#8-32)	UNC	2A 3A	.0009 .0000	.1631 .1640	.1571 .1580	-- --	.1428 .1437	.1399 .1415	.0029 .0022	.1248 .1257	2B 3B	.130 .1300	.139 .1389	.1437 .1437	.1475 .1465	.0038 .0028	.1640 .1640
.190-32 (#10-32)	UNF	2A 3A	.0009 .0000	.1891 .1900	.1831 .1840	-- --	.1688 .1697	.1658 .1674	.0030 .0023	.1508 .1517	2B 3B	.156 .1560	.164 .1641	.1697 .1697	.1736 .1726	.0039 .0029	.1900 .1900
.250-28	UNF	1A 2A 3A	.0010 .0010 .0000	.2490 .2490 .2500	.2392 .2425 .2435	-- -- --	.2258 .2258 .2268	.2208 .2225 .2243	.0050 .0033 .0025	.2052 .2052 .2062	1B 2B 3B	.2110 .2110 .2110	.2200 .2200 .2190	.2268 .2268 .2268	.2333 .2311 .2300	.0065 .0043 .0032	.2500 .2500 .2500
.3125-24	UNF	1A 2A 3A	.0011 .0011 .0000	.3114 .3114 .3125	.3006 .3042 .3053	-- -- --	.2843 .2843 .2854	.2788 .2806 .2827	.0055 .0037 .0027	.2603 .2603 .2614	1B 2B 3B	.2670 .2670 .2670	.2770 .2770 .2754	.2854 .2854 .2854	.2925 .2902 .2890	.0071 .0048 .0036	.3125 .3125 .3125
.375-24	UNF	1A 2A 3A	.0011 .0011 .0000	.3739 .3739 .3750	.3631 .3667 .3678	-- -- --	.3468 .3468 .3479	.3411 .3430 .3450	.0057 .0038 .0029	.3228 .3228 .3239	1B 2B 3B	.3300 .3300 .3300	.3400 .3400 .3372	.3479 .3479 .3479	.3553 .3528 .3516	.0074 .0049 .0037	.3750 .3750 .3750
.4375-20	UNF	1A 2A 3A	.0013 .0013 .0000	.4362 .4362 .4375	.4240 .4281 .4294	-- -- --	.4037 .4037 .4050	.3974 .3995 .4019	.0063 .0042 .0031	.3749 .3749 .3762	1B 2B 3B	.3830 .3830 .3830	.3950 .3950 .3916	.4050 .4050 .4050	.4131 .4104 .4091	.0081 .0054 .0041	.4375 .4375 .4375
.500-20	UNF	1A 2A 3A	.0013 .0013 .0000	.4987 .4987 .5000	.4865 .4906 .4919	-- -- --	.4662 .4662 .4675	.4598 .4619 .4643	.0064 .0043 .0032	.4374 .4374 .4378	1B 2B 3B	.4460 .4460 .4460	.4570 .4570 .4537	.4675 .4675 .4675	.4759 .4731 .4717	.0084 .0056 .0042	.5000 .5000 .5000
.5625-18	UNF	1A 2A 3A	.0014 .0014 .0000	.5611 .5611 .5625	.5480 .5524 .5538	-- -- --	.5250 .5250 .5264	.5182 .5205 .5230	.0068 .0045 .0034	.4929 .4929 .4943	1B 2B 3B	.5020 .5020 .5020	.5150 .5150 .5106	.5264 .5264 .5264	.5353 .5323 .5308	.0089 .0059 .0044	.5625 .5625 .5625
.625-18	UNF	1A 2A 3A	.0014 .0014 .0000	.6236 .6236 .6250	.6105 .6149 .6163	-- -- --	.5875 .5875 .5889	.5805 .5828 .5854	.0070 .0047 .0035	.5554 .5554 .5568	1B 2B 3B	.5650 .5650 .5650	.5780 .5780 .5730	.5889 .5889 .5889	.5980 .5949 .5934	.0091 .0060 .0045	.6250 .6250 .6250

16-15

TABLE II (continued)
MIL-S-7742 Selected Series
Standard Series Limits of Size - Unified Screw Threads

NOMINAL SIZE & THREADS PER INCH	SERIES DESIGNA- TION	CLASS	ALLOW- ANCE	EXTERNAL									INTERNAL							MAJOR DIAM- ETER MIN
				MAJOR DIAMETER LIMITS			*PITCH DIAMETER LIMITS			MINOR DIAM- ETER		CLASS	MINOR DIAM- ETER LIMITS		*PITCH DIAMETER LIMITS				
				MAX	MIN	MIN	MAX	MIN	TOL				MIN	MAX	MIN	MAX	TOL		
				INCHES	INCHES	INCHES	INCHES	INCHES	INCHES	INCHES	INCHES		INCHES	INCHES	INCHES	INCHES	INCHES	INCHES	INCHES
.750-16	UNF	1A	.0015	.7485	.7343	- -	.7079	.7004	.0075		.6718	1B	.6820	.6960	.7094	.7192	.0098	.7500
		2A	.0015	.7485	.7391	- -	.7079	.7029	.0050		.6718	2B	.6820	.6960	.7094	.7159	.0065	.7500
		3A	.0000	.7500	.7406	- -	.7094	.7056	.0038		.6733	3B	.6820	.6908	.7094	.7143	.0049	.7500
.875-14	UNF	1A	.0016	.8734	.8579	- -	.8189	.8108	.0081		.7858	1B	.7980	.8140	.8286	.8392	.0106	.8750
		2A	.0016	.8734	.8631	- -	.8270	.8216	.0054		.7858	2B	.7980	.8140	.8286	.8356	.0070	.8750
		3A	.0000	.8750	.8647	- -	.8286	.8245	.0041		.7874	3B	.7980	.8068	.8286	.8339	.0053	.8750
1.000-12	UNF	1A	.0018	.9982	.9810	- -	.9441	.9353	.0088		.8960	1B	.9100	.9280	.9459	.9573	.0114	1.0000
		2A	.0018	.9982	.9868	- -	.9441	.9382	.0059		.8960	2B	.9100	.9280	.9459	.9535	.0076	1.0000
		3A	.0000	1.0000	.9886	- -	.9459	.9415	.0044		.8978	3B	.9100	.9198	.9459	.9516	.0057	1.0000
1.125-12	UNF	1A	.0018	1.1232	1.1060	- -	1.0691	1.0601	.0090	1	.0210	1B	1.0350	1.0530	1.0709	1.0826	.0117	1.1250
		2A	.0018	1.1232	1.1118	- -	1.0691	1.0631	.0060	1	.0210	2B	1.0350	1.0530	1.0709	1.0787	.0078	1.1250
		3A	.0000	1.1250	1.1136	- -	1.0709	1.0664	.0045	1	.0228	3B	1.0350	1.0448	1.0709	1.0768	.0059	1.1250
1.250-12	UNF	1A	.0018	1.2482	1.2310	- -	1.1941	1.1849	.0092	1	.1460	1B	1.1600	1.1780	1.1959	1.2079	.0120	1.2500
		2A	.0018	1.2482	1.2368	- -	1.1941	1.1879	.0062	1	.1460	2B	1.1600	1.1780	1.1959	1.2039	.0080	1.2500
		3A	.0000	1.2500	1.2386	- -	1.1959	1.1913	.0046	1	.1478	3B	1.1600	1.1698	1.1959	1.2019	.0060	1.2500
1.375-12	UNF	1A	.0019	1.3731	1.3559	- -	1.3190	1.3096	.0094	1	.2709	1B	1.2850	1.3030	1.3209	1.3332	.0123	1.3750
		2A	.0019	1.3731	1.3617	- -	1.3190	1.3127	.0063	1	.2709	2B	1.2850	1.3030	1.3209	1.3291	.0082	1.3750
		3A	.0000	1.3750	1.3636	- -	1.3209	1.3162	.0047	1	.2728	3B	1.2850	1.2948	1.3209	1.3270	.0061	1.3750
1.500-12	UNF	1A	.0019	1.4981	1.4809	- -	1.4440	1.4344	.0096	1	.3959	1B	1.4100	1.4280	1.4459	1.4584	.0125	1.5000
		2A	.0019	1.4981	1.4867	- -	1.4440	1.4376	.0064	1	.3959	2B	1.4100	1.4280	1.4459	1.4542	.0083	1.5000
		3A	.0000	1.5000	1.4886	- -	1.4459	1.4411	.0048	1	.3978	3B	1.4100	1.4198	1.4459	1.4522	.0063	1.5000
1.750-12	UN	2A	.0018	1.7482	1.7368	- -	1.6941	1.6881	.0060	1	.6460	2B	1.6600	1.678	1.6959	1.7037	.0078	1.7500
		3A	.0000	1.7500	1.7386	- -	1.6959	1.6914	.0045	1	.6478	3B	1.6600	1.6698	1.6959	1.7017	.0058	1.7500
2.000-12	UN	2A	.0018	1.9982	1.9868	- -	1.9441	1.9380	.0061	1	.8960	2B	1.910	1.928	1.9459	1.9538	.0079	2.0000
		3A	.0000	2.0000	1.9886	- -	1.9459	1.9414	.0045	1	.8978	3B	1.9100	1.9198	1.9459	1.9518	.0059	2.0000
2.250-12	UN	2A	.0018	2.2482	2.2368	- -	2.1941	2.1880	.0061	2	.1460	2B	2.160	2.178	2.1959	2.2038	.0079	2.2500
		3A	.0000	2.2500	2.2386	- -	2.1959	2.1914	.0045	2	.1478	3B	2.1600	2.1698	2.1959	2.2018	.0059	2.2500
2.500-12	UN	2A	.0019	2.4981	2.4867	- -	2.4440	2.4378	.0062	2	.3959	2B	2.410	2.428	2.4459	2.4540	.0081	2.5000
		3A	.0000	2.5000	2.4886	- -	2.4459	2.4413	.0046	2	.3978	3B	2.4100	2.4198	2.4459	2.4519	.0060	2.5000
2.750-12	UN	2A	.0019	2.7481	2.7367	- -	2.6940	2.6878	.0062	2	.6459	2B	2.660	2.678	2.6959	2.7040	.0081	2.7500
		3A	.0000	2.7500	2.7386	- -	2.6959	2.6913	.0046	2	.6478	3B	2.6600	2.6698	2.6959	2.7019	.0060	2.7500
3.000-12	UN	2A	.0019	2.9981	2.9867	- -	2.9440	2.9377	.0063	2	.8959	2B	2.910	2.928	2.9459	2.9541	.0082	3.0000
		3A	.0000	3.0000	2.9886	- -	2.9459	2.9412	.0047	2	.8978	3B	2.9100	2.9198	2.9459	2.9521	.0062	3.0000
3.250-12	UN	2A	.0019	3.2481	3.2367	- -	3.1940	3.1877	.0063	3	.1459	2B	3.160	3.178	3.1959	3.2041	.0082	3.2500
		3A	.0000	3.2500	3.2386	- -	3.1959	3.1912	.0047	3	.1478	3B	3.1600	3.1698	3.1959	3.2021	.0062	3.2500
3.500-12	UN	2A	.0019	3.4981	3.4867	- -	3.4440	3.4376	.0064	3	.3959	2B	3.410	3.428	3.4459	3.4543	.0084	3.5000
		3A	.0000	3.5000	3.4886	- -	3.4459	3.4411	.0048	3	.3978	3B	3.4100	3.4198	3.4459	3.4522	.0063	3.5000

TABLE II (continued)
MIL-S-7742 Selected Series
Standard Series Limits of Size - Unified Screw Threads

NOMINAL SIZE & THREADS PER INCH	SERIES DESIGN- ATION	CLASS	ALLOW- ANCE	EXTERNAL MAJOR DIAMETER LIMITS		*PITCH DIAMETER LIMITS			MINOR DIAM- ETER	CLASS	INTERNAL MINOR DIAM- ETER LIMITS		*PITCH DIAMETER LIMITS			MAJOR DIAM- ETER MIN
				MAX INCHES	MIN INCHES	MAX INCHES	MIN INCHES	TOL. INCHES	INCHES		MIN INCHES	MAX INCHES	MIN INCHES	MAX INCHES	TOL. INCHES	INCHES
3.750-12	UN	2A	.0019	3.7481	3.7367	3.6940	3.6876	.0064	3.6459	2B	3.660	3.678	3.6959	3.7043	.0084	3.7500
		3A	.0000	3.7500	3.7386	3.6959	3.6911	.0048	3.6478	3B	3.6600	3.6698	3.6959	3.7022	.0063	3.7500
4.000-12	UN	2A	.0020	3.9980	3.9866	3.9439	3.9374	.0065	3.8958	2B	3.910	3.928	3.9459	3.9544	.0085	4.0000
		3A	.0000	4.0000	3.9886	3.9459	3.9410	.0049	3.8978	3B	3.9100	3.9198	3.9459	3.9523	.0064	4.0000
4.250-12	UN	2A	.0020	4.2480	4.2366	4.1939	4.1874	.0065	4.1458	2B	4.160	4.178	4.1959	4.2044	.0085	4.2500
		3A	.0000	4.2500	4.2386	4.1959	4.1910	.0049	4.1473	3B	4.1600	4.1698	4.1959	4.2023	.0064	4.2500
4.500-12	UN	2A	.0020	4.4980	4.4866	4.4439	4.4374	.0065	4.3953	2B	4.410	4.428	4.4459	4.4544	.0085	4.5000
		3A	.0000	4.5000	4.4886	4.4459	4.4410	.0049	4.3973	3B	4.4100	4.4198	4.4459	4.4523	.0064	4.5000
4.75-12	UN	2A	.0020	4.7480	4.7366	4.6939	4.6872	.0067	4.6458	2B	4.660	4.673	4.6959	4.7046	.0087	4.7500
		3A	.0000	4.7500	4.7386	4.6959	4.6909	.0050	4.6478	3B	4.6600	4.6698	4.6959	4.7025	.0066	4.7500
5.000-12	UN	2A	.0020	4.9980	4.9866	4.9439	4.9372	.0067	4.8958	2B	4.910	4.928	4.9459	4.9546	.0087	5.0000
		3A	.0000	5.0000	4.9886	4.9459	4.9409	.0050	4.8978	3B	4.9100	4.9198	4.9459	4.9525	.0066	5.0000
5.250-12	UN	2A	.0020	5.2480	5.2366	5.1939	5.1872	.0067	5.1458	2B	5.160	5.178	5.1959	5.2046	.0087	5.2500
		3A	.0000	5.2500	5.2386	5.1959	5.1909	.0050	5.1478	3B	5.1600	5.1698	5.1959	5.2025	.0066	5.2500
5.500-12	UN	2A	.0020	5.4980	5.4866	5.4439	5.4409	.0067	5.3958	2B	5.410	5.428	5.4459	5.4546	.0087	5.5000
		3A	.0000	5.5000	5.4886	5.4459	5.4409	.0050	5.3978	3B	5.4100	5.4198	5.4459	5.4525	.0066	5.5000
5.750-12	UN	2A	.0021	5.7479	5.7365	5.6938	5.6869	.0069	5.6457	2B	5.660	5.678	5.6959	5.7049	.0090	5.7500
		3A	.0000	5.7500	5.7386	5.6959	5.6907	.0052	5.6478	3B	5.6600	5.6698	5.6959	5.7026	.0067	5.7500
6.000-12	UN	2A	.0021	5.9979	5.9865	5.9438	5.9369	.0069	5.8957	2B	5.910	5.928	5.9459	5.9549	.0090	6.0000
		3A	.0000	6.0000	5.9886	5.9459	5.9407	.0052	5.8978	3B	5.9100	5.9198	5.9459	5.9526	.0067	6.0000

16

16-17

Table III
Dimensions for Drawing Representation of Pipe Threads

NOMINAL PIPE SIZE	THREAD SIZE & PITCH	PIPE OUTSIDE DIA	PIPE INSIDE DIA	DIMENSION A			DIMENSION B		DIMENSION C
				THREAD DESIGNATION			THREAD DESIGNATION		
				NPT ANPT	NPTF	PTF-SAE SHORT	NPSF NPSI (Note 2)	NPT ANPT NPTF	PTF-SAE SHORT	(Note 3)
1/16	.062-27	.3125	---	.400	.385	.364	.312	.390	.327	.271
1/8	.125-27	.405	.269	.401	.384	.365	.312	.392	.329	.291
1/4	.250-18	.540	.364	.587	.555	.533	.469	.595	.495	.367
3/8	.375-18	.675	.493	.599	.561	.546	.500	.601	.501	.407
1/2	.500-14	.840	.622	.782	.751	.713	.656	.782	.651	.534
3/4	.750-14	1.050	.824	.801	.758	.732	.656	.794	.663	.553
1	1.00-11.5	1.315	1.049	.963	.916	.878	.656	.985	.823	.661
1-1/4	1.25-11.5	1.660	1.380	.983	.943	.898		1.008	.847	.681
1-1/2	1.50-11.5	1.900	1.610	.983	.957	.898		1.025	.864	.681
2	2.00-11.5	2.375	2.067	.999	.972	.914		1.058	.897	.697
2-1/2	2.50-8	2.875	2.469	1.366	1.435	1.369		1.571	1.335	1.057
3	3.00-8	3.500	3.068	1.500	1.519	1.454		1.634	1.398	1.141

NOTES:
1. Values shown are for drawing representation only, see applicable document for design values.
2. NPSF and NPSI are straight, internal threads and value shown is minimum used for design.
3. Value shown for Dimension C based on three threads wrench make-up allowance. When showing 2.50-8 NPT or 3.00-8 NPT, deduct .125 from Dimension C to compensate for the two thread normal wrench make-up allowance for these threads.

A = Length of internal thread, including imperfect thread, to point of vanish.
B = Length of external thread, including imperfect thread, to point of vanish.
C = Length of engaged thread, wrench tight.

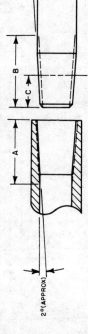

Table IV

Aeronautical National Taper Pipe Thread (ANPT) Data
Ref: MIL-P-7105

Nominal Pipe Size	Thread Size & Pitch	Internal Thread				External Thread 45° Chamfer
		Tap Hole Diameter Min	Tap Hole Diameter Max	Depth Min	$90° \pm 5°$ CSK DIA $^{+.03}_{-.00}$	
1/16	.062-27	.233	.239	.609	.312	.03-.05
1/8	.125-27	.327	.334	.625	.406	.03-.05
1/4	.250-18	.421	.428	.812	.562	.04-.07
3/8	.375-18	.561	.570	.843	.688	.04-.07
1/2	.500-14	.686	.695	1.062	.875	.05-.08
3/4	.750-14	.889	.900	1.062	1.062	.05-.08
1	1.000-11.5	1.124	1.137	1.312	1.312	.06-.09
1-1/4	1.250-11.5	1.467	1.480	1.312	1.641	.06-.09
1-1/2	1.500-11.5	1.702	1.715	1.343	1.906	.06-.09
2	2.000-11.5	2.186	2.196	1.343	2.500	.06-.09
2-1/2	2.500-8	2.592	2.602	1.875	2.906	.08-.11
3	3.00-8	3.217	3.227	1.937	3.531	.08-.11

TABLE V Fine Thread Series
MIL-S-8879 Selected Series
Standard Series Limits of Size - Unified Screw Threads
Equivalent Size Numbers For Reference Only

BASIC SIZE (NOM-INAL)	THDS PER INCH	EXTERNAL THREAD - UNJF CLASS 3A								INTERNAL THREAD - UNJF CLASS 3B							INTERNAL THREAD - UNJF CLASS 3BG					
		MAJOR DIAMETER		PITCH DIAMETER		MINOR DIAMETER		ROOT RADIUS		MINOR DIAMETER		PITCH DIAMETER		MAJOR DIA	MINOR DIAMETER		PITCH DIAMETER		MAJOR DIA			
		MIN	MAX	MIN	MAX	MIN	MAX	MIN	MAX	MIN	MAX	MIN	MAX	MIN	MIN	MAX	MIN	MAX	MIN			
0.0600 (#0-80)	80	0.0568	0.0600	0.0506	0.0519	0.0435	0.0456	0.0019	0.0023	0.0479	0.0511	0.0519	0.0536	0.0600	0.0489	0.0521	0.0529	0.0456	0.0610			
0.1120 (#4-48)	48	0.1075	0.1120	0.0967	0.0985	0.0849	0.0880	0.0031	0.0038	0.0917	0.0971	0.0985	0.1008	0.1120	0.0927	0.0981	0.0995	0.1018	0.1130			
0.1380 (#6-40)	40	0.1329	0.1380	0.1198	0.1218	0.1057	0.1092	0.0038	0.0045	0.1137	0.1202	0.1218	0.1243	0.1380	0.1147	0.1212	0.1228	0.1253	0.1390			
0.1640 (#8-36)	36	0.1585	0.1640	0.1439	0.1460	0.1282	0.1320	0.0042	0.0050	0.1370	0.1442	0.1460	0.1487	0.1640	0.1380	0.1452	0.1470	0.1497	0.1650			
0.1900 (#10-32)	32	0.1840	0.1900	0.1674	0.1697	0.1497	0.1539	0.0047	0.0056	0.1596	0.1675	0.1697	0.1726	0.1900	0.1626	0.1705	0.1727	0.1756	0.1930			
0.2500	28	0.2435	0.2500	0.2243	0.2268	0.2041	0.2088	0.0054	0.0064	0.2152	0.2229	0.2268	0.2300	0.2500	0.2182	0.2259	0.2298	0.2330	0.2530			
0.3125	24	0.3053	0.3125	0.2827	0.2854	0.2591	0.2644	0.0063	0.0075	0.2719	0.2799	0.2854	0.2890	0.3125	0.2749	0.2829	0.2884	0.2920	0.3155			
0.3750	24	0.3678	0.3750	0.3450	0.3479	0.3214	0.3268	0.0063	0.0075	0.3344	0.3418	0.3479	0.3516	0.3750	0.3374	0.3448	0.3509	0.3546	0.3780			
0.4375	20	0.4294	0.4375	0.4019	0.4050	0.3736	0.3797	0.0075	0.0090	0.3888	0.3970	0.4050	0.4091	0.4375	0.3918	0.4000	0.4080	0.4121	0.4405			
0.5000	20	0.4919	0.5000	0.4643	0.4675	0.4360	0.4422	0.0075	0.0090	0.4513	0.4591	0.4675	0.4717	0.5000	0.4543	0.4621	0.4705	0.4747	0.5030			
0.5625	18	0.5538	0.5625	0.5230	0.5264	0.4916	0.4983	0.0083	0.0100	0.5084	0.5166	0.5264	0.5308	0.5625	0.5114	0.5196	0.5294	0.5338	0.5655			
0.6250	18	0.6163	0.6250	0.5854	0.5889	0.5540	0.5608	0.0083	0.0100	0.5709	0.5788	0.5889	0.5934	0.6250	0.5739	0.5818	0.5919	0.5964	0.6280			
0.7500	16	0.7406	0.7500	0.7056	0.7094	0.6702	0.6778	0.0094	0.0113	0.6892	0.6977	0.7094	0.7143	0.7500	0.6922	0.7007	0.7124	0.7173	0.7550			
0.8750	14	0.8647	0.8750	0.8245	0.8286	0.7841	0.7925	0.0107	0.0129	0.8055	0.8152	0.8286	0.8339	0.8750	0.8085	0.8182	0.8316	0.8369	0.8780			
1.0000	12	0.9886	1.0000	0.9415	0.9459	0.8944	0.9038	0.0125	0.0150	0.9189	0.9298	0.9459	0.9516	1.0000	0.9219	0.9339	0.9489	0.9546	1.0030			
1.1250	12	1.1136	1.1250	1.0664	1.0709	1.0192	1.0288	0.0125	0.0150	1.0439	1.0539	1.0709	1.0768	1.1250	1.0469	1.0569	1.0739	1.0798	1.1280			
1.2500	12	1.2386	1.2500	1.1913	1.1959	1.1442	1.1538	0.0125	0.0150	1.1689	1.1789	1.1959	1.2019	1.2500	1.1719	1.1819	1.1989	1.2049	1.2550			
1.3750	12	1.3636	1.3750	1.3162	1.3209	1.2690	1.2788	0.0125	0.0150	1.2939	1.3039	1.3209	1.3270	1.3750	1.2969	1.3069	1.3239	1.3300	1.3780			
1.5000	12	1.4886	1.5000	1.4411	1.4459	1.3940	1.4038	0.0125	0.0150	1.4189	1.4289	1.4459	1.4522	1.5000	1.4219	1.4319	1.4489	1.4552	1.5030			

MIL-S-8879

TABLE VI Twelve Thread Series

	EXTERNAL THREAD – 12 UNJ CLASS 3A ROOT RADIUS 0.0125 MIN 0.0150 MAX							INTERNAL THREAD – 12 UNJ CLASS 3B							INTERNAL THREAD – 12 UNJ CLASS 3BG					
BASIC SIZE	MAJOR DIAMETER		PITCH DIAMETER		MINOR DIAMETER			MINOR DIAMETER		PITCH DIAMETER		MAJOR DIA		MINOR DIAMETER		PITCH DIAMETER		MAJOR DIA	
	MIN	MAX	MIN	MAX	MIN	MAX		MIN	MAX	MIN	MAX	MIN		MIN	MAX	MIN	MAX	MIN	
1.6250	1.6136	1.6250	1.5665	1.5709	1.5194	1.5288		1.5439	1.5539	1.5709	1.5736	1.6250		1.5469	1.5569	1.5739	1.5796	1.6280	
1.7500	1.7386	1.7500	1.6914	1.6959	1.6442	1.6538		1.6689	1.6789	1.6959	1.7017	1.7500		1.6719	1.6819	1.6989	1.7047	1.7530	
1.8750	1.8636	1.8750	1.8164	1.8209	1.7692	1.7788		1.7939	1.8039	1.8209	1.8267	1.8750		1.7969	1.8069	1.8239	1.8297	1.8780	
2.0000	1.9886	2.0000	1.9414	1.9459	1.8942	1.9038		1.9189	1.9289	1.9459	1.9518	2.0000		1.9219	1.9319	1.9489	1.9548	2.0030	
2.2500	2.2386	2.2500	2.1914	2.1959	2.1442	2.1538		2.1689	2.1789	2.1959	2.2018	2.2500		2.1719	2.1819	2.1989	2.2048	2.2530	
2.5000	2.4886	2.5000	2.4413	2.4459	2.3942	2.4038		2.4189	2.4289	2.4459	2.4519	2.5000		2.4219	2.4319	2.4489	2.4549	2.5030	
2.7500	2.7386	2.7500	2.6913	2.6959	2.6442	2.6538		2.6689	2.6789	2.6959	2.7019	2.7500		2.6719	2.6819	2.6989	2.7049	2.7530	
3.0000	2.9886	3.0000	2.9412	2.9459	2.8940	2.9038		2.9189	2.9289	2.9459	2.9521	3.0000		2.9219	2.9319	2.9489	2.9551	3.0030	
3.2500	3.2386	3.2500	3.1912	3.1959	3.1440	3.1538		3.1689	3.1789	3.1959	3.2021	3.2500		3.1719	3.1819	3.1989	3.2051	3.2530	
3.5000	3.4886	3.5000	3.4411	3.4459	3.3940	3.4038		3.4189	3.4289	3.4459	3.4522	3.5000		3.4219	3.4319	3.4489	3.4552	3.5030	
3.7500	3.7386	3.7500	3.6911	3.6959	3.6440	3.6538		3.6689	3.6789	3.6959	3.7022	3.7500		3.6719	3.6819	3.6989	3.7052	3.7530	
4.0000	3.9886	4.0000	3.9410	3.9459	3.8938	3.9038		3.9189	3.9289	3.9459	3.9523	4.0000		3.9219	3.9319	3.9489	3.9553	4.0030	
4.2500	4.2386	4.2500	4.1910	4.1959	4.1438	4.1538		4.1689	4.1789	4.1959	4.2023	4.2500		4.1719	4.1819	4.1989	4.2053	4.2530	
4.5000	4.4886	4.5000	4.4410	4.4459	4.3938	4.4038		4.4189	4.4289	4.4459	4.4523	4.5000		4.4219	4.4319	4.4489	4.4553	4.5030	
4.7500	4.7386	4.7500	4.6909	4.6959	4.6438	4.6538		4.6689	4.6789	4.6959	4.7025	4.7500		4.6719	4.6819	4.6989	4.7055	4.7530	
5.0000	4.9886	5.0000	4.9409	4.9459	4.8938	4.9038		4.9189	4.9289	4.9459	4.9525	5.0000		4.9219	4.9319	4.9489	4.9555	5.0030	
5.2500	5.2386	5.2500	5.1909	5.1959	5.1438	5.1538		5.1689	5.1789	5.1959	5.2025	5.2500		5.1719	5.1819	5.1989	5.2055	5.2530	
5.5000	5.4886	5.5000	5.4409	5.4459	5.3938	5.4038		5.4189	5.4289	5.4459	5.4525	5.5000		5.4219	5.4319	5.4489	5.4555	5.5030	
5.7500	5.7386	5.7500	5.6907	5.6959	5.6436	5.6538		5.6689	5.6789	5.6959	5.7026	5.7500		5.6719	5.6819	5.6989	5.7056	5.7530	
6.0000	5.9886	6.0000	5.9407	5.9459	5.8936	5.9038		5.9189	5.9289	5.9459	5.9526	6.0000		5.9219	5.9319	5.9489	5.9556	6.0030	

16

16-21

16.5 METRIC THREADS

16.5.1 BASIC THREAD DESIGNATION —

16.5.1.1 ISO Metric Threads are designated by the letter "M" followed by the NOMINAL SIZE in millimetres, and the PITCH in millimetres, separated by the sign "X".
  Example: M16X1.5
Above designation format is followed for all thread series.

16.5.1.2 Coarse Pitch ISO Metric Threads may be designated by only the letter "M" and the NOMINAL SIZE in millimetres.
  Example: M16
This is a 16 millimetre diameter, 2 millimetre pitch ISO metric thread. Although the ISO standards use the above designations for coarse pitch, USA practice has been to include the pitch symbol even for the coarse pitch series. The inclusion of the pitch symbol should not create any problems, and will serve to avoid confusion.

16.5.2 TOLERANCE SYSTEM — The ISO Metric Screw Thread Tolerance System provides for allowances and tolerances defined by tolerance grades, tolerance positions and tolerance classes briefly defined as follows:

16.5.2.1 TOLERANCE GRADE — Basically, there are three Metric Tolerance Grades recommended by ISO (International Organization for Standardization): Grades 4, 6 and 8 which reflect the SIZE of the tolerance.

  a.) Grade 6 Tolerance is the closest ISO recommendation to our Unified Class 2A and 2B fits and is most frequently used as this grade is recommended for "medium" quality and normal lengths of engagement-or General Purpose Threads. Grade 4 Tolerance is closest to our unified class 3A and 3B fits.

  b.) Tolerances below Grade 6 are smaller than Grade 6 and recommended for "fine" quality or short lengths of engagement.

  c.) Tolerances above Grade 6 are larger than Grade 6 and recommended for "coarse" quality or long lengths of engagement.

16.5.2.2 TOLERANCE POSITION — ISO has established "amounts of allowance" by a series of tolerance position symbols, as follows:

External Threads (Bolts):
  small "e" = large allowance
  small "g" = small allowance
  small "h" = no allowance

Internal Threads (Nuts):
  Large "G" = small allowance
  Large "H" = no allowance

The above symbols are used after the Tolerance Grade such as; 6g which designates a "Medium" Tolerance Grade with small allowance for an external thread.

16.5.2.3 TOLERANCE CLASSES — ISO Tolerance classes of fit are determined by selecting one of the three qualities, (Fine, Medium or Coarse) combined with one of the three lengths of engagement, Short (S), Normal (N) or Long (L) and applying the proper allowance. TOLERANCE POSITIONS "g" for external threads and "H" for internal threads are preferred.

16.5.3 THREAD CALLOUT ON DRAWINGS — A complete designation for an ISO Metric screw thread comprise of:
  a.) Basic designation including nominal size.
  b.) Pitch
  c.) Tolerance Grade for Pitch Diameter
  d.) Tolerance Position for Pitch Diameter
  e.) Tolerance Grade for Crest Diameter
  f.) Tolerance Position for Crest Diameter

EXTERNAL THREAD

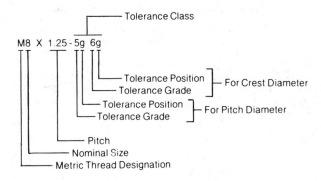

INTERNAL THREAD

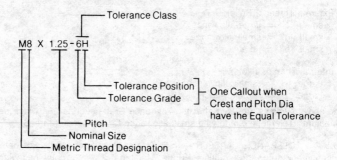

16.5.4 LENGTH OF ENGAGEMENT DESIGNATIONS — Where considered necessary, the length of engagement group symbol may be added to the tolerance class designation.

M20 X 2 — 7g 6g L
                  └── Length of Engagement Group Symbol

16.5.5 COMMERCIAL THREADS — For General thread (See tables 1 and 2) applications, the tolerance class recommended for:
1.) External threads is 6g
2.) Internal threads is 6H
This compares approximately with 2A & 2B class for inch threads.

METRIC SCREW THREADS

Limiting Dimensions of Standard Series Threads for Commercial Screws, Bolts and Nuts (inches)

Nominal Size Diam. (mm)	Pitch P (mm)	Basic Thread Designation	Tol Class	Allowance	External Thread (Bolt)							Tol Class	Internal Thread (Nut)					
					Major Diameter		Pitch Diameter			Minor Diameter			Minor Diameter		Pitch Diameter			Major Dia
					Max	Min	Max	Min	Tol	Max a	Min b		Min	Max	Min	Max	Tol	Min
1.6	0.35	M1.6	6g	0.0008	0.0622	0.0589	0.0533	0.0509	0.0024	0.0453	0.0419	6H	0.0481	0.0520	0.0541	0.0574	0.0033	0.0630
1.8	0.35	M1.8	6g	0.0008	0.0701	0.0668	0.0611	0.0588	0.0023	0.0531	0.0498	6H	0.0560	0.0598	0.0620	0.0652	0.0032	0.0709
2	0.4	M2	6g	0.0009	0.0779	0.0743	0.0677	0.0652	0.0025	0.0586	0.0549	6H	0.0617	0.0661	0.0686	0.0720	0.0034	0.0788
2.2	0.45	M2.2	6g	0.0009	0.0858	0.0819	0.0743	0.0716	0.0027	0.0640	0.0601	6H	0.0675	0.0723	0.0752	0.0788	0.0033	0.0867
2.5	0.45	M2.5	6g	0.0009	0.0976	0.0938	0.0861	0.0834	0.0027	0.0759	0.0719	6H	0.0793	0.0841	0.0870	0.0906	0.0036	0.0985
3	0.5	M3	6g	0.0009	0.1173	0.1132	0.1045	0.1016	0.0029	0.0931	0.0889	6H	0.0969	0.1023	0.1054	0.1092	0.0038	0.1182
3.5	0.6	M3.5	6g	0.0009	0.1369	0.1321	0.1216	0.1183	0.0033	0.1079	0.1030	6H	0.1123	0.1185	0.1225	0.1268	0.0043	0.1378
4	0.7	M4	6g	0.0009	0.1566	0.1512	0.1387	0.1352	0.0034	0.1227	0.1173	6H	0.1277	0.1347	0.1396	0.1442	0.0046	0.1575
4.5	0.75	M4.5	6g	0.0010	0.1762	0.1708	0.1571	0.1536	0.0035	0.1400	0.1345	6H	0.1452	0.1526	0.1580	0.1626	0.0046	0.1772
5	0.8	M5	6g	0.0010	0.1959	0.1900	0.1754	0.1717	0.0037	0.1572	0.1513	6H	0.1628	0.1706	0.1764	0.1812	0.0048	0.1969
6	1	M6	6g	0.0012	0.2351	0.2282	0.2096	0.2052	0.0044	0.1868	0.1797	6H	0.1936	0.2028	0.2107	0.2165	0.0058	0.2363
7	1	M7	6g	0.0011	0.2745	0.2675	0.2489	0.2446	0.0043	0.2262	0.2191	6H	0.2330	0.2422	0.2500	0.2559	0.0059	0.2756
8	1.25	M8	6g	0.0012	0.3138	0.3056	0.2818	0.2773	0.0045	0.2535	0.2454	6H	0.2617	0.2721	0.2810	0.2892	0.0062	0.3150
8	1	M8x1	6g	0.0011	0.3139	0.3069	0.2883	0.2840	0.0043	0.2656	0.2584	6H	0.2724	0.2816	0.2894	0.2952	0.0058	0.3150
10	1.5	M10	6g	0.0013	0.3924	0.3832	0.3540	0.3489	0.0051	0.3199	0.3102	6H	0.3298	0.3415	0.3554	0.3624	0.0070	0.3937
10	1.25	M10x1.25	6g	0.0012	0.3925	0.3843	0.3606	0.3560	0.0046	0.3322	0.3241	6H	0.3404	0.3508	0.3618	0.3680	0.0062	0.3937
12	1.75	M12	6g	0.0014	0.4711	0.4607	0.4263	0.4205	0.0058	0.3865	0.3758	6H	0.3979	0.4110	0.4277	0.4355	0.0078	0.4725
12	1.25	M12x1.25	6g	0.0012	0.4713	0.4630	0.4393	0.4342	0.0051	0.4109	0.4023	6H	0.4192	0.4295	0.4405	0.4475	0.0070	0.4725
14	2	M14	6g	0.0016	0.5496	0.5387	0.4985	0.4923	0.0062	0.4530	0.4412	6H	0.4660	0.4807	0.5001	0.5083	0.0082	0.5512
14	1.5	M14x1.5	6g	0.0013	0.5493	0.5407	0.5115	0.5061	0.0054	0.4774	0.4677	6H	0.4873	0.4990	0.5129	0.5203	0.0074	0.5512
16	2	M16	6g	0.0016	0.6284	0.6175	0.5772	0.5710	0.0062	0.5318	0.5199	6H	0.5447	0.5594	0.5788	0.5871	0.0083	0.6300
16	1.5	M16x1.5	6g	0.0014	0.6286	0.6194	0.5903	0.5849	0.0054	0.5561	0.5465	6H	0.5660	0.5777	0.5916	0.5990	0.0074	0.6300
18	2.5	M18	6g	0.0017	0.7070	0.6939	0.6430	0.6364	0.0066	0.5862	0.5725	6H	0.6022	0.6198	0.6448	0.6535	0.0087	0.7087
18	1.5	M18x1.5	6g	0.0013	0.7074	0.6982	0.6690	0.6636	0.0054	0.6349	0.6252	6H	0.6448	0.6565	0.6704	0.6777	0.0073	0.7087
20	2.5	M20	6g	0.0018	0.7857	0.7726	0.7218	0.7152	0.0066	0.6649	0.6513	6H	0.6809	0.6985	0.7235	0.7322	0.0087	0.7875
20	1.5	M20x1.5	6g	0.0014	0.7861	0.7769	0.7477	0.7423	0.0054	0.7136	0.7039	6H	0.7235	0.7352	0.7491	0.7565	0.0074	0.7875
22	2.5	M22	6g	0.0018	0.8644	0.8513	0.8005	0.7939	0.0066	0.7437	0.7300	6H	0.7597	0.7773	0.8023	0.8110	0.0087	0.8662
22	1.5	M22x1.5	6g	0.0014	0.8648	0.8556	0.8265	0.8211	0.0054	0.7924	0.7827	6H	0.8023	0.8140	0.8278	0.8352	0.0074	0.8662
24	3	M24	6g	0.0020	0.9429	0.9283	0.8662	0.8584	0.0078	0.7980	0.7817	6H	0.8171	0.8366	0.8682	0.8785	0.0103	0.9449
24	2	M24x2	6g	0.0016	0.9433	0.9324	0.8922	0.8856	0.0066	0.8467	0.8345	6H	0.8597	0.8744	0.8938	0.9025	0.0087	0.9449
27	3	M27	6g	0.0019	1.0611	1.0464	0.9843	0.9765	0.0078	0.9161	0.8999	6H	0.9352	0.9548	0.9863	0.9966	0.0103	1.0630
27	2	M27x2	6g	0.0016	1.0614	1.0505	1.0103	1.0037	0.0066	0.9648	0.9526	6H	0.9778	0.9925	1.0119	1.0206	0.0087	1.0630
30	3.5	M30	6g	0.0022	1.1790	1.1623	1.0895	1.0812	0.0083	1.0099	0.9917	6H	1.0320	1.0539	1.0917	1.1026	0.0109	1.1812
30	2	M30x2	6g	0.0016	1.1796	1.1686	1.1284	1.1218	0.0066	1.0829	1.0707	6H	1.0959	1.1106	1.1300	1.1387	0.0087	1.1812
33	3.5	M33	6g	0.0022	1.2971	1.2804	1.2076	1.1993	0.0083	1.1280	1.1099	6H	1.1501	1.1720	1.2098	1.2207	0.0109	1.2993
33	2	M33x2	6g	0.0016	1.2977	1.2867	1.2465	1.2399	0.0066	1.2011	1.1888	6H	1.2140	1.2287	1.2481	1.2568	0.0087	1.2993
36	4	M36	6g	0.0025	1.4149	1.3963	1.3126	1.3039	0.0087	1.2217	1.2017	6H	1.2469	1.2704	1.3151	1.3268	0.0117	1.4174
36	3	M36x3	6g	0.0020	1.4154	1.4007	1.3386	1.3309	0.0077	1.2705	1.2542	6H	1.2895	1.3091	1.3406	1.3510	0.0104	1.4174
39	4	M39	6g	0.0025	1.5330	1.5144	1.4307	1.4220	0.0087	1.3398	1.3198	6H	1.3650	1.3885	1.4332	1.4449	0.0117	1.5355
39	3	M39x3	6g	0.0020	1.5335	1.5188	1.4568	1.4490	0.0078	1.3886	1.3723	6H	1.4076	1.4272	1.4587	1.4691	0.0104	1.5355

a Design form.
b Required for high strength applications where rounded root is specified.

Table I

METRIC SCREW THREADS

Limiting Dimensions of Standard Series Threads for Commercial Screws, Bolts and Nuts (mm)

Nominal Size Diam.	Pitch P	Basic Thread Designation	Tol Class	Allowance	External Thread (Bolt)							Tol Class	Internal Thread (Nut)					Major Dia Min
					Major Diameter		Pitch Diameter			Minor Diameter			Minor Diameter		Pitch Diameter			
					Max	Min	Max	Min	Tol	Max[a]	Min[b]		Min	Max	Min	Max	Tol	
1.6	0.35	M1.6	6g	0.019	1.581	1.496	1.354	1.291	0.063	1.151	1.063	6H	1.221	1.321	1.373	1.458	0.085	1.600
1.8	0.35	M1.8	6g	0.019	1.781	1.696	1.554	1.491	0.063	1.351	1.263	6H	1.421	1.521	1.573	1.658	0.085	1.800
2	0.4	M2	6g	0.019	1.981	1.886	1.721	1.654	0.067	1.490	1.394	6H	1.567	1.679	1.740	1.830	0.090	2.000
2.2	0.45	M2.2	6g	0.020	2.180	2.080	1.888	1.817	0.071	1.628	1.525	6H	1.713	1.838	1.908	2.003	0.095	2.200
2.5	0.45	M2.5	6g	0.020	2.480	2.380	2.188	2.117	0.071	1.928	1.825	6H	2.013	2.138	2.208	2.303	0.095	2.500
3	0.5	M3	6g	0.020	2.980	2.874	2.655	2.580	0.075	2.367	2.256	6H	2.459	2.599	2.675	2.775	0.100	3.000
3.5	0.6	M3.5	6g	0.021	3.479	3.354	3.089	3.004	0.085	2.742	2.614	6H	2.850	3.010	3.110	3.222	0.112	3.500
4	0.7	M4	6g	0.022	3.978	3.838	3.523	3.433	0.090	3.119	2.979	6H	3.242	3.422	3.545	3.663	0.118	4.000
4.5	0.75	M4.5	6g	0.022	4.478	4.338	3.991	3.901	0.090	3.558	3.414	6H	3.688	3.878	4.013	4.131	0.118	4.500
5	0.8	M5	6g	0.024	4.976	4.826	4.456	4.361	0.095	3.994	3.841	6H	4.134	4.334	4.480	4.605	0.125	5.000
6	1	M6	6g	0.026	5.974	5.794	5.324	5.212	0.112	4.747	4.563	6H	4.917	5.153	5.350	5.500	0.150	6.000
7	1	M7	6g	0.026	6.974	6.794	6.324	6.212	0.112	5.747	5.563	6H	5.917	6.153	6.350	6.500	0.150	7.000
8	1.25	M8	6g	0.028	7.972	7.760	7.160	7.042	0.118	6.439	6.231	6H	6.647	6.912	7.188	7.348	0.160	8.000
8	1	M8x1	6g	0.026	7.974	7.794	7.324	7.212	0.112	6.747	6.563	6H	6.917	7.153	7.350	7.500	0.150	8.000
10	1.5	M10	6g	0.032	9.968	9.732	8.994	8.862	0.132	8.127	7.879	6H	8.376	8.676	9.026	9.206	0.180	10.000
10	1.25	M10x1.25	6g	0.028	9.972	9.760	9.160	9.042	0.118	8.439	8.231	6H	8.647	8.912	9.188	9.348	0.160	10.000
12	1.75	M12	6g	0.034	11.966	11.701	10.829	10.679	0.150	9.819	9.543	6H	10.106	10.441	10.863	11.063	0.200	12.000
12	1.25	M12x1.25	6g	0.028	11.972	11.760	11.160	11.028	0.132	10.439	10.217	6H	10.647	10.912	11.188	11.368	0.180	12.000
14	2	M14	6g	0.038	13.962	13.682	12.663	12.503	0.160	11.508	11.204	6H	11.835	12.210	12.701	12.913	0.212	14.000
14	1.5	M14x1.5	6g	0.032	13.968	13.732	12.994	12.854	0.170	12.127	11.879	6H	12.376	12.676	13.026	13.216	0.190	14.000
16	2	M16	6g	0.038	15.962	15.682	14.663	14.503	0.160	13.508	13.204	6H	13.835	14.210	14.701	14.913	0.212	16.000
16	1.5	M16x1.5	6g	0.032	15.968	15.732	14.994	14.854	0.170	14.127	13.879	6H	14.376	14.676	15.026	15.216	0.190	16.000
18	2.5	M18	6g	0.042	17.958	17.623	16.334	16.164	0.170	14.891	14.541	6H	15.294	15.744	16.376	16.600	0.224	18.000
18	1.5	M18x1.5	6g	0.032	17.968	17.732	16.994	16.854	0.170	16.127	15.879	6H	16.376	16.676	17.026	17.216	0.190	18.000
20	2.5	M20	6g	0.042	19.958	19.623	18.334	18.164	0.170	16.891	16.541	6H	17.294	17.744	18.376	18.600	0.224	20.000
20	1.5	M20x1.5	6g	0.032	19.968	19.732	18.994	18.854	0.140	18.127	17.879	6H	18.376	18.676	19.026	19.216	0.190	20.000
22	2.5	M22	6g	0.042	21.958	21.623	20.334	20.164	0.170	18.891	18.541	6H	19.294	19.744	20.376	20.600	0.224	22.000
22	1.5	M22x1.5	6g	0.032	21.968	21.732	20.994	20.854	0.140	20.127	19.879	6H	20.376	20.676	21.026	21.216	0.190	22.000
24	3	M24	6g	0.048	23.952	23.577	22.003	21.803	0.200	20.271	19.855	6H	20.752	21.252	22.051	22.316	0.265	24.000
24	2	M24x2	6g	0.038	23.962	23.682	22.663	22.493	0.170	21.508	21.194	6H	21.835	22.210	22.701	22.925	0.224	24.000
27	3	M27	6g	0.048	26.952	26.577	25.003	24.803	0.200	23.271	22.855	6H	23.752	24.252	25.051	25.316	0.265	27.000
27	2	M27x2	6g	0.038	26.962	26.682	25.663	25.493	0.170	24.508	24.194	6H	24.835	25.210	25.701	25.925	0.224	27.000
30	3.5	M30	6g	0.053	29.947	29.522	27.674	27.462	0.212	25.653	25.189	6H	26.211	26.771	27.727	28.007	0.280	30.000
30	2	M30x2	6g	0.038	29.962	29.682	28.663	28.493	0.170	27.508	27.194	6H	27.835	28.210	28.701	28.925	0.224	30.000
33	3.5	M33	6g	0.053	32.947	32.522	30.674	30.462	0.212	28.653	28.189	6H	29.211	29.771	30.727	31.007	0.280	33.000
33	2	M33x2	6g	0.038	32.962	32.682	31.663	31.493	0.170	30.508	30.194	6H	30.835	31.210	31.701	31.925	0.224	33.000
36	4	M36	6g	0.060	35.940	35.465	33.342	33.118	0.224	31.033	30.521	6H	31.670	32.270	33.402	33.702	0.300	36.000
36	3	M36x3	6g	0.048	35.952	35.577	34.003	33.803	0.200	32.271	31.855	6H	32.752	33.252	34.051	34.316	0.265	36.000
39	4	M39	6g	0.060	38.940	38.465	36.342	36.118	0.224	34.033	33.521	6H	34.670	35.270	36.402	36.702	0.300	39.000
39	3	M39x3	6g	0.048	38.952	38.577	37.003	36.803	0.200	35.271	34.855	6H	35.752	36.252	37.051	37.316	0.265	39.000

[a] Design form.
[b] Required for high strength applications where rounded root is specified.

Table 2

16.5.6   AEROSPACE THREADS — For Aerospace thread (See table 3) applications, the tolerance class recommended for.
1.) External threads is 4 h 6 h.
2.) Internal threads for sizes 1 thru 5 mm is 4H6H
Internal threads for sizes 6 mm and larger is 4H5H. This compares approximately with 3A & 3B class for inch threads.

16.5.6.1   EXTERNAL THREAD ROOT CONTOUR — The root of the external thread shall have a controlled radius as defined by the Society of Automotive Engineers (SAE) Aerospace Standard   ASI338 for use in the aerospace industry.

16.5.6.2   DESIGNATION OF CONTROLLED ROOT RADIUS THREAD —

EXTERNAL THREAD

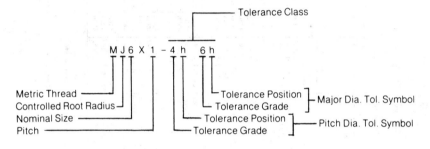

INTERNAL THREAD

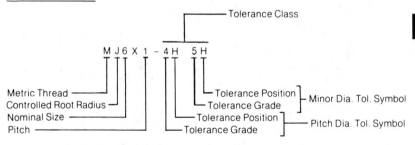

16.5.7   AEROSPACE PREFERRED DIAMETER-PITCH COMBINATIONS— Table 3 lists the preferred diameter-pitch combinations for use in the aerospace industry for metric module bolts and nuts. For additional aerospace Metric Screw Thread Standard and Special Sizes see Society of Automotive Engineers (SAE) Aerospace Standard AS1370 and ANSI Y14.6aM.

16.5.8   Other product information may also be conveyed by the ISO metric thread designations. Complete specifications and product limits may be found in the ISO Recommendations or in the B1 report "ISO Metric Screw Threads".

METRIC SCREW THREADS

Limiting Dimensions of Standard Series Threads for Aerospace Screws, Bolts and Nuts (mm)

Nominal Size Dia	Pitch P	Basic Thread Designation	Tol Class	External Thread (Bolt) Major Diameter Max	Min	Pitch Diameter Max	Min	Tol	Minor Diameter Max	Min	Tol Class	Internal Thread (Nut) Minor Diameter Min	Max	Pitch Diameter Min	Max	Tol	Major Diameter Min	Max
1.6	0.35	MJ1.6	4h6h	1.600	1.515	1.373	1.333	0.040	1.196	1.135	4H6H	1.259	1.359	1.373	1.426	0.053	1.600	1.704
2	0.4	MJ2	4h6h	2.000	1.905	1.740	1.698	0.042	1.538	1.472	4H6H	1.610	1.722	1.740	1.796	0.056	2.000	2.114
2.5	0.45	MJ2.5	4h6h	2.500	2.400	2.208	2.163	0.045	1.981	1.908	4H6H	2.062	2.187	2.208	2.268	0.060	2.500	2.625
3	0.5	MJ3	4h6h	3.000	2.894	2.675	2.627	0.048	2.422	2.344	4H6H	2.513	2.653	2.675	2.738	0.063	3.000	3.135
3.5	0.6	MJ3.5	4h6h	3.500	3.375	3.110	3.057	0.053	2.807	2.718	4H6H	2.915	3.075	3.110	3.181	0.071	3.500	3.657
4	0.7	MJ4	4h6h	4.000	3.860	3.545	3.489	0.056	3.191	3.093	4H6H	3.318	3.498	3.545	3.620	0.075	4.000	4.176
5	0.8	MJ5	4h6h	5.000	4.850	4.480	4.420	0.060	4.076	3.967	4H6H	4.221	4.421	4.480	4.560	0.080	5.000	5.195
6	1	MJ6	4h6h	6.000	5.820	5.350	5.279	0.071	4.845	4.713	4H5H	5.026	5.216	5.350	5.445	0.095	6.000	6.239
7	1	MJ7	4h6h	7.000	6.820	6.350	6.279	0.071	5.845	5.713	4H5H	6.026	6.216	6.350	6.445	0.095	7.000	7.239
8	1	MJ8	4h6h	8.000	7.820	7.350	7.279	0.071	6.845	6.713	4H5H	7.026	7.216	7.350	7.445	0.095	8.000	8.239
10	1.25	MJ10	4h6h	10.000	9.788	9.188	9.113	0.075	8.557	8.406	4H5H	8.782	8.994	9.188	9.288	0.100	10.000	10.280
12	1.25	MJ12	4h6h	12.000	11.788	11.188	11.103	0.085	10.557	10.396	4H5H	10.782	10.994	11.188	11.300	0.112	12.000	12.292
14	1.5	MJ14	4h6h	14.000	13.764	13.026	12.936	0.090	12.268	12.087	4H5H	12.539	12.775	13.026	13.144	0.118	14.000	14.335
16	1.5	MJ16	4h6h	16.000	15.764	15.026	14.936	0.090	14.268	14.087	4H5H	14.539	14.775	15.026	15.144	0.118	16.000	16.335
18	1.5	MJ18	4h6h	18.000	17.764	17.026	16.936	0.090	16.268	16.087	4H5H	16.539	16.775	17.026	17.144	0.118	18.000	18.335
20	1.5	MJ20	4h6h	20.000	19.764	19.026	18.936	0.090	18.268	18.087	4H5H	18.539	18.775	19.026	19.144	0.118	20.000	20.335
22	1.5	MJ22	4h6h	22.000	21.764	21.026	20.936	0.090	20.268	20.087	4H5H	20.539	20.775	21.026	21.144	0.118	22.000	22.335
24	2	MJ24	4h6h	24.000	23.720	22.701	22.595	0.106	21.691	21.463	4H5H	22.051	22.351	22.701	22.841	0.140	24.000	24.429
27	2	MJ27	4h6h	27.000	26.720	25.701	25.595	0.106	24.691	24.463	4H5H	25.051	25.351	25.701	25.841	0.140	27.000	27.429
30	2	MJ30	4h6h	30.000	29.720	28.701	28.595	0.106	27.691	27.463	4H5H	28.051	28.351	28.701	28.841	0.140	30.000	30.429
33	2	MJ33	4h6h	33.000	32.720	31.701	31.595	0.106	30.691	30.463	4H5H	31.051	31.351	31.701	31.841	0.140	33.000	33.429
36	2	MJ36	4h6h	36.000	35.720	34.701	34.595	0.106	33.691	33.463	4H5H	34.051	34.351	34.701	34.841	0.140	36.000	36.429
39	2	MJ39	4h6h	39.000	38.720	37.701	37.595	0.106	36.691	36.463	4H5H	37.051	37.351	37.701	37.841	0.140	39.000	39.429

a Design form.
b Required for high strength applications where rounded root is specified.

Table 3

NOTES

NOTES

Section 17 — ELECTRICAL AND ELECTRONIC DRAFTING

17.1 SCOPE — This section establishes standard methods for preparing electrical and electronic drawings.

17.2 APPLICABLE DOCUMENTS

MIL-W-5088	Wiring, Aircraft, Installation of
MIL-W-8160	Wiring, Guided Missile, Installation of General Specification for
MIL-P-55110	Printed Wiring Boards
DOD-STD-100	Engineering Drawing Practices
MIL-STD-275	Printed Wiring for electronic Equipment
MIL-STD-454	Standard General Requirements for Electronic Equipment
MIL-STD-681	Identification Coding and Application of Hook Up and Lead Wire
ANSI/IPC-T-50	Terms and Definitions for interconnecting and packaging electrical circuits.
ANSI Y14.5	Dimensioning and Tolerancing
ANSI Y14.15	Electrical and Electronics Diagrams
ANSI Y32.2 (IEEE 315)	Graphic Symbols for Electrical and Electronics Diagrams
ANSI Y32.14	Graphic Symbols for Logic Diagrams
ANSI Y32.16	Reference Designations for Electrical and Electronics Parts and Equipments

17.3 DEFINITIONS (For Complete Coverage, See MIL-STD-429)

17.3.1 ACTIVE ELECTRONIC PARTS — Parts capable of controlling the flow of electrons without mechanical adjustment. This includes vacuum tubes and semi-conductor devices but excludes switches and variable resistors.

17.3.2 BLOCK DIAGRAM — Shows circuit information in a more simplified form than the Single Line Diagram. It represents the circuit functions by the means of single lines and rectangular blocks without using graphical symbols or reference designations. (See Figure 2)

17.3.3 CABLE ASSEMBLY — A cable of a definite continuous length, having one or more ends processed or terminated in fittings which provide for connection to other items.

17.3.4 CIRCUIT CARD ASSEMBLY — A printed wiring board with separately manufactured parts such as resistors, capacitors, etc. mounted on it. It may or may not have plug-in connecting facilities either attached to the board or integral etched, printed or deposited contacts. (See Figure 15)

17.3.5 COLOR CODED WIRING — The utilization of a different color or combination of colors on the insulation of conductors to identify them.

17.3.6 COMPONENT BOARD ASSEMBLY — A terminal board upon which separately manufactured component parts have been mounted, e.g., resistors, capacitors, etc. (Boards with printed electrical conductor paths and components or paths only are not a part of this assembly — See 17.3.38 and 17.3.41)

17.3.7 CONDUCTIVE PATTERN — A design formed from any electrically conductive material, on an insulating base.

17.3.8 DISCRETE ELECTRONIC PARTS — A separate and distinct device, e.g. transistor, diode, resistor, etc.

17.3.9 ELECTRICAL AND ELECTRONIC REFERENCE DESIGNATIONS — Combinations of letters and numbers used to identify and locate items on diagrams and assemblies and for relating items in a set. Reference designations are not intended to replace other identifications such as part numbers.

17.3.10 ELECTRICAL AND ELECTRONIC SYMBOLS — Graphic representations of electrical or electronic parts. They are used in single line diagrams, schematics, or, if applicable, on connection or wiring diagrams.

17.3.11 ELECTRICAL DRAWINGS — Related to power distribution, relays, motors, generators, etc.

17.3.12 ELECTRICAL SCHEMATIC DIAGRAM — A drawing showing, by means of graphic symbols, the electrical connections and functions of a circuit arrangement. A schematic diagram does not show the size, shape or location of the component devices or parts.

17.3.13 ELECTRONIC DRAWINGS — Related to electrical circuits in vacuum or gas filled tubes and to solid state devices such as transistors, diodes, etc.

17.3.14 ELEMENT (OF A MICROCIRCUIT OR INTEGRATED CIRCUIT) — A constituent of the microcircuit or integrated circuit that contributes directly to its operation. (A discrete part incorporated into a microcircuit becomes an element of the microcircuit.)

17.3.15 ETCHING — A process wherein a printed pattern is formed by chemical or chemical and electrolytic removal of the unwanted portion of conductive material bonded to a base.

17.3.16 FILM INTEGRATED CIRCUIT — An integrated circuit consisting of elements which are films formed in situ upon an insulating substrate.

17.3.17 GRID — An orthogonal network of equidistant lines providing the basis for an incremental location system.

17.3.18 HARNESS ASSEMBLY — Two or more insulated conductors grouped into a bundle or harness, held together by lacing, or other similar binding.

17.3.19 HIGHWAY SYSTEM — A method of diagramming in which wires are grouped into a single path, or "Highway," to conserve space and simplify the diagram.

17.3.20 HYBRID MICROCIRCUIT — A microcircuit consisting of elements which are a combination of the film circuit type and the semiconductor types or a combination of one or both of the types with discrete parts.

17.3.21 INDEXING HOLES — Holes placed in a printed circuit base material to enable the base to be positioned accurately for processing. (Indexing holes may or may not be on the finished board.)

17.3.22 INDEXING NOTCHES — Notches placed in the edge of printed circuit base material to enable the base to be positioned accurately for processing. (Indexing notches may or may not be on the finished board.)

17.3.23 INTEGRATED CIRCUITS — Devices in which a number of electronic elements (both active and passive) are diffused as patterns into a wafer to function as a complete circuit.

17.3.24 INTERCONNECTION DIAGRAM — A drawing that shows only external wiring connections between units, sets, groups and systems.

17.3.25 LAND — Use the term "terminal area."

17.3.26 LOGIC DIAGRAM — A diagram that depicts by logic symbols and supplementary notations the details of signal flow and control, but not necessarily the point-to-point wiring existing in a system of two-state devices.

17.3.27 MASTER ARTWORK — An accurately scaled configuration which is used to produce the master pattern.

17.3.28 MASTER DRAWING — A drawing showing the dimensional limits or grid location applicable to any or all parts of a printed wiring or printed circuit including the base.

17.3.29 MASTER PATTERNS — A line-to-line scale pattern (production negative) which is used to produce the printed wiring or printed circuit within the accuracy specified on the master drawing.

17.3.30   MICROCIRCUIT — A small circuit having a high equivalent circuit element density, which is considered as a single part composed of interconnected elements on or within a single substrate to perform an electronic circuit function. (This excludes printed wiring boards, circuit card assemblies and modules composed exclusively of discrete electronic parts.)

17.3.31   MICROCIRCUIT MODULE — An assembly of microcircuits or an assembly of microcircuits and discrete parts, designed to perform one or more electronic circuit functions, and constructed such that for the purposes of specification testing, commerce, and maintenance, it is considered indivisible.

17.3.32   MICROELECTRONICS — That area of electronic technology associated with or applied to the realization of electronic systems from extremely small electronic parts or elements.

17.3.33   MONOLITHIC INTEGRATED CIRCUIT — An integrated circuit consisting of elements formed in situ on or within a semiconductor substrate with at least one of the elements formed within the substrate.

17.3.34   MULTICHIP MICROCIRCUIT — A microcircuit consisting of elements formed on or within two or more semiconductor chips which are separately attached to substrate.

17.3.35   PASSIVE ELECTRONIC PARTS — Parts that are non-active, and do not produce power or signal gain.

17.3.36   PLATED THROUGH HOLE — An interfacial or interlayer connection formed by deposition of conductive material on the sides of hole through the base.

17.3.37   PRINTED CIRCUIT — A pattern composed of printed wiring and printed parts all formed on a common base. Also see MIL-STD-429 for additional terms and definitions not included herein.

17.3.38   PRINTED CIRCUIT BOARD — A completely processed conductive pattern or patterns and printed parts.

17.3.39   PRINTED COMPONENT PART — A component part in printed form, such as a printed inductor, resistor, capacitor, transmission line, etc.

17.3.40   PRINTED CONTACT — A portion of printed wiring used for the purpose of providing electrical connection by pressure contact.

17.3.41   PRINTED WIRING ASSEMBLY — A wiring board upon which separately manufactured component parts have been added.

17.3.42   PRINTED WIRING BOARD — A completely processed conductive pattern or patterns.

17.3.43   REGISTER — The relative position of one or more printed wiring patterns with respect to their desired locations on the base material.

17.3.44 RUNNING (WIRE) LIST — A book form drawing consisting of tabular data and instructions required to establish wiring connections within or between units of an equipment, or between equipments, sets or assemblies of a system. A running (wire) list is a form of interconnection diagram or wiring diagram.

17.3.45 SINGLE LINE DIAGRAM — Represents the circuit by means of single lines and simplified graphical symbols. A typical single line diagram is shown in Figure 1.

17.3.46 SUBSTRATE (OF A MICROCIRCUIT OR INTEGRATED CIRCUIT) — The supporting material upon or within which the elements of a microcircuit or integrated circuit are fabricated or attached.

17.3.47 TERMINAL AREA — A portion of a printed-circuit or printed wiring used for making electrical connections to the conductive pattern.

17.3.48 TERMINAL BOARD — An item usually consisting of insulating material which is designed specifically for, or on which are mounted terminals such as screws, solder lugs, solder studs, etc. It does not include parts such as resistors, capacitors, etc.

17.3.49 TERMINAL BOARD ASSEMBLY — Terminal boards, two or more, which are mounted on a common surface or to each other.

17.3.50 TERMINAL PAD — Use the term "terminal area."

17.3.51 THIN FILM CIRCUITS — Consist of parts, interconnecting conductor paths and provision for electrically connecting discrete active parts, normally deposited as thin films on a substrate (usually glass).

17.3.52 WELDED MODULE — A package of separately manufactured parts stacked between two or more insulating wafers with metal ribbons welded to the leads of the components rather than soldered interconnections. Finished product is normally encapsulated.

17.3.53 WIRE DESTINATION — A reference designation, with or without terminal identification, placed to the right of and above the wire number or color code to indicate the component to which the wire is routed.

17.3.54 WIRE NUMBERS — Numbers or combinations of numbers and letters sequentially assigned to individual conductors on wiring diagrams for the purpose of identifying them. Wire numbers are not intended to replace other identification such as part numbers.

17.3.55 WIRING DIAGRAM — A drawing showing electrical connections of an installation or its component devices or parts. It may show internal or external connections or both, and contains the detail necessary to make or trace connections that are involved. Wiring diagrams usually show the general physical arrangement of the component devices or parts.

17.4 ELECTRICAL AND ELECTRONIC SCHEMATIC DIAGRAMS

17.4.1 These are engineering reference documents and are not used for fabrication. The schematic diagram shows by means of graphic symbols and connecting lines the electrical connection and functions of a specific circuit arrangement. It facilitates tracing the circuit and its functions without regard to the actual size, shape or location of the components. the components are identified by reference designations and electrical values. (See Figure 3)

17.4.1.1 GRAPHIC SYMBOLS DRAFTING PRACTICES

17.4.1.1.1 Symbols shall comply with ANSI Y32.2.

17.4.1.1.2 The position of a symbol on a drawing does not alter its meaning.

17.4.1.1.3 A symbol may be drawn to any proportionate size or line thickness to suit reproduction and commensurate with drawing size.

17.4.1.1.4 Electrically operated devices are shown in a position with the power off. Where otherwise, a note should so indicate.

17.4.1.1.5 The terminal symbol (0) may be added for attachment of conductors to any of the part symbols.

17.4.1.1.6 Parts of a symbol, such as a relay, may be separate on a circuit diagram provided designations are given to show the relationship of parts.

17.4.1.1.7 Switches are shown in a position with no operating force applied. Where there are two or more switch positions with no operating force applied or where switches are operated by a mechanical device, such as air pressure, a note is added to indicate the switch position.

17.4.1.1.8 Symbols are drawn in heavier weight lines than lines joining symbols.

17.4.1.2 REFERENCE DESIGNATIONS DRAFTING PRACTICES

17.4.1.2.1 Reference designations shall comply with ANSI Y32.16.

17.4.1.2.2 A complete reference designation is a combination of letters and numbers which identifies a part, subassembly, or unit of a set on diagrams, drawings, parts, lists, technical manuals, etc. The letters in a reference designation identify the class of item such as resistor, coil, electron tube or subassembly. The number differentiates between parts or subassemblies of the same class as illustrated below.

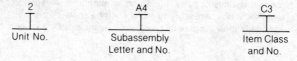

2A4C3 is CAPACITOR 3 of Subassembly A4 in Unit 2

17.4.1.2.3   Partial Reference designations may be used in accordance with ANSI Y32.16.

17.4.1.2.4   A reference designation is not an abbreviation for the name of an item.

17.4.1.2.5   Reference designations initially should be assigned with the lowest number of each designation item class in the upper left hand corner of a schematic diagram and proceed from left to right in top to bottom order on the diagram. Numbers within each class shall start with 1 and run consecutively. (See Figure 3)

17.4.1.2.6   Additional information such as type designation, input or output, etc., may be added as shown in the example below:

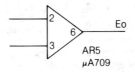

17.4.1.3   COMPONENT NUMERICAL VALUES

17.4.1.3.1   The values of capacitance, inductance and resistance are designated by whole numbers, or whole numbers and decimals. For example:

Use .092 μF (microfarad) not 92 000 pF (picofarads)

5 mH (millihenry) not 0.005 H (henrys)

2.5 KΩ (kilohms) not 2500 (ohms)

17.4.1.3.2   The comma is omitted in values of four or more digits. Repetition of component values should be eliminated through the use of general notes:

RESISTANCE IN OHMS, UNLESS OTHERWISE SPECIFIED.

CAPACITANCE IN PICOFARADS, UNLESS OTHERWISE SPECIFIED.

CAPACITANCE IN MICROMICROFARADS AND RESISTANCE IN OHMS, UNLESS OTHERWISE SPECIFIED.

ALL RESISTOR VALUES ARE IN OHMS ± _____% AND ARE _____WATT, UNLESS OTHERWISE SPECIFIED.

ALL CAPACITOR VALUES ARE IN MICROFARADS ± _____% AND. ARE_____VDC, UNLESS OTHERWISE SPECIFIED.

17.4.1.4   ADDITIONAL PART INFORMATION — May be included next to the graphical symbols or as a note or tabulated in a chart on a schematic diagram. Resistors may specify wattage ratings and tolerance, capacitors working voltage and polarity, inductances direct current resistance or impedance, transformers voltage and current rating of each winding, and switches and connectors voltage and current rating.

17.4.1.5   AN EQUIPMENT LIST — May be shown for reference only. If used, it is not to appear in the same area as a parts list. Except in special cases when it is deemed useful, schematics should omit equipment lists.

17.5 WIRING DIAGRAM

17.5.1 WIRING DIAGRAM — Shows general physical arrangement of the parts; reference designations identical with the schematic except parts in sockets are prefixed with "X"; and lines representing the wiring between parts. It may cover internal or external connections or both and contains the necessary detail to make or trace the connections that are involved. It is a supplementary document to the assembly drawing, containing all wire connection information, but does not have a parts list. The wiring diagram may be shown on the assembly drawing if practical. (See Figure 4)

17.5.1.1 PART REPRESENTATION — Parts on the wiring diagram are shown by a physical outline of the part, suggestive of its appearance, but confined to bare essentials and terminal identification (as viewed from wiring side). (See Figure 4)

17.5.1.2 PART TERMINAL IDENTIFICATION — Wiring diagrams show detailed terminal identification of each part. If the terminals on the part are identified, these identifications are used. If the terminals on the part are not identified, sufficient detail is shown to permit ready identification of all terminals. Where the schematic diagram identifies terminals, the wiring diagram will show identical terminal identification.

17.5.1.2.1 The leads of transistors, diodes, electrolytic capacitors, other semiconductor devices and batteries shall have their identification of polarity indicated on the wiring diagram to aid in terminal identification.

17.5.2 REFERENCE DESIGNATIONS

17.5.2.1 REFERENCE DESIGNATIONS — Placed on the wiring diagram below or along side the component symbol, preferably to the right side.

17.5.2.2 SOCKETS OR FUSE HOLDERS — Identified by a composite designation of letter "X" and the designation identifying the associated part. For example, the socket for electron tube V6 would be identified XV6. Only the designations for the sockets or fuse holders are given and not those for the parts which plug into them.

17.5.3 CONDUCTOR IDENTIFICATION

17.5.3.1 WIRE IDENTIFICATION — Wires are identified on wiring diagrams by one of three methods as explained in the following paragraphs. Only one method will be utilized on a group of drawings prepared for a specific project.

17.5.3.1.1 COLOR CODE IDENTIFICATION — To be in accordance with MIL-STD-681 may be used on either point-to-point or highway type diagrams. System I, II, III, or IV may be selected for use; but once a system has been selected, it will be utilized for all drawings in the set. A note on the drawing will invoke the standard and the system selected. (See Figure 4)

17.5.3.1.2 WIRE NUMBERS — Consisting of sequentially assigned numerals starting with "1", or preferably, sequentially assigned numerals with a suffix letter may be used on either point-to-point or highway type diagrams. If numerals only are used, a new number is assigned to each wire segment regardless of potential. If the preferred method is used, all wires of the same potential (i.e., having common connection) are assigned the same number with a different suffix letter, starting with "A", for each wire segment. The number is changed and a new suffix sequence is started when the circuit passes through an active or passive electrical component. (See Figure 5)

17.5.3.1.3 WIRE NUMBERS (per MIL-W-5088 or MIL-W-8160) — Used on either point-to-point or highway type diagrams. This method provides a significant numbering system identifying circuit function, unit number, etc., and should be utilized only when a contractual or other specific requirement exists. A note on the drawing will invoke this specification when it is used for wire numbering.

17.5.3.2 WIRE NUMBER OR COLOR CODE — Place in a break in the conductor line and the wire destination (indicated by reference designation) is placed to the right and above the wire number or color coding. Wire destinations are required when the highway system is used, and lines going into highways must curve in the direction of the component to which they are routed. All wires including jumpers are identified. (See Figure 4)

17.5.4 WIRE TABLE — A tabulation of all wires including jumpers. It may be included on diagrams but is required only when all wires are being assigned dash numbers of the mechanical assembly drawing. When shown, a wire table lists the wire dash number, wire identification, and wire size as a minimum. (See Figure 4)

17.5.4.1 EQUIPMENT LIST may be shown on a wiring diagram if it conforms to the requirements specified in paragraph 17.4.1.5.

17.6 ELECTRICAL OR ELECTRONIC ASSEMBLY DRAWING

17.6.1 ELECTRICAL OR ELECTRONIC ASSEMBLY DRAWING — Meets all the requirements of an assembly drawing as described in Section 4. In addition, it must also include or reference the schematic and the wiring diagram or wire list in the general notes. (See Figure 6)

17.7 ELECTRICAL OR ELECTRONIC DETAIL DRAWING

17.7.1 ELECTRICAL OR ELECTRONIC DETAIL DRAWING — Meets all the requirements of a mechanical detail drawing as described in Section 4. (See Figure 7)

17.8 CABLE ASSEMBLY DRAWING — An assembly drawing containing all of the necessary information to fabricate a finished cable. The wiring diagram is usually an integral part of the drawing. A parts list is included and reference designations required to be marked on components are shown. Single usage cables have mating connector reference designations marked in parentheses on each connector. (See Figure 8)

17.9 WIRING HARNESS DRAWING — All dimensions necessary are shown to define the harness form and termination points, or grid system, in lieu of dimensions. The drawing shall also include a wire data tabulation of wire numbers, circuit reference designations, color codes, lengths, parts list, and other data, as necessary. Included in note form should be instructions or references thereto, for the preparation and installation of the harness, associated schematic diagram, and the wiring diagram. (See Figure 9)

17.10 INTERCONNECTION DIAGRAM

17.10.1 Units are shown as phantom lined rectangular boxes omitting all internal circuitry. Interconnecting wiring or harness assemblies are shown in wiring diagram fashion. All part subassemblies and wiring harness are identified by referencing their part numbers. (See 17.3.24 and Figure 10.)

17.10.2 The interconnection diagram is acceptable as a final assembly drawing for some contracts. When it is contractually acceptable, the drawing is not a reference document and must meet all other requirements for an assembly drawing.

17.11 PRINTED CIRCUITS AND PRINTED WIRING

17.11.1 The following drawings which comply with MIL-STD-1495 are required to develop and document the design of a circuit card assembly: master artwork, board detail, or master drawing, assembly drawing and schematic diagram. The master drawing and the master pattern drawing may be separate sheets of the same drawing.

17.11.1.1 A layout may be necessary to determine component and circuit positioning, but is not a required drawing. The layout is normally drawn to a 4:1 or 2:1 scale on a stable base material such as matte surface Mylar drawing film. It may be preprinted with appropriate size grid spacing or used with a gridded underlay.

17.11.1.2 Once the location of the components and the routing of the conductors are determined on the layout, a printed circuit/wiring board master drawing is drawn to the same scale and on the same standard grid spacing. It is an accurate drawing of the conductor pattern and shows all the printed conductors, terminal areas, edges of the board, mounting and indexing holes (see Figures 12 and 13) and all conductive areas that will appear on the finished board. Reference designations are to be shown if they are etched.

17.11.2 The master pattern is a precise scale pattern which is used to produce the printed wiring or printed circuit within the accuracy specified on the master drawing. It is a contact reproducible of the original tape master prepared on a stable base material. (See Figure 11)

17.11.2.1 REQUIREMENTS FOR MASTER PATTERN — The master artwork from which the master pattern is made is prepared on stable material using black pressure sensitive tape and pre-cut shapes (MS16912, Figure 16) for the entire conductive pattern must conform with MIL-P-55110 and MIL-STD-275. Identification markings to be etched on the board and information on the drawing other than conductive pattern may be done with ink. Separate views are prepared for each side of double-sided boards. Grids spaced at .100, .050 or .025, either preprinted on the drawing material or on an underlay, are utilized. All terminal areas, mounting holes, test points, etc. are located on grid intersections or are dimensioned from a hole that is on a grid intersection. Board edges are located on grid lines or dimensioned from holes that are on grid lines. Unless otherwise specified on the master drawing (board detail), the accuracy for locating terminal areas on the tape master must be such that the resulting pattern on the printed wiring board is in true position within .010 diameter. Trim marks for the board outline are approximately .125 wide with the inside edge representing the trim line and are normally shown only at the corners. In all cases, regardless of size and board complexity, where the drawings are to comply with DOD-STD-100, two (2) reduction dimensions will be required on the master artwork drawing for the purpose of reproduction to full scale. For further definition of trammel points, see paragraph 17.11.4.1. Three indexing holes or notches are required and are used to locate the board for drilling holes. On double-sided boards the indexing holes or notches must match within .005 inch. Two indexing holes (or notches) are located on a line parallel to the long dimension of the board and the third is located on a line perpendicular to the first line and in line with one of the first two holes. Indexing holes are located as far apart as practical and with different dimensions between holes in the two perpendicular directions.

17.11.2.2 APPLICATION OF TAPE — The pressure sensitive black tape for the conductive pattern is applied without tension to minimize creepage. Tape should be overlapped to avoid gaps which would cause discontinuity in the conductor. Bends 90° or greater with bend radii smaller than those in the tabulation below are to be made with precut radius tape.

Tape Width on Drawing	Recommended Bend Radius for 90° Bends
.125	1.00
.187	2.00
.250	3.00
.3125	4.00
.375	5.00
.4375	6.00

17.11.2.2.1 CONDUCTORS HAVING EXTERIOR AND INTERIOR CORNERS LESS THAN 90 DEGREES INCLUDED ANGLE — All conductors having exterior and interior corners less than 90 degrees included angle shall be rounded and filleted as shown below.

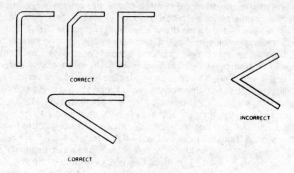

Conductor corners.

17.11.2.2.2 LENGTH OF THE CONDUCTOR BETWEEN VARIOUS TERMINAL AREAS — The length of the conductor between various terminal areas shall be held to a minimum, consistent with other design requirements as shown below.

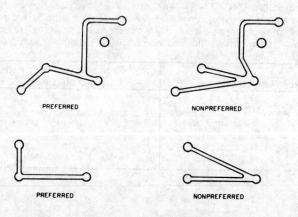

Conductor pattern (shortest distance).

17.11.2.2.3 MINIMUM SPACING — The minimum spacing between conductors, between conductor patterns, and between conductive materials (such as conductive markings or mounting hardware) and conductors shall be in accordance with tables I, II, or III. The minimum spacing between the conductive patterns and the edge of the printed wiring board or the printed wiring board guides shall be 0.025 inch; however, this is not applicable to edge board connectors or ground planes. Larger spacings shall be used whenever possible.

17.11.2.2.4 The minimum spacing between conductors on uncoated printed wiring boards at sea level to altitudes up to and including 10,000 feet shall be in accordance with table I.

TABLE I. *Conductor spacing (uncoated printed wiring boards)*
(sea level through 10,000 feet).

Voltage between conductors DC or AC peak (volts)	Minimum spacing	
0-150	0.025 inch (0.64 mm)	
151-300	0.050 inch (1.27 mm)	
301-500	0.100 inch (2.54 mm)	
Greater than 500	0.0002 inch per volt	(0.00508 mm per volt)

17.11.2.2.5 The minimum spacing between conductors on uncoated printed wiring boards shall be in accordance with table II when the printed wiring boards are subjected to reduced pressure equivalent to altitudes greater than 10,000 feet.

TABLE II. *Conductor spacing (uncoated printed wiring boards)*
(over 10,000 feet).

Voltage between conductors DC or AC peak (volts)	Minimum spacing	
0-50	0.025 inch (0.64 mm)	
51-100	0.060 inch (1.52 mm)	
101-170	0.125 inch (3.18 mm)	
171-250	0.250 inch (6.35 mm)	
251-500	0.500 inch (12.7 mm)	
Greater than 500	0.001 inch per volt	(0.025 mm per volt)

17.11.2.2.6 The minimum spacing between conductors on coated printed wiring boards shall be as indicated in table III.

TABLE III. *Conductor spacing (conformal coated printed wiring boards) (applicable to all altitudes).*

Voltage between conductors DC or AC peak (volts)	Minimum spacing	
0-30	0.010 inch (0.25 mm)	
31-50	0.015 inch (0.38 mm)	
51-100	0.020 inch (0.51 mm)	
101-300	0.030 inch (0.76 mm)	
301-500	0.060 inch (1.52 mm)	
Greater than 500	0.00012 inch per volt	(0.00305 mm per v

17.11.2.3 CARE OF MASTER ARTWORK — After the necessary contact reproducible copies of the tape master have been prepared, the master artwork should be stored flat if it is being retained. The drawing number may be removed and it may be stored for future use in a similar design or for possible revision to the same drawing.

17.11.2.4 CONDUCTOR PATTERNS — Show as viewed from the pattern side. Double-sided boards will be identified as "Side 1" and "Side 2." The reference designation of the assembly, trademarks, assembly part numbers and other desirable information may be shown on the master artwork if it is to be etched on the board. Assembly reference designations, part identification and trademarks are located so they will be visible when assembled in the unit.

17.11.2.5 PLATED THROUGH HOLES — These require a terminal area on both sides of the board. Large conductive areas, required for shielding or other reasons, must be broken up into a gridwork or other pattern that will make the conductive pattern continuous. The maximum unbroken conductor area should be no larger than can be covered by a .500 diameter circle at 1:1. When boards utilize a portion of the conductor pattern as male contacts to mate with a connector receptacle, the contacts must be extended at least .030 inch beyond the board outline and be joined together outside the board outline by one continuous strip of conductor material .125 inch wide at 4:1 to aid in plating.

17.11.2.6 TERMINAL AREA DIAMETER FOR UNSUPPORTED HOLES (not plated through) — A minimum of .040 diameter larger than the diameter of the hole in the terminal area is required. For plated through holes, the terminal area must be a minimum of .020 diameter larger than the diameter of the hole. Larger terminal areas are desirable. It should be kept in mind that after considering all manufacturing tolerances, the minimum acceptable width of the annular ring of conductive material around an unsupported hole is ⁱ5, and for plated through holes, the minimum width is .005.

17.11.2.7 CHANGES TO A MASTER ARTWORK DRAWING — Made only by direct drawing change utilizing a drawing change notice (DCN).

17.11.3 MASTER DRAWING — A master drawing is one that shows the dimensional limits or grid location applicable to any or all parts of a printed circuit or printed wiring, including the base. All permanently attached hardware such as nonremovable terminals, eyelets, soldered clips or sockets, etc. are included on this drawing. Electronic parts and removable hardware are not included. (See Figure 14)

17.11.3.1 GENERAL REQUIREMENTS, MASTER DRAWING — The master drawing in accordance with MIL-STD-275 establishes the size and shape of the board, the size and location of all holes therein, the shape and arrangement of both conductor and nonconductor patterns or elements, with separate views of each side of double sided or multilayer boards. Any pattern features not controlled by hole sizes and locations must be dimensioned. All locations are dimensioned by use of a modular grid system with grid spacing of .100, .050, or .025 inch, in that order of preference. The accuracy of the resulting hole pattern on the printed wiring board shall have all centers located within .005 inch radius of the positional tolerance indicated by the grid location or dimensioned location. Dimensioning and tolerancing practices shall be in accordance with ANSI Y14.5.

17.11.3.2 A master drawing consists of the following: It shows the finished board with all permanently attached hardware. It contains the general notes and a parts list.

17.11.3.3 The master drawing may be prepared from a transparency of the master artwork drawing.

17.11.3.4 Hole locations not shown on the master artwork drawing are specified on the master drawing. Holes may be coded for size with a tabulation showing the size for each code letter. The diameter of an unsupported hole must not exceed the diameter of the lead to be inserted by more than .028 diameter unless the lead is crimped and soldered to the terminal area. The inside diameter of plated through holes, after plating, must not exceed the lead diameter by more than .035. The diameter of holes for eyelets must not exceed the outside diameter of the eyelet by more than .010. There must be a separate hole for each lead, part terminal and end of wire jumper. The number of different size holes should be kept to the minimum consistent with the above. Hole information is normally specified on the conductor side of the board.

17.11.3.5 When component reference designations are required, they are marked or etched on the component side of the board. If they are to be etched, they are shown on the master artwork drawing.

17.11.3.6 Whenever the drawings are to comply with **DOD-STD-100**, workmanship standards and detail requirements for the printed circuit or printed wiring board are specified in the notes on sheet 1. A note similar to the following is recommended: "COMPLETED BOARD TO BE PER MIL-P-55110." Those drawings not required to meet **DOD-STD-100** need not reference the workmanship standard note; however, it may be included at the discretion of the project engineer. Other notes generally required are: (1) a note referencing **DOD-STD-100**; (2) part number marking; (3) reference designation marking; (4) reference to the schematic diagram; and (5) a note stating that hole coding is not to appear on the finished board.

17.11.3.7 The dimension of the grid size used for location is provided by showing a small section of grid work with horizontal and vertical spacing dimensions or the grid spacing is specified in a general note. Overall board dimensions are shown on sheet 1. Other dimensions, as necessary, are shown.

17.11.3.8 The parts list shows the board as -2 and the -1 assembly consists of the -2 board and all permanently attached (not removable) hardware.

17.11.4 TABLES AND DRAFTING PRACTICES FOR ETCHED BOARDS

17.11.4.1 In all cases, regardless of size and board complexity, where the drawings are to comply with DOD-STD-100, two (2) dimensions will be required on the master artwork drawing for the purpose of reproduction to full scale. These will be shown one (1) in the horizontal direction and one (1) in the vertical direction. Those drawings not required to meet DOD-STD-100 may have a minimum of one dimension on the master artwork drawing for the purpose of reproduction to full scale. This dimension may be in the horizontal or vertical position, whichever distance is the greater. All boards, regardless of DOD-STD-100, that are 6'' x 6'' or greater will have two reduction dimensions on master artwork drawing to assure accuracy. The reduction dimension is the full scale distance between the vertical or horizontal cross hairs of the targets.

17.11.4.2 All of the conductive pattern must be dense uniform black with smooth edges, applied with pressure sensitive tape.

17.11.4.3 All letters, numerals, and characters must be dense uniform black with smooth edges, applied with ink, paste-on's, or transfer decals.

17.11.4.4 When standoff terminals are permanently attached to the board, it is recommended that MIL-STD-454, Requirement 5, be invoked by a general note on the master drawing, sheet 1, to cover the requirements for part attachment. The same note should be invoked on electronic assembly drawings to cover the requirements for attaching electronic components to the board or to terminals.

17.11.4.5 The title for the master drawing should be Printed Wiring (or Circuit) Board, as applicable, followed by the noun which describes the function of the completed board, i.e., amplifier, multiplexer, etc. and any required modifiers.

17.11.4.6 The application of tape and drafting practices utilized in the preparation of the master artwork drawing must conform with the requirements of MIL-P-55110 and MIL-STD-275.

17.11.4.7 TABLES APPLICABLE TO PRINTED WIRING (OR CIRCUIT) BOARDS.

Table IV

Maximum Allowable Tolerance, Full Scale,
for Locations on Master Artwork

Feature	True Position Tolerance Zone
Location of any target with respect to another	.005 Dia
Terminal areas with respect to grids	.010 Dia
Mismatch targets and terminal areas, side 1 to side 2	.005 Dia

Table V

Width of Lines and Height of Characters, Full Scale

Characters and Lines	Minimum
Lines (other than circuit)	.015 wide
Component Reference Designations	.062 high
Board and Assy Identification Number	.125 high
Terminal Numbers & Contact Designations	.062 high

17.12 LOGIC DIAGRAMS

17.12.1 These are engineering reference documents and are not used for fabrication. The logic diagram shows by means of logic symbols and supplementary notations the details of signal flow and control. (See Figure 17)

17.12.1.1 Graphic Symbols Drafting Practices

17.12.1.1.1 Symbols shall comply with ANSI Y32.14.

17.12.1.1.2 The position of a symbol on a drawing does not alter its meaning.

17.12.1.1.3 A symbol may be drawn to any proportionate size or line thickness to suit reproduction and commensurate with drawing size.

17.13 DRAWING CALLOUT — GENERAL NOTES

17.13.1   A general note is used to establish application requirements that are not otherwise covered by an existing specification, or other acceptable document, or when space will not permit entry of all necessary information.

17.13.2   When requirements are established by a general note, the note is limited to concise statements.

17.13.3   If lengthy or complex data is required to provide adequate information that is not covered by an existing document, the responsible designer initiates action for the preparation of an appropriate company specification.

17.13.4   GENERAL NOTES — The following notes illustrate the type of general note coverage intended to be included on the engineering drawing. They may be varied to suit individual requirements. (See Section 7)

17.13.4.1   GENERAL APPLICATION NOTE

⚠ PERMISSIBLE TO USE (alternate part number and code identification number, when required) IN PLACE OF (parts list part number).

17.13.4.2   ELECTRICAL NOTES

MARK PER _____ WITH .12 HIGH REFERENCE DESIGNATIONS LOCATED APPROXIMATELY AS SHOWN.

MAKE SILK SCREEN USING REPRODUCIBLES PER MIL-D-8510 TYPE _____ .

THIS DRAWING TO BE USED IN CONJUNCTION WITH

   WIRING DIAGRAM _____
   SCHEMATIC DIAGRAM _____
   ASSEMBLY DRAWING _____
   INTERCONNECTION DIAGRAM _____

SOLDER CONNECTIONS PER _____

WIRE NUMBER CODING PER _____

ELECTRICAL BONDING PER _____

WIRE COLOR CODING PER _____

CONNECTORS TO BE CLOSED DURING HANDLING AND STORAGE.

REMOVE DUST CAP ONLY AS NECESSARY.

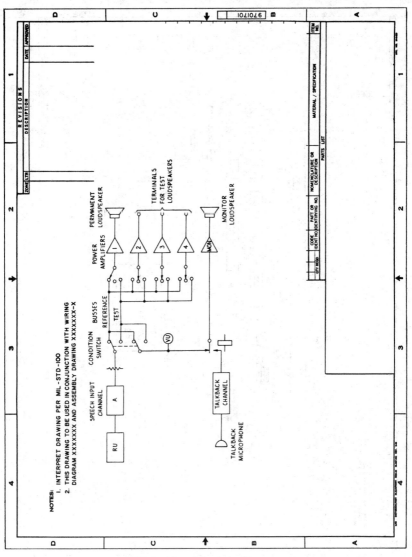

Single Line Schematic Diagram
Figure 1

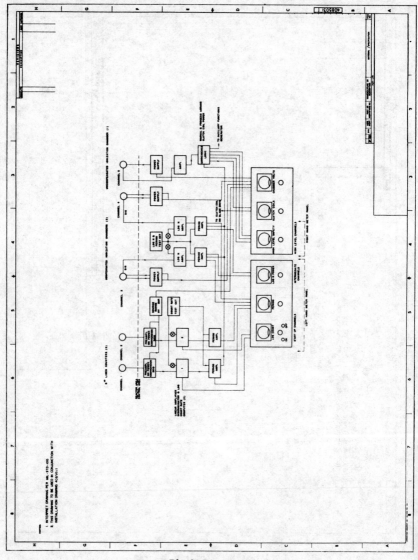

Block Diagram
Figure 2

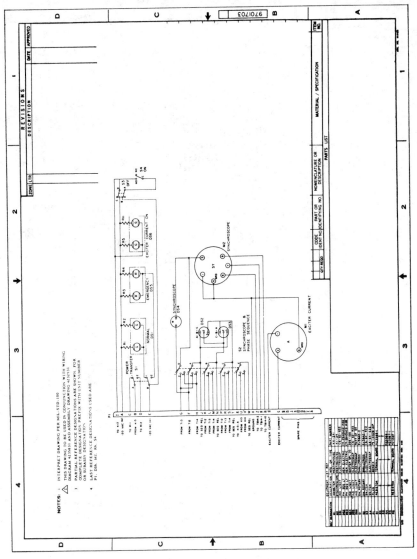

Schematic Diagram
Figure 3

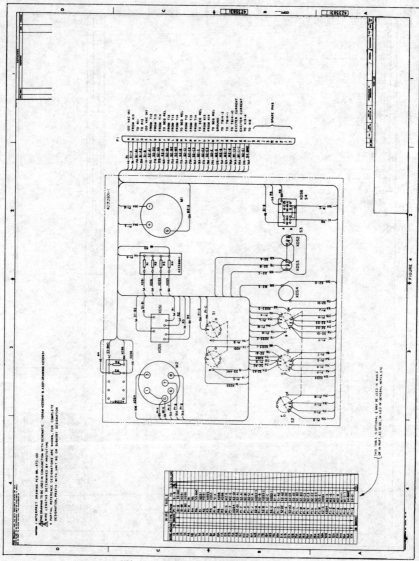

Wiring Diagram - Highway Type
Figure 4

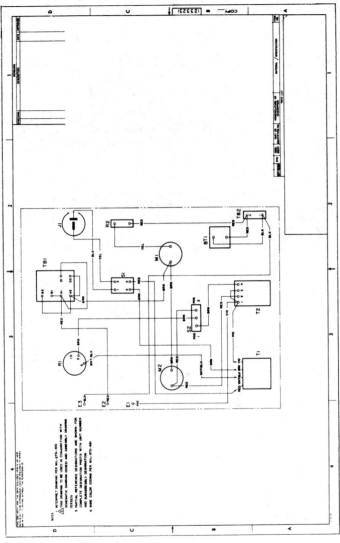

Wiring Diagram - Point-to-Point
Figure 5

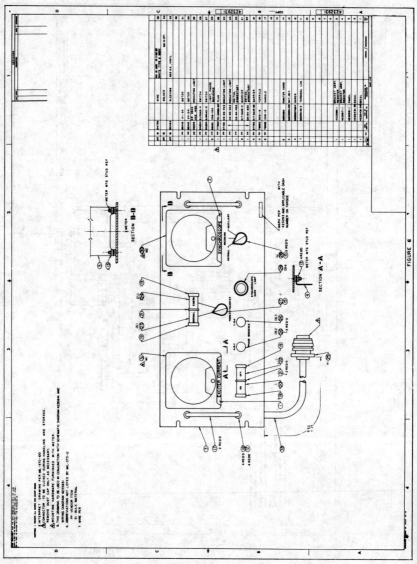

Electronic Assembly Drawing
Figure 6

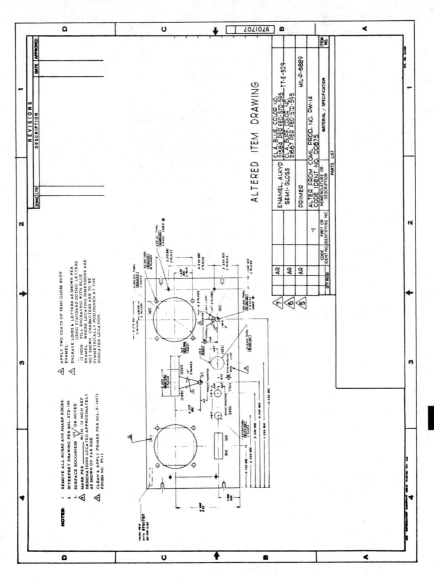

Electronic Detail Drawing
Figure 7

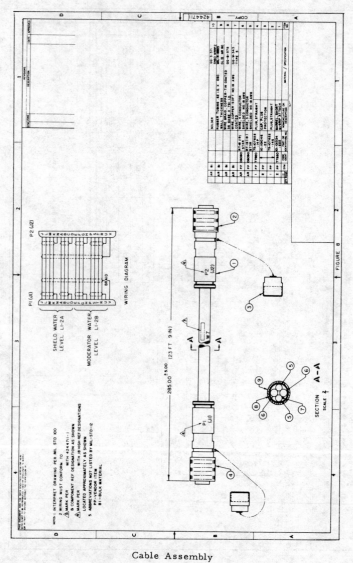

Cable Assembly
Figure 8

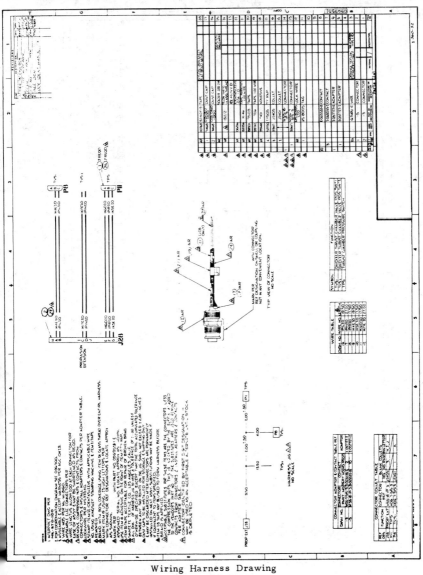

Wiring Harness Drawing
Figure 9

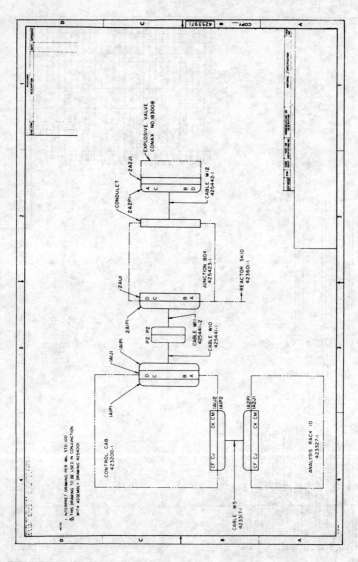

Interconnection Diagram Drawing
Figure 10

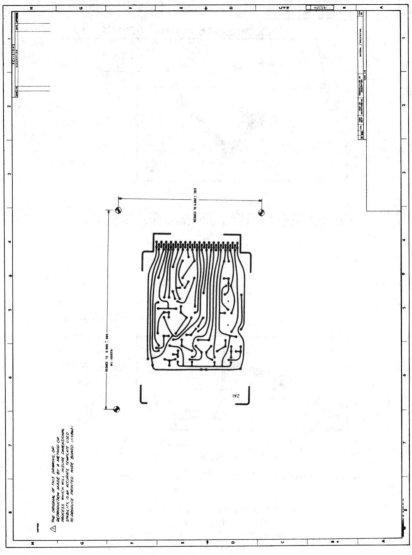

Master Pattern Drawing
Figure 11

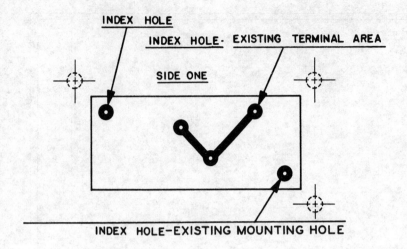

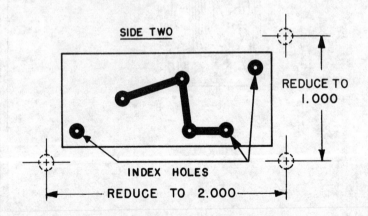

Two-Sided Printed Wiring Board,
Method of Registering
Using Index Holes Within Board Area
Figure 12

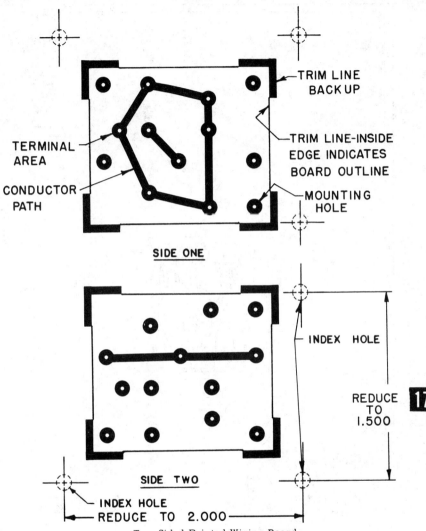

Two-Sided Printed Wiring Board
Method of Registering
With Index Holes Outside Board Area
Figure 13

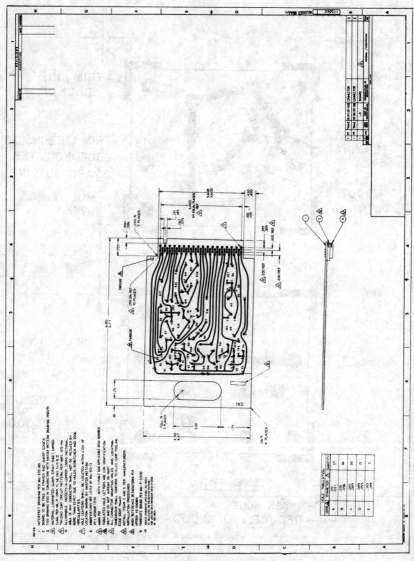

Master Drawing (Board Detail)
Figure 14

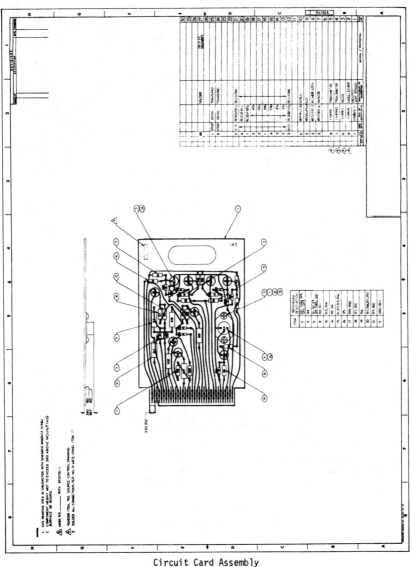

Circuit Card Assembly
Figure 15

PREFERRED   NON-PREFERRED

Areas on each   The radius
side of dotted   R should be
line should be   from two to   The solder fillet will
approximately   three times   be non-symmetrical.
equal.   the hole
                       diameter.

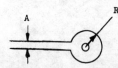

                                         Solder will
                                         flow away
                                         from fillet.

The dimension 'A' shall be 1/3
to 2/3 of 'R' (.040 inch min.)

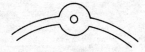   

Uniform conductor pattern
around hole.   Solder fillet will
                                         be non-symmetrical.

Terminal Area Patterns for Soldered Connections
Figure 16

PREFERRED   NON-PREFERRED

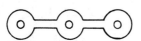   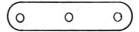

All solder fillets
will be symmetrical.

Outside solder fillets
will be non-symmetrical.

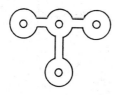

All solder fillets
will be good.

Conductor in center hole
will pull solder from
outside conductors

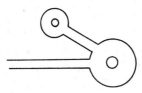   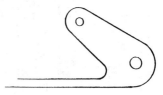

Solder will bleed
(flow) evenly.

Solder will bleed (flow)
toward large hole.

Figure 16 (continued)

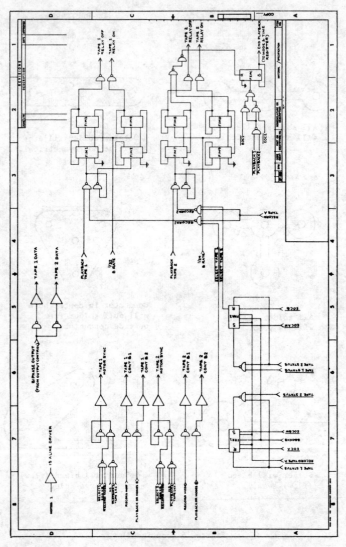

Logic Diagram
Figure 17

NOTES

NOTES

Section 18 — DRAWING CHANGE PROCEDURE

18.1 SCOPE — This section establishes the methods of revising engineering drawings. These procedures apply only to drawings that have been released.

18.2 APPLICABLE DOCUMENTS
        DOD-STD-100   Engineering Drawing Practices
        ANSI Y14.5   Dimensions and Tolerancing

18.3 DEFINITIONS (Applicable to drawing change procedures)

18.3.1 ADVANCE DOCUMENT CHANGE NOTICE (ADCN) is an advance notice of approved revisions that will be incorporated on the drawing at a later date.

18.3.2 DIRECT DRAWING CHANGE — A change incorporated on to the original vellum at the time it is authorized and recorded on a DCN.

18.3.3 DOCUMENT CHANGE NOTICE (DCN) is a record of revisions that have been incorporated on a drawing, and the authorization for those changes.

18.3.4 DOCUMENT CHANGE REQUEST is a document used to officially request drawing changes.

18.3.5 DRAWING CHANGES are additions, deletions and/or revisions to any information appearing on the drawing.

18.3.6 CHANGE OF DRAWING STATUS

18.3.6.1 INACTIVATE A DRAWING–The process of removing a drawing from active status. Inactive drawings may be those no longer used or those superseded by other documents, and are not used for fabrication. Vellums or microfilm records are retained and may be reactivated for use when required.

18.3.6.2 OBSOLETE A DRAWING — An obsolete drawing is one which has been replaced by another drawing of the same number, and shall be conspicuously marked: 'OBSOLETE—REDRAWN WITH (OR WITHOUT) CHANGE'. The note shall be located in the general area of the title block, on sheet 1 only of a book form drawing and on every sheet of multi-sheet drawings with 0.25 inch high characters. The replacement drawing shall be assigned the next revision letter and all previous entries in the Revision Block shall be removed and 'REDRAWN WITH CHANGE' entered, followed by a desription of the changes.

18.3.6.3 CANCELLED DRAWINGS — A cancelled drawing is one which has been removed from the system and the number retired, never to be used again. The drawing shall be conspicuously marked: 'CANCELLED—REPLACED BY _____ or NOT REPLACED' with 0.25 inch high characters located in the general area of the title block, on sheet 1 only of a book form drawing and on every sheet of multi-sheet drawings. A new change letter is required to cancel a drawing using the next letter available.

18.3.7 INCORPORATE AN ADCN is to transfer information from a released ADCN to a drawing.

18.3.8 INTERCHANGEABILITY is a condition which permits the direct interchange or substitution of a component for another without modification or selection of the component. Upon interchange of components, the same physical and functional characteristics of the original components are assured. (See Figure 1)

18.3.8.1 FUNCTIONALLY INTERCHANGEABLE PARTS are items equivalent in characteristics of operation, performance, durability, serviceability, structural strength, material, protective finish and safety.

18.3.8.2 PHYSICALLY INTERCHANGEABLE PARTS are those equivalent parts capable of being installed, removed or replaced without sustaining damage, or causing misalignment, damage to, or interference with adjoining parts or portions of the complete assembly.

18.3.9 PRE-RELEASED ADCN is the same as an ADCN but requires initially only the approval of the responsible design activity and is used for a more rapid release of changes where an emergency condition exists and for a specific period.

18.4 GENERAL CHANGE PROCEDURES

18.4.1 REVISION METHODS — Revisions shall be made by erasure, crossing out, addition of information, or by redrawing.

18.4.2 CHANGE IN DIMENSIONS — In general, any change in a dimension of a part should be made to scale. However, it is permissible to leave the delineation of the part unchanged when the new proportion of the part is not noticeably different from the original one. If the change to scale is not made, the practice outlined in ANSI Y14.5 shall be followed. If the drawing is redrawn, delineation shall be made to scale.

18.4.3 CHANGE AUTHORIZATION AND FORMAT — The preprinted ADCN-DCN form is used to authorize and record drawing changes. The form is prepared per Figure 2 and the applicable change condition in Figure 6. When additional sheets are required, use the drawing change notice continuation sheet shown in Figure 3. Typing or legible hand lettering is permissible.

18.4.3.1 CHANGE REQUEST — The responsible engineer has the primary responsibility for initiating document change requests or for approving document change requests originated outside the project. On receipt of an approved document change request and/or drafting work request, the drafting organization determines the correct procedure for implementing the change. (See 18.6.1 and 18.7.1)

18. 4. 4 REVISION IDENTIFICATION — The revision status of a drawing is identified by an upper case letter. The first change issued (ADCN or DCN) is identified by the letter "A". Each successive change uses the next letter of the alphabet in sequence, except the letters "I", "O", "Q", "S", "X", and "Z", are never used. Upon exhaustion of the alphabet, revisions are identified by letters "AA", "AB", "AC", etc., then "BA", "BB", etc. Revision letters shall not exceed two characters. Initial release of a drawing does not constitute need for a revision letter and the revision block shall be left blank.

18.4.5 APPROVAL SIGNATURES — Approval signatures on a DCN or ADCN are equal to those on the drawing affected by the change except as noted in Figure 6.

18.4.6 LIMITATION ON DRAWINGS AFFECTED — An ADCN-DCN is issued against one drawing only.

18.4.7 INCORPORATION OF REVISIONS ON DRAWINGS — All revisions incorporated on drawings are authorized by an ADCN-DCN unless otherwise specifically excepted in this section. For exceptions, see paragraph 18.4.8 and Item 13 of Figure 6.

18.4.8 UNDOCUMENTED CHANGES — The following drawing corrections or additions may be made without recording on an ADCN or DCN: similar to data, required approval signatures (except checked), weight information, misspelled words, missing arrowheads, next assembly. Such changes are normally made when incorporating other authorized changes, and will be checked by an authorized checker.

18.4.9 DOCUMENTED CHANGES — Each change on an ADCN-DCN is assigned an item number and is referenced to its drawing location by a zone symbol or other applicable reference, i.e., P/L, T/B, G/N, etc. Changes made on the field of the drawing are not identified by revision symbols. Where several changes are contained within a small area of one zone, they may be recorded under one item number. (See Figure 4)

18.4.10 CHANGES IN GENERAL NOTES — A general note may be changed if the original purpose is maintained. Example: Note 5 is for sandblasting. The note contents may be changed providing its general intent "Sandblasting" is maintained. When the old note is removed, the note number cannot be reused for a "Heat Treat" note.

18.4.10.1 REMOVING GENERAL NOTES — When a general note has been removed, the word "REMOVED" shall appear in its place. (See Figure 4)

18.4.10.2 RELISTING GENERAL NOTES — On drawings that have been revised, redrawn, etc. the general notes are listed in their original numerical order.

18.4.11 CHANGES TO SECURITY CLASSIFIED DRAWINGS — When an ADCN-DCN affects a classified drawing, and the data contained on the ACDN-DCN is unclassified, the statement below appears on the ACDN-DCN.

This document is downgraded to Unclassified when detached from parent drawing.

The appropriate security classification is placed at the top and bottom of the ADCN-DCN. See Figure 2.

18.4.12 SECURITY CLASSIFICATION REVISION — Security classification revisions, upgrading or downgrading, are made by a direct drawing change only. When the security classification is revised, the existing classification is not removed from the drawing, but is lined out, and the new classification added (if any). (See Figure 5) When redrawing drawings with previously revised security classifications, or downgrading on redraw, the old lined out classification is not repeated on the new drawing. (See Figure 6, Item 12)

18.4.13 REVISION OF DESIGN ACTIVITY CODE IDENTIFICATION NUMBER — When design responsibility is transferred from one design activity to another, the drawing number and original design activity code identification will not be changed. The new design activity may add their code identification to the drawing near the title block: "DESIGN ACTIVITY TRANSFERRED TO (Code Ident.No.) (Date)."

18.4.14 REVISING SEPARATE PARTS LIST — When the parts list is separate from the drawing, it is revised in the same manner as a drawing. There is no requirement for a drawing and its separate parts list to have the same change letter. See Section 9.

18.4.14.1 REMOVING DASH NUMBERED ITEMS — When a part identified by a dash number is removed from a drawing or parts list, the word "REMOVED" is inserted in the description column. The dash number is left in the part number column and the item number is left in the item number column (if item numbers were listed) and all other information on that line is deleted.

18.4.15 REVISION DATE — Dates shown in revision blocks shall be specified numerically — year, month, day (e.g., 75-04-14).

18.4.16 REVISION BLOCK

18.4.16.1 CONTENTS — The revision block contains the change letter in the LETTER COLUMN, DESCRIPTION and DATE revision was released. The zone column may be left blank, since zones are recorded on the DCN. See Figure 4.

18.4.16.2 APPROVAL SIGNATURES — The approval signature block will be signed by an authorized drawing checker. See Figure 6.

18.4.16.3 DESCRIPTION IN REVISION BLOCK — The statement placed in the description column shall be selected from those samples listed in Figure 6.

18.4.17 PROCESSING REVISED DRAWINGS — All revised drawings are resubmitted for approval signatures and necessary reproduction in accordance with individual Division release procedures.

18.5 REVISION TO MULTI-SHEET DRAWINGS — Concurrent changes affecting any or all sheets are identified by the same revision letter on all affected sheets and on sheet 1. The sequence of revision letters applies to the drawing as a whole and not to individual sheets.

18.5.1 REVISION BLOCK ENTRIES FOR MULTI-SHEET DRAWINGS — When a change is made on any or all sheets, the latest change letter is entered in the change letter column of all affected sheets and on sheet 1 whether it is affected or not. The revision letter does not advance for any sheets not affected by the change. Changes are recorded in the revision description column of each affected sheet in accordance with the applicable change condition in Figure 7.

18.5.2 REVISION STATUS BLOCK FOR MULTI-SHEET DRAWINGS — Sheet 1 of multi-sheet drawings will include a tabulation, as shown in Section 5, to indicate the revision status of each sheet in the group. When a change is made on any or all sheets in the group, the latest change letter is entered in the revision status block for all sheets affected and for sheet 1 whether it is affected by the change or not. The revision letter of sheets not affected by the change does not advance, but the last applicable revision letter is entered in the appropriate column of the revision status block.

18.6 ADVANCE DOCUMENT CHANGE NOTICE (ADCN)

18.6.1 GENERAL PROCEDURES — Upon receipt of an approved Document Change Request, the drafting group will prepare and issue an ADCN or a DCN, as applicable.

18.6.2 MAXIMUM NUMBER ALLOWED — The maximum number of ADCN's permitted to accumulate prior to incorporation is five (5). Multi-sheet ADCNs are limited to a maximum of two (2) sheets. All outstanding ADCNs are incorporated prior to final submittal of the drawing to the customer.

18.6.3 DESCRIPTION OF REVISION — A brief description of each change is written in a manner that will provide adequate information for incorporation at a later date, i.e., .250 WAS .375; REMOVED G/N 1 WAS "REMOVE ALL BURRS AND SHARP EDGES"; ADDED G/N 6 "DIMENSION APPLY AFTER PLATING".

18.6.4 CHANGES TO RELEASED ADCNs — After release an ADCN is considered an integral part of the subject drawing, and therefore cannot be cancelled, but may be revised in part or in entirety by another ADCN or DCN. Each item of an ADCN or DCN that revises an item on a previously released but unincorporated ADCN references the previous ADCN and the specific item number. This is accomplished by stating "REF ITEM_____OF _____CHANGE ADCN" at the end of the new item's revision description. When the "WAS" condition for the superseding ADCN or DCN is shown, it will be the "IS" condition from the superseded ADCN.

18.6.5 PROCESSING COMPLETED ADCNs — The ADCN original is submitted for check with a copy of the latest change drawing and all outstanding, unincorporated ADCNs. After receiving all required approvals, the original is processed through the release group and a copy is attached to all prints of the subject drawing. After an ADCN is incorporated, the original or microfilm is retained as a record of the change.

18.6.6 PRERELEASING AN ADCN — When a critical time element exists, the prerelease of an ADCN may be initiated by the design activity to authorize a change, pending release of the fully approved ADCN. A prereleased ADCN is supplied in a composite form.

18.6.6.1 APPROVAL SIGNATURES — All procedures governing the use of an ADCN apply with the exception of the initially required approval signatures. Only the responsible design activity approval is required to authorize implementation.

18.6.6.2 PRERELEASED COPY — The prereleased (hard) copy is used as the working document, remaining in effect until the specified "Void Date", or until replaced by the fully approved ADCN. The period of effectivity shall be established by each Division. Changes to a prereleased copy are accomplished only by a superseding DCN, ADCN, or another prereleased ADCN.

18.6.6.3 ORIGINAL COPY — The original ADCN is immediately processed by the responsible engineer. Release and distribution of the fully approved ADCN, without revision, is mandatory, and automatically supersedes the prereleased copy. Minor changes that do not affect fabrication or design are permitted.

18.6.6.4 IDENTIFICATION — The original ADCN and all copies are identified with the identical drawing number and change letter.

18.7 DOCUMENT CHANGE NOTICE (DCN)

18.7.1 GENERAL PROCEDURES — Upon receipt of an approved Document Change Request, the drafting group will prepare and issue a DCN or ADCN, as applicable. When a DCN is prepared, all outstanding ADCNs are incorporated.

18.7.2 DESCRIPTION OF REVISION — A brief description of each change is written, i.e., REMOVED G/N 1 was REMOVE BURRS AND SHARP EDGES; ADDED G/N 6; .250 WAS .375, etc. It is not necessary to show the "IS" condition for a change on a DCN.

18.7.3 PROCESSING COMPLETED DCNs — The DCN original, when completely incorporated, is submitted with the drawing vellum, a print of the drawing prior to change, and a print of each outstanding ADCN for check. The checked DCN original becomes the "Change Record" for the related drawing change, and is processed in accordance with individual Division procedures.

18.8 DOCUMENT CHANGE REQUEST

18.8.1 GENERAL PROCEDURE — The Document Change Request form (see Figure 8) is used to request drawing changes. It may be initiated by personnel outside the project or within the project.

18.8.2 PROCESSING — After a document change request has been prepared, the original is forwarded to the responsible design activity for action. The design activity completes the applicable disposition block, etc., and returns a copy to the person requesting the change. The original is delivered to the drafting group for implementation.

TYPES OF CHANGES	CHANGE DOCUMENTS	
	ADCN OR DCN	DOCUMENT CHANGE REQUEST
Within limits of physical and functional interchangeability	Retain same drawing number and suffix number	Constitutes a request only until approved by the responsible design activity. May be initiated by personnel outside or within the design activity. When approved, it is implemented by drafting with the issue of an ADCN or DCN.
Functional interchangeability affected.	Retain same drawing number with a new suffix number, or a new drawing number may be issued.	
Physical interchangeability affected.	Assign a new drawing number. (On tabulated drawings, a new suffix number may be assigned.)	

Item Identification and Change Document Usage

Figure 1

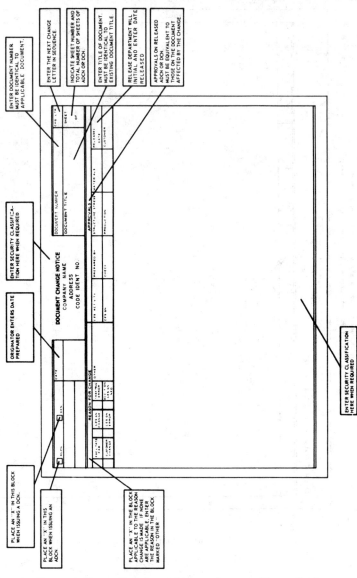

Document Change Notice and Advance Document Change Notice
Figure 2

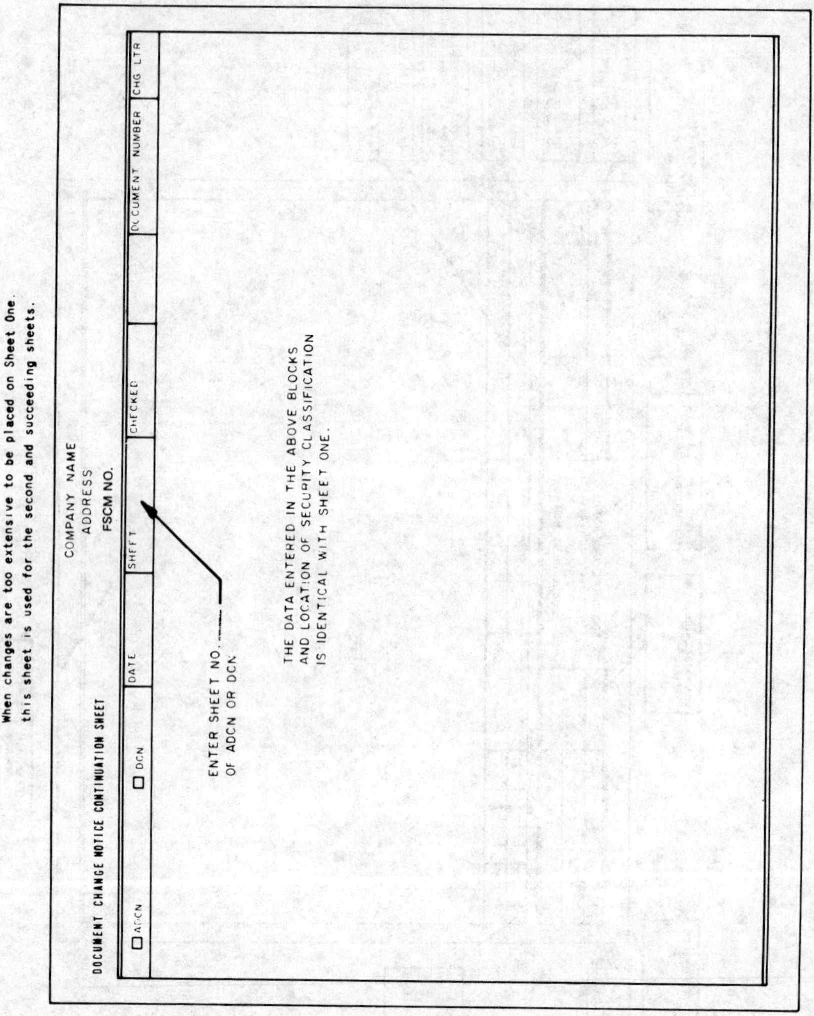

When changes are too extensive to be placed on Sheet One, this sheet is used for the second and succeeding sheets.

DOCUMENT CHANGE NOTICE CONTINUATION SHEET			COMPANY NAME ADDRESS FSCM NO.		
☐ ADCN	☐ DCN	DATE	SHEET	CHECKED	DOCUMENT NUMBER	CHG LTR

ENTER SHEET NO.
OF ADCN OR DCN

THE DATA ENTERED IN THE ABOVE BLOCKS
AND LOCATION OF SECURITY CLASSIFICATION
IS IDENTICAL WITH SHEET ONE.

Document Change Notice Continuation Sheet

Figure 3

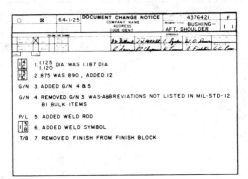

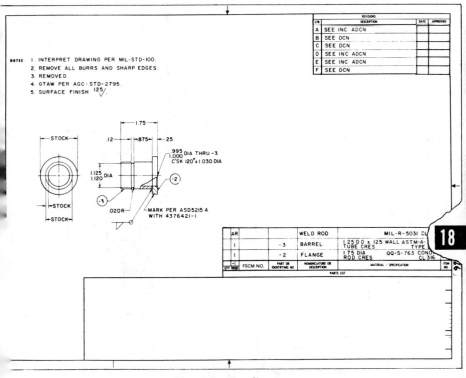

Direct Drawing Change
Figure 4

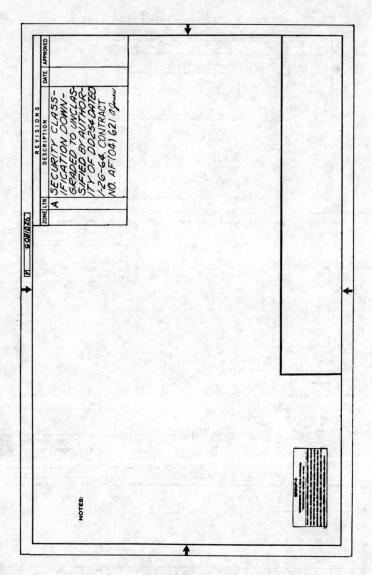

Change in Drawing Security Classifications

Figure 5

18-10

This figure simulates the revision block of an actual drawing except in some examples the same entries are alternates for the previous entry (or entries)

Release Group enters release date of revised drawing & on DCN when applicable

Authorized checker signs & enters date of signature in approval column & on DCN when applicable

ITEM	CHANGE CONDITION	REVISION BLOCK			REMARKS
		LTR	DESCRIPTION	DATE / APPROVED	
1	Incorporating ADCNs	C	SEE INC A, B & C ADCN		All changes authorized by the released ADCN(s) are incorporated. If any approved changes are not incorporated or if additional changes are made a DCN must be issued, see item 3.
2	Direct drawing change without outstanding ADCNs	D	SEE D DCN		Record all changes made to the drawing except as noted in 18.4.6.
3	Direct drawing change with outstanding ADCNs	H	SEE INC E, F & G ADCN & H DCN		All outstanding ADCNs must be incorporated at this time. Any ADCN items that are changed or not incorporated must be noted on the DCN. (See 18.6.4).
4	Drawing redrawn or reproduced without change, no outstanding ADCNs	-	Old drawing REPLACED WITHOUT CHANGE BY REV L		See footnote 1, 2 & 3. A DCN is not issued and the revision letter does not advance.
		L	New drawing REPLACES REV L DATED 5-21-68 (enter old revision letter & date old revision was released) WITHOUT CHANGE		
5	Drawing redrawn or reproduced without change, but outstanding ADCNs incorporated	-	Old drawing REPLACED WITHOUT CHANGE BY REV N		See footnote 1, 2 & 3 A DCN is not issued and the revision letter on the new drawing advances to the latest outstanding ADCN. The incorporation of ADCNs is not considered additional changes.
		N	New drawing REPLACES REV L WITHOUT CHANGE. SEE INC M & N ADCN		
6	Drawing redrawn or reproduced with change, no outstanding ADCNs	-	Old drawing REPLACED WITH CHANGE BY REV M		See footnote 1, 2 & 3 A DCN is issued in accordance with Item 2.
		M	New drawing REPLACES REV L WITH CHANGE. SEE M DCN		
7	Drawing redrawn or reproduced with change and outstanding ADCNs incorporated	-	Old drawing REPLACED WITH CHANGE BY REV P		See footnote 1, 2 & 3 All outstanding ADCNs must be incorporated at this time. Any ADCN items that are changed or not incorporated must be noted on the DCN (See 18.6.4)
		P	New drawing REPLACES REV L WITH CHANGE. SEE INC M & N ADCN & P DCN		

1. New approval dates and signatures equivalent to the replaced dwg are required in the T/B of the redrawn or reproduced dwg. The redrawing draftsman letters his name and the date the redraw was started in the T/B and enters the original release date in the T/B release date block. Previously lined out security classifications and previous revision history are not carried forward to the new drawing. If a security classification change occurs simultaneous with redraw, both drawings, will carry the statement shown in item 13 when applicable to both drawings. The new drawing will carry only the latest classification.
2. Print the word "SUPERSEDED" in .25 high letters near the title block on the old drawing. (The old drawing original may be destroyed after microfilming)
3. If the original vellum of the old drawing has been lost, the DCN or revision block on the drawing will state "ORIGINAL VELLUM LOST."
4. When the drawing has not been signed by any of the drawing review organizations or when signed by Check only, the DCN may be omitted and the revision block statement "SEE _____ DCN" replaced with the applicable DCN statement.

Revision Block Description
Figure 6

ITEM	CHANGE CONDITION	REVISION BLOCK				REMARKS
		LTR	DESCRIPTION	DATE	APPROVED	
8	Inactivate a drawing, no superseding document, no outstanding ADCNs	R	SEE R DCN			The drawing advances to the next change letter and a DCN is issued stating "INACTIVE FOR FUTURE DESIGN." This same statement is placed near the T/B in .25 high letters. The drawing original may be destroyed after release and microfilming. See footnote 4.
9	Inactivate a drawing, no superseding document, outstanding ADCNs incorporated or not incorporated	T	ADCNs incorporated SEE INC R & S ADCN & T DCN			
		T	ADCNs not incorporated SEE T DCN. R & S ADCN RETAINED FOR RECORD & NOT INC			
10	Superseded drawing		Use applicable entry of item 9 or 10			The statements in the "REMARKS" column of item 9 & 10 apply except replace "INACTIVE FOR FUTURE DESIGN" with INACTIVE FOR FUTURE DESIGN, SUPERSEDED BY _____ (enter document no.)"
11	Reinstating an inactivated drawing	U	ADCNs previously incorporated SEE U DCN			The drawing advances to the next change letter and a DCN is issued stating "DRAWING REACTIVATED". The inactivation statement near the T/B is removed and all unincorporated ADCNs are incorporated. See footnote 4.
		U	ADCNs not previously incorporated SEE INC R & S ADCN & U DCN			
12	Security Classification change	V	SECURITY CLASSIFIC-ATION DOWNGRADED (or UPGRADED) TO (new security classification) BY AUTHORITY OF (see remarks column) DATED (date of authorizing document) CONTRACT (applicable contract number number or PROJECT designation) (Signature of authorized classifying officer) a. jones			The drawing advances to the next change letter, but a DCN is not issued. If a DCN is issued for additional changes, the statement "REVISED SECURITY CLASSIFICATION, SEE REVISION BLOCK" is included on the DCN. The authority may be a "DD254" (Security Requirements Check List) or a Company or Customer project letter (to implement interim action). Existing classifications on the drawing are lined out and the new classification, if any, added. See footnote 1 if reclassification occurs simultaneous with redraw.
13	Transfer of drawings from outside Company Facilities	Z	SEE Z DCN			Prior to accepting these drawings, a letter from the old design activity must be on file with the Release Group stating that the design activity has been transferred to Aerojet-General Corp. The drawings and any outstanding change notices are microfilmed prior to making any changes. After microfilming, the drawing advances to the next change letter and a DCN is issued listing all changes. The statement "DESIGN ACTIVITY TRANSFERRED TO _____ (enter new code ident no.)" will appear on the DCN and near the drawing T/B. Any other changes required to ensure the drawings meet the minimum Facility and/or contract requirements shall be

Figure 6 (continued)

ITEM	CHANGE CONDITION	REVISION BLOCK				REMARKS
		LTR	DESCRIPTION	DATE	APPROVED	
13	Transfer of drawings from outside Company Facilities (continued)					made at this time, including the incorporation of all outstanding change notices, the removal of any old design activity restrictive markings and the addition of the appropriate Company legend. The drawing is checked by all applicable drawing review organizations and their signatures on the DCN signify drawing acceptance. All old design activity specifications, standards, etc, referenced on the drawings must be on file with the Release Group and, wherever allowed by the old design activity, have the activity transfer statement added. If the old design activity makes the above changes prior to shipment, the drawings shall be microfilmed and prints sent to all applicable drawing review organizations. If the drawings meet the minimum requirements, no change will be required.
14	Transfer of drawings from one Company Facility to another or to another Company	AA	Example 1 SEE INC Z ADCN & AA DCN			Preferred Method The old design activity shall incorporate all outstanding ADCNs and issue a DCN. The statement "DESIGN ACTIVITY TRANSFERRED TO ____ (enter new code ident no.)" will appear on the DCN and near the drawing T/B, (See Example 1). The DCN may be omitted if the proper authority has issued written instructions listing the drawing numbers to be transferred, (See Example 2). The drawings and DCNs will be microfilmed prior to shipment of the drawings. The new design activity shall remicrofilm the drawings and submit prints to all drawing review organizations who will ensure the drawings meet the minimum Facility and/or contract requirements. Existing signatures will be honored. However, if additional signatures are required and there are no other changes, the statement "ADDED ADDITIONAL SIGNATURES" shall be entered in the revision block and the drawing shall advance to the next change letter. Alternate Method When approved by the new design activity, the old design activity may leave active ADCNs outstanding (See Example 3) or, when necessary, may transfer the drawings and outstanding ADCNs without any changes. However, if the drawings are transferred without any change, the new design activity must make the changes in accordance with Example 1 or 2 immediately after receipt and microfilming.
		AA	Example 2 SEE INC Z ADCN DESIGN ACTIVITY TRANSFERRED TO ____ (enter new design activity code ident no.)			
		AA	Example 3 DESIGN ACTIVITY TRANSFERRED TO ____ (enter new design activity code ident no.). Z ADCN OUTSTANDING			
15	Changes to Tooling drawings.	A	Preferred Method SEE A DCN (or SEE INC A ADCN)			When the drawing has approval signatures other than Design, Design Activity or Check, a DCN (or ADCN) must be issued recording the changes. The ADCN or DCN must have signatures equivalent to those on the affected drawing
		A	Alternate Methods 1. .50 WAS .38 2. ADDED NOTE 4 *a. jones* or			When the drawing has no approval signatures other than Design, Design Activity or Check, the responsible engineer has the option of recording the changes in the revision block or adding the statement shown. In either case the drawing advances to the next change letter and the engineer signs at the end of the revision block notation. The "APPROVAL" column is signed by an authorized checker only if the drawing has been checked.
		A	DWG REVISED. RECORD ONLY *a. jones*			

Figure 6 (continued)

Note: This figure simulates the revision blocks and "REVISION STATUS OF SHEETS" block of an actual multisheet drawing except some entries are alternates for the previous entry (or entries).

ITEM	CHANGE CONDITION AND CHANGE NOTICE (S)	DWG SH NO.	ADDITIONS AND CHANGES TO REVISION BLOCKS				REMARKS
1.	Incorporating ADCN(s) & DCN A Change ADCN ZONE ITEM [1][8/H] 1. REMOVED G/N; WAS "SANDBLAST PER___." [1][3/B] 2. 3.750 DIA WAS 3.870 DIA [2][3/A] 3. .125 WAS .125 .135 WAS .130 [4][6/C] 4. .125 R WAS .250 R ZONE ITEM B CHANGE DCN [4][3/C] 1. 3.40 WAS 3.25 [4][8/D] 2. 3.40 REF WAS 3.25 REF	1	LTR	DESCRIPTION		DATE	APPROVED	See Figure 6, items 1 thru 4 for similar changes and the requirements for the changes. See footnotes 1, 2 & 3.
			B	SEE INC A ADCN & B DCN				
			B - A B REV REV STATUS 4 3 2 1 SHEET OF SHEETS					
		2	LTR	DESCRIPTION		DATE	APPROVED	
			A					
		4	LTR	DESCRIPTION		DATE	APPROVED	
			B					
2.	Sheet(s), except sheet 1, redrawn or reproduced without change, no outstanding ADCNs	1	LTR	DESCRIPTION		DATE	APPROVED	The requirements of Figure 6, item 5 apply, except new approval signatures are not required. The slash in the "DATE" column of the old sheet(s) is replaced with the date for the revision letter shown. (The date can be obtained from sheet 1) The "REV STATUS OF SHEETS" block is not altered. See footnotes 1 & 3 A DCN is not issued
			B	SEE INC A ADCN & B DCN				
			-	SHEET 2 REPLACES REV A DATED 1-5-68 WITHOUT CHANGE				
			Old sheet 2					
		2	LTR	DESCRIPTION		DATE	APPROVED	
			A					
			-	REPLACED BY REV A WITHOUT CHANGE				
			New sheet 2					
			LTR	DESCRIPTION		DATE	APPROVED	
			A					
3.	Sheet(s), except sheet 1, redrawn or reproduced without change, but outstanding ADCN(s) incorporated. ZONE ITEM C Change ADCN [4][5/D] 1. .750 WAS .812 [4][6/B] 2. 3.00 WAS 3.12	1	LTR	DESCRIPTION		DATE	APPROVED	The requirements of Figure 6, item 6 apply, except new approval signatures are not required. The slash in the "DATE" column of the old sheet(s) is replaced with the date for the revision letter shown. (The date may be obtained from sheet 1). The "REV STATUS OF SHEETS" block is updated to reflect the latest change letter of each revised sheet. See footnotes 1, 2 & 3.
			B	SEE INC A ADCN & B DCN				
			C	SHEET 2 REPLACES REV A DATED 1-5-68 WITHOUT CHANGE. SHEET 4 REPLACES REV B WITHOUT CHANGE. SEE INC C ADCN				
			C - A C REV REV STATUS 4 3 2 1 SHEET OF SHEETS					
		2	See item 2					
			Old sheet 4					A DCN is not issued
		4	LTR	DESCRIPTION		DATE	APPROVED	
			B					
			-	REPLACED WITHOUT CHANGE BY REV C				
			New sheet 4					
			LTR	DESCRIPTION		DATE	APPROVED	
			C					

1. Normally, only sheet 1 and the affected sheet (s) will be removed from storage. After revision and release only the revised sheet (s) should be remicrofilmed and replaced in the active microfilm file.
2. The revision letter for each sheet and the "REV STATUS OF SHEETS" block advance only to the change letter affecting that sheet.
3. The change letter block in the right hand margin of each sheet will carry the same change letter as the revision block for that sheet.

Changes to Multisheet Drawings

Figure 7

18-14

ITEM	CHANGE CONDITION AND CHANGE NOTICE (S)	DWG SH NO.	ADDITIONS AND CHANGES TO REVISION BLOCKS	REMARKS
4.	Sheet 1 redrawn or reproduced without change, no outstanding ADCNs	1	Old sheet 1 **LTR / DESCRIPTION / DATE / APPROVED** B / SEE INC A ADCN & B DCN — / SHEET 1 REPLACED WITHOUT CHANGE BY REV B New sheet 1 **LTR / DESCRIPTION / DATE / APPROVED** B / SHEET 1 REPLACES REV B DATED 1-5-68 WITHOUT CHANGE	The requirements of Figure 6, item 5 apply. The "REV STATUS OF SHEETS" block on the old sheet is not altered and is repeated on the new sheet. See footnotes 1 & 3. If other sheets are redrawn or reproduced without change at this time, add entries shown in item 2. A DCN is not issued
5.	Sheet 1 redrawn or reproduced without change, but outstanding ADCNs incorporated C Change ADCN (See Item 3)	1 4	Old sheet 1 **LTR / DESCRIPTION / DATE / APPROVED** B / SEE INC A ADCN & B DCN — / SHEET 1 REPLACED WITHOUT CHANGE BY REV C New sheet 1 **LTR / DESCRIPTION / DATE / APPROVED** C / SHEET 1 REPLACES REV B WITHOUT CHANGE SEE INC C ADCN See sheet 4 in item 1 or 3 for applicable entries	The requirements of Figure 6, item 6 apply. The "REV STATUS OF SHEETS" block on the old sheet 1 is not altered but is updated to reflect the latest change letter of each sheet on the new sheet. See footnotes 1 & 3. If other sheets are redrawn or reproduced with or without change at this time, add entries shown in item 2 or 3 as applicable.
6.	Sheets redrawn or reproduced with change ZONE ITEM C Change DCN [1 B/G] 1. ADDED G/N 12 [2 C/E] 2. REVISED MARKING CALLOUT [4 D/F] 3. .812 DIA WAS .750 DIA A on sheet 2, B on sheet 4	1 2 & 4	Sheet 1 when not replaced **LTR / DESCRIPTION / DATE / APPROVED** B / SEE INC A ADCN & B DCN C / SHEET 2 REPLACES REV A WITH CHANGE. SHEET 4 REPLACES REV B WITH CHANGE. SEE C DCN Old sheet 1 when replaced **LTR / DESCRIPTION / DATE / APPROVED** B / SEE INC A ADCN & B DCN — / SHEETS 1, 2 & 4 REPLACED WITH CHANGE BY REV C New sheet 1 **LTR / DESCRIPTION / DATE / APPROVED** C / SHEETS 1 & 4 REPLACE REV B WITH CHANGE. SHEET 2 REPLACES REV A WITH CHANGE. SEE C DCN C / — / C / C / REV / REV STATUS 4 / 3 / 2 / 1 / SHEET / OF SHEETS Old sheet 2 & 4 **LTR / DESCRIPTION / DATE / APPROVED** — / REPLACED WITH CHANGE BY REV C New sheet 2 & 4 **LTR / DESCRIPTION / DATE / APPROVED** C	The requirements of Figure 6, item 7 & 8 apply, except new approval signatures are not required unless sheet 1 is replaced. The "REV STATUS OF SHEETS" block on the old sheet 1 is not altered but is updated to reflect the latest change letter of each sheet on the new sheet 1. See footnote 1 & 3.

Figure 7 (continued)

ITEM	CHANGE CONDITION AND CHANGE NOTICE (S)	DWG SH NO.	ADDITIONS AND CHANGES TO REVISION BLOCKS	REMARKS
7.	Inactivating, superseding or reactivating all sheets of a multisheet drawing			The revision block entries and the requirements of Figure 6, items 9 thru 12 apply. However, only sheet 1 is changed if (1) there are no outstanding ADCNs or (2) the ADCNs affect sheet 1 only and are incorporated or (3) the ADCNs are left outstanding.
8.	Inactivating or adding a sheet (s) to a multisheet drawing ZONE ITEM C Change DCN 1. SHEETS 2 & 4 INACTIVE FOR FUTURE DESIGN 1/B/G 2. ADDED G/N 13 3. ADDED SHEET 5 A on sheet 2, B on sheet 4	1 2 & 4 5	LTR / DESCRIPTION / DATE / APPROVED B / SEE INC A ADCN & B DCN C / SHEETS 2 & 4 INACTIVE FOR FUTURE DESIGN. ADDED SHEET 5. SEE C DCN INACTIVE SHEETS C / - / - / - / C / REV SHEET / REV STATUS OF SHEETS 5 / 4 / 3 / 2 / 1 Sheet 1 "SHEET" block SHEET 1 OF /13\ G/N /13\ will read: "SEE REV STATUS OF SHEETS BLOCK" Inactivated sheets LTR / DESCRIPTION / DATE / APPROVED C / INACTIVE FOR FUTURE DESIGN New sheet (s) LTR / DESCRIPTION / DATE / APPROVED C	The requirements of Figure 6, item 9 & 10 apply, except if an item on an outstanding ADCN affects the inactivated sheet and is not incorporated the statement "ITEM ___ & ___ OF ___ ADCN NOT INC" shall be added to the revision block of the inactivated sheet. Once a sheet has been inactivated that sheet number shall never be reused except if the same sheet is reactivated. The "REV STATUS OF SHEETS" block will always carry the notation "INACTIVE" for that sheet unless the same sheet is reactivated. See footnotes 1, 2 & 3.
9.	Reinstating an inactivated sheet D Change DCN ZONE ITEM 1. SHEET 4 REACTIVATED	1 4	LTR / DESCRIPTION / DATE / APPROVED B / SEE INC A ADCN & B DCN C / SHEETS 2 & 4 INACTIVE FOR FUTURE DESIGN ADDED SHEET 5 SEE C DCN D / SHEET 4 REACTIVATED SEE D DCN INACTIVE SHEET C / D / - / - / D / REV SHEET / REV STATUS OF SHEETS 5 / 4 / 3 / 2 / 1 Reactivated sheet LTR / DESCRIPTION / DATE / APPROVED D	The requirements of Figure 6, item 12 apply. All unincorporated ADCN items affecting the reactivated sheet shall be incorporated. The Inactivation statement in the revision block (and near the T/B) shall be removed from the reactivated sheet.

Figure 7 (continued)

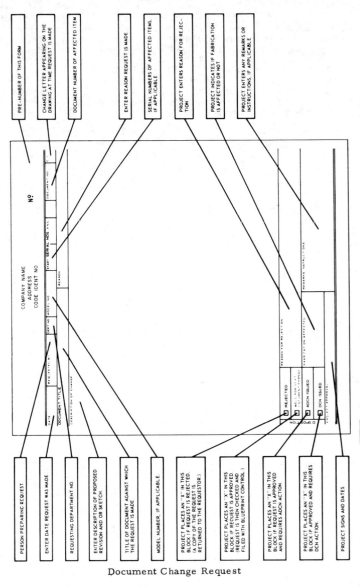

Document Change Request

Figure 8

18.9 CHANGES REQUIRING NEW IDENTIFICATION — Whenever a part's performance or durability is affected, or a replacement is not interchangeable or when a replacement part is limited in it's performance for the part it replaces or the part has been altered or selected, the part must be reidentified.

18.10 ASSEMBLIES THAT CONTAIN NON-INTERCHANGEABLE PARTS — When repairable assemblies contain a non-interchangeable part, the part number reidentification of the non-interchangeable part, of it's next assembly and all the progressively higher assemblies shall be reidentified up to and including the assembly where interchangeability is re-established.

NOTES

Section 19 — PROTECTIVE COATINGS

19.1 SCOPE — This section establishes the methods for specifying protective coating requirements.

19.2 APPLICABLE DOCUMENTS
    MIL-STD-171   Finishing of Metal and Wood Surfaces
    FED-STD-595   Color (Requirements for Individual Color Chips)

19.3 DEFINITIONS

19.3.1 PROTECTIVE COATING — A film applied to the base material of an item to protect the item from corrosion, abrasion, erosion and other forms of deterioration.

19.3.1.1 PERMANENT PROTECTIVE COATING — A protective coating used to preserve an item during its designed service life.

19.3.1.2 TEMPORARY PROTECTIVE COATING — A strippable protective coating used to preserve an item during fabrication, handling and storage. This type of coating is not an engineering requirement and is not specified on the engineering drawing.

19.3.2 CHEMICAL COATING — A superficial layer of a metallic compound produced by the chemical or electrochemical treatment of a surface, e.g., chromate, oxide and anodize films.

19.3.3 ORGANIC COATING — A film of organic material, e.g., paint, lacquer, enamel, varnish, primer, etc.

19.3.4 METALLIC COATING — A film of metal or metal alloy deposited by chemical, electrochemical or other processes, e.g., electroless plate, electroplate, hot dip, spray plate, vapor plate, etc.

19.3.5 VITREOUS COATING — A film of fused silicates or glasses, porcelain enamel or ceramic.

19.3.6 LUBRICANT COATING — A fluid or dry film used to reduce surface friction. The fluid films are produced by petroleum or synthetic base oils, waxes and greases. The dry films are produced by inert mineral substances, such as graphite or molybdenum disulphide.

19.3.7 PRESERVATIVE — A temporary protective coating. (See 19.3.1.2)

19.3.8 PASSIVATION — A surface cleaning treatment that immunizes corrosion resistant steel surfaces against rust by the removal of all free iron particles from the surface. This is not a protective coating but produces the same result by preventing deterioration of the surface.

19.3.9 CONFORMAL COATING — A coating that conforms to the configuration of the object coated for protection from environmental exposure, electrical shorts and mechanical damage. It is also used as a mechanical support for electrical parts.

19.4 DESIGN PRACTICE — A protective coating may consist of one or more types and provide protection from corrosion and/or abrasion. In addition, it may be used for decorative purposes and, as a color-coding, may be used for supplementary identification. The type of coating is determined by design requirements and/or service conditions in which the item is used. The responsible designer shall specify the suitable protective coating.

19.5 GENERAL REQUIREMENTS

19.5.1 When engineering drawings define items to be plated or coated, the applicability of dimensions (and surface roughness, when applicable) before or after coating is specified on the drawing by one of the general notes listed in paragraph 19.7.2.

19.5.2 Due to the variation in specification coverage, careful consideration is to be given to completely specify the protective coating. Some specifications are for the material only; others are for the application of a material or materials; and still others cover the requirements for both the material and its application.

19.5.3 When chemical pretreatment or metallic coats are included in the specification requirements of a multiple coat protective coating, it is not necessary to call out the elements separately since they will be adequately controlled by the overall specification requirements.

19.5.4 It is not necessary that a detail part, of an inseparable assembly, receive a final coat of organic protective coating prior to assembly; it may be prime-coated and receive additional prime and/or finishing coats after assembly.

19.5.5 The order of preference for specifying the protective coating requirements on the engineering drawing is: (1) finish block, and (2) general notes.

19.6 FINISH BLOCK ENTRIES

19.6.1 When a specification adequately controls the application or process, the finish block is to contain an entry similar to one of the following examples:
Example 1: MIL-STD-171, FINISH NO. 1.1.3.1
Example 2: Where the number of coats are not controlled by the specification:
   MIL-X-XXXX
   (X) COATS
Example 3: Where the thickness is not controlled by the specification:
   MIL-STD-XXX, FINISH CODE NO._____, TYPE_____, FILM_____,
   FINAL FILM .0004 TO .0006 THK.
Example 4: Where the specification, application or process is called out in the general notes:
   SEE NOTE ⚠

19.7 DRAWING NOTE CALLOUTS

19.7.1 When space within the finish block will not permit the entry of all the necessary data or when it is more practical, a delta note is used to specify the finish requirements. The delta note is cross-referenced to the finish block and is to be composed as stated in Section 7. For Example:

⚠ ANODIZE PER_____, TYPE_____, CLASS_____.
⚠ NICKEL PLATE PER_____, FINISH NO._____, ITEM 5 ONLY.
⚠ PRIME PER_____, 2 COATS.
⚠ FINISH PER MIL-STD-XXX, FINISH CODE NO._____, TYPE_____, FILM_____, FINAL FILM .0004 TO .0006 THK.

19.7.2 When plating or coating (and surface roughness, when applicable) is a requirement on the drawing, a general note is specified as follows:

DIMENSIONS (AND SURFACE ROUGHNESS, when applicable) APPLY _____.
    A — BEFORE COATING
    B — BEFORE PLATING
    C — AFTER COATING
    D — AFTER PLATING

19.7.3 Where finish does not apply to all items on the drawing:

⚠ MIL-X-XXXX, TYPE_____, CLASS_____, ETC. (AS APPLICABLE) ITEM 5

19.8 PARTS LIST ENTRIES

19.8.1 Parts list entries are required for materials used to produce organic coatings and some of the other types of protective coatings. Each material required, except those used for in-process thinning, reducing, cleaning, etc., is entered in the parts list as stated in Section 9.

19.8.2 When the application specification for organic coatings controls the materials without reference to other material specifications, the application specification is to be entered in both the finish block and parts list. However, when the application specification references other specifications for the control of the material, the material specifications are entered in the parts list and the applicable specification entered in the finish block.

19.8.3 The parts list will always specify the final color of an organic coating. The entry may be by class, type, etc., of (a) the applicable specification, (b) reference to a material specification called out in the application specification or, (c) when not controlled by either the application or material specification, by reference to a separate document such as FED-STD-595, ANA BULLETIN 166, etc.

19.8.4 A detail part is not considered an assembly when a parts list entry is required to qualify the coating material requirements.

19.9 REFERENCE DATA

19.9.1 The following bibliography provides general information on the characteristics and application of the various types of protective coatings.

Materials Handbook, Brady, George S., McGraw-Hill Book Co., Inc., New York, N.Y.

Protective Coatings for Metals, Burns, R. M., Reinhold Publishing Corp., New York, N.Y.

Metal Finishing Guidebook, Metals and Plastic Publications, Inc., Westwood, New Jersey.

Electroplating Engineering Handbook, Graham, A. Kenneth, Editor, Reinhold Publishing Corp., New York, N.Y.

NOTES

Section 20 — ABBREVIATIONS

20.1 SCOPE — This section contains the requirements for and a list of abbreviations used on drawings.

20.2 APPLICABLE DOCUMENTS

MIL-STD-12 — Abbreviations for Use on Drawings, Specifications, Standards, and in Technical Documents

20.3 DEFINITIONS

20.3.1 ABBREVIATION — An abbreviation is a shortened form of a word, expression or phrase used when essential in the conservation of space.

20.4 DRAWING APPLICATION

20.4.1 Abbreviations of short words are omitted except those established by long-standing practice.

20.4.2 Periods are used with abbreviations which spell whole words to provide clarity and avoid misinterpretation. (e.g., Figure — FIG.)

20.4.3 Upper case letters are used except where the use of lower case letters has been established by long practice.

20.4.4 Abbreviations for Chemical Elements or Compounds do not necessarily agree with the chemical symbols, for example: Copper: COP (Cu), Cadmium: CAD (Cd).

20.5 RULES FOR USE

20.5.2 Use an abbreviation of a word combination as such; do not separate for use singly. Single abbreviations may be combined when necessary. i.e. Air Blast = AB

20.5.3 Spaces between word combination abbreviations may be filled with a hyphen (-) for clarity, particularly on drawings with free hand lettering. i.e. USAF-AL = United States Air Force Air Lock

20.5.4 Use the same abbreviation for all tenses, the possessive case, and the singular and plural forms of a given word.

20.5.5 If a drawing uses an abbreviation that does not appear in this DRM (or in MIL-STD-12), it must be decoded on the drawing, directly or by reference.

Term	Abbreviation
Abbreviate	ABBR
Absolute	ABS
Accelerate	ACCEL
Accordance with	A/W
Accumulate	ACCUM
Acme Screw Thread	ACME
Addendum	ADD.
Adhesive	ADH
Adjacent	ADJ
Aerodynamic	AERODYN
Aeronautic	AERO
Aeronautical national taper pipe threads	ANPT
Aerospace Information Report (SAE)	AIR
Aerospace material specification (SAE)	AMS
Aerospace recommended practices (SAE)	ARP
Aerospace Standard (SAE)	AS
Air Force	AF
Air Force-Navy	AN
Air Force-Navy Aeronautical	ANA
Air Force-Navy Design	AND.
Air-to-Air	A-A
Air-to-air missile	AAM
Air-to-ground	A-G
Air-to-surface missile	ASM
Air-to-underwater missile	AUM
Alignment	ALIGN.
Allowance	ALLOW.
Alloy	ALY
Alternate	ALTN
Alternating current	AC
Alternating current volts	VAC
Alternator	ALTNTR
Altitude	ALT
Aluminum	AL
Amber	AMB
Ambient	AMB
American	AMER
American Steel Wire Gage	ASWG
American Wire Gage	AWG
Ammeter	AMM
Amount	AMT
Amperage	AMP
Ampere	A
Amplifier	AMPL
Amplitude modulation	AM.
And	&
Anhydrous	ANHYD
Anneal	ANL
Annunciator	ANN
Anodize	ANDZ
Antenna	ANT.
Antilogarithm	ANTILOG
Apparatus	APPAR
Appendix	APPX
Application	APPL
Approved	APVD
Approximate	APPROX
April	APR
Architecture	ARCH
Armature	ARM.
Army	A
Arrangement	ARR
Article	ART.
Asbestos	ASB
As required	AR
Assemble	ASSEM
Assembly	ASSY
Associate(s)	ASSOC
Association	ASSN
Atmosphere	ATM
Atomic	AT.
Atomic weight	ATWT
Attention	ATTN
Audible	AUD
Audio frequency	AF
August	AUG
Authorize	AUTH
Automatic	AUTO.
Automatic frequency control	AFC
Automatic gain control	AGC
Automatic phase control	APC
Automatic volume control	AVC
Automatic zero set	AZS
Automotive	AUTOM

Term	Abbreviation
Auxiliary	AUX
Auxiliary power unit	APU
Auxiliary switch (breaker) normally closed	ASC
Auxiliary switch (breaker) normally open	ASO
Average	AVG
Avoirdupois	AVDP
Babbit	BAB
Back pressure	BP
Baffle	BAF
Balance	BAL
Ball bearing	BBRG
Base line	BL
Basic	BSC
Battery	BAT.
Bearing	BRG
Bending moment	M
Between centers	BC
Black	BLK, BK
Blue	BLU, BL
Board	BD
Boiling Point	BP
Bolt Circle	BC
Booster	BSTR
Both faces	BF
Both sides	BS
Bottom	BOT
Bottom face	BF
Bottoming	BOTMG
Bracket	BRKT
Brake horsepower	BHP
Branch	BR
Brazing	BRZG
Brinell hardness	BH
Brinell hardness number	BHN
British thermal unit	BTU
Bronze	BRZ
Brown	BRN, BR
Brown and Sharpe (wire gage)	B&S
Building	BLDG
Bulkhead	BHD
Bulletin	BULL.
Bureau	BU
Bushing	BSHG
Buttock line	BL
Butt line	BL
By (used between dimensions)	X
Cadmium	CAD.
Calculate	CALC
Calibrate	CAL
Candlepower	CP
Cantilever	CANTIL
Cap Screw	CAP SCR
Capacitance	C
Capacitor	CAP.
Capacity	CAP.
Carbon	C
Carbon molybdenum steel	CMOS
Carbon steel	CS
Carbon tetrachloride	CBN TET
Carburize	CARB
Case Harden	CH
Casting	CSTG
Cast iron	CI
Cast-iron pipe	CIP
Cast steel	CS
Catalogue	CAT.
Cathode	CATH
Cathode-ray tube	CRT
Center	CTR
Center line	CL
Center of gravity	CG
Center of section	CS
Centrifugal	CNTFGL
Centrifugal force	CF
Ceramic	CER
Chamfer	CHAM
Change notice	CN
Channel	CHAN
Chassis	CHAS
Check	CHK
Check valve	CV
Chemical	CHEM
Chemically pure	CP
Chrome	CRM
Chrome molybdenum	CR MOLY
Chrome vanadium	CR VAN

Term	Abbreviation
Chromium	CHR
Circuit	CKT
Circuit breaker	CB
Circular	CIRC
Circular pitch	CP
Circumference	CRCMF
Class	CL
Classification	CLASS.
Clearance	CL
Clevis	CLV
Clockwise	CW
Coaxial	COAX.
Coefficient	COEF
Cognizant	COG.
Cold-drawn	CD
Cold-drawn steel	CDS
Cold finished	CF
Cold finished steel	CFS
Cold-punched	CP
Cold-rolled	CR
Cold-rolled steel	CRS
Cologarithm	COLOG
Color code	CC
Commercial	COML
Commercial quality	CQ
Commutator	COMM
Company	CO
Complete	COMPL
Composition	CMPSN
Compound	CMPD
Compressor	CPRSR
Concentric	CNCTRC
Condenser	COND
Conductivity	CNDCT
Conductor	CNDCT
Conduit	CND
Confidential	CONF
Connector	CONN
Consist of	C/O
Console	CSL
Constant	CONST
Construction	CONSTR
Container	CNTNR
Contaminated	CONTAM
Continued	CONT
Contour	CTR
Contract(or)	CONTR
Contract change notice	CCN
Contractor furnished equipment	CFE
Control	CONT
Converter	CONV
Conveyor	CNVR
Coordinate	COORD
Copper	COP.
Corner	COR
Corner bead	COR BD
Corporation	CORP
Correct	CORR
Corrosion	CRSN
Corrosion-resistant	CRE
Corrosion-resistant steel	CRES
Corrugate	CORR
Cosecant	CSC
Cosine	COS
Contangent	COT.
Counterbalance	CBAL
Counterbore	CBORE
Counterclockwise	CCW
Counterdrill	CDRILL
Counterdrill other side	CDRILLO
Countersink	CSK
Countersink other side	CSKO
Countersunk head	CSKH
Counterweight	CTWT
Coupling	CPLG
Crankpin	CPIN
Crankshaft	CSHAFT
Cross arm	XARM
Cross connection	XCONN
Cross section	XSECT
Cross head	CRSHD
Cryogenics	CRYOG
Crystal	XTAL
Cubic	CU
Cubic centimeter	CC
Cubic feet	CU FT
Cubic feet per minute	CFM

Term	Abbreviation
Cubic feet per second	CFS
Cubic inch	CU IN.
Cubic meter	CU M
Cubic millimeter	CU MM
Cubic yard	CU YD
Current	CUR.
Cutoff	CO
Cutter	CTR
Cutting	CTG
Cyanide	CYN
Cycle	CY
Cycles per minute	CPM
Cycles per second (see Hertz)	
Cylinder, Cylindrical	CYL
Dated	DTD
Datum	DAT
Decalcomania	DECAL
December	DEC
Decimal	DEC
Decrease	DECR
Dedendum	DED
Deep-drawn	DD
Defense	DEF
Deflect	DEFL
Department	DEPT
Design	DSGN
Detail	DET
Developed horsepower	DHP
Deviation	DEVN
Diagonal	DIAG
Diagram	DIAG
Diameter	DIA
Diameter bolt circle	DBC
Diametral pitch	DP
Diaphragm	DIAPH
Differential	DIFF
Dimension	DIM.
Direct current	DC
Direct-current volts	VDC
Direct-current working volts	VDCW
Direction	DIR
Disassemble	DISASSM
Disassembly	DISASSY
Disc Grind	DG

Term	Abbreviation
Discharge	DISCH
Disconnect	DISC.
Distance	DIST
Distributor	DISTR
Ditto	DO.
Division	DIV
Document	DOC.
Double	DBL
Double acting	DBL ACT.
Double contact	DC
Double Pole	DP
Double-pole double throw	DPDT
Double-pole single throw	DPST
Doubler	DBLR
Double throw	DT
Dovetail	DVTL
Dowel	DWL
Down	DN
Dozen	DOZ
Drafting	DFTG
Drawing requirement manual	DRM
Drawing	DWG
Drawing list	DL
Drawn	DR
Drill	DR
Drill rod	DR
Drive fit	DF
Drop forge	DF
Duplicate	DUP
Dynamic	DYN
Dynamometer	DYNMT
Each	EA
Each face	EF
East	E
Eccentric	ECC
Effective	EFF
Efficiency	EFF
Elastic limit	EL
Electric	ELEC
Electronic	ELEK
Electronics	ELEX
Electromotive Force	EMF
Element	ELEM
Elevate	ELEV

Term	Abbreviation
Elevation	EL
Elongation	ELONG
Emitter	EMTR
Emergency	EMER
Enclose, enclosure	ENCL
Engine	ENG
Engineer	ENGR
Engineering	ENGRG
Engineering change notice	ECN
Engineering change order	ECO
Engineering change proposal	ECP
Engineering change request	ECR
Engineering order	EO
Engineering work order	EWO
English	ENG
Envelope	ENV
Equally spaced	EQ SP
Equipment	EQPT
Equivalent	EQUIV
Estimate	EST
Et cetera	ETC
Evaporate, evaporator	EVAP
Example	EX
Exhaust	EXH
Exhaust gas temperature	EGT
Existing	EXST
Experiment, experimental	EXP
Extension	EXT
Exterior	EXT
External pipe thread	EPT
Extra fine (threads)	EF
Extra heavy	XHVY
Extra strong	XSTR
Extremely-high frequency	EHF
Extrude	EXTD
Fabricate	FAB
Face to face	F TO F
Fahrenheit	F
Fairing	FAIR.
Farad	F
Far side	FS
Fastener	FSTNR
Fast Operating (relay)	FO
Fast release (relay)	FR
February	FEB
Federal	FED
Federal Stock Number (Now: NSN)	FSN
Feet per minute	FPM
Feet per second	FPS
Field	FLD
Figure	FIG.
Filament	FIL
Fillet	FIL
Fillister head	FILH
Final assembly	FA
Finish	FNSH
Finish all over	FAO
Finish one side	F1S
Finish two sides	F2S
Fitting	FTG
Fixture	FXTR
Flange	FLG
Flat bar	FB
Flat fillister head	FFILH
Flat head	FLH
Flat pattern	F/P
Flexible	FLEX.
Flight	FLT
Fluid	FL
Fluorescent	FLUOR
Fluorine	F
Foot	FT
Foot-candle	FC
Footing	FTG
Foot-pound (work)	FT LB
Foot-pound-second (system)	FPS
Force	F
For example	EG
Forged steel	FST
Forging	FORG
Forward	FWD
Foundation	FDN
Foundry	FDRY
Four conductor	4/C
Four pole	4P
Four-way	4 WAY
Free cutting brass	FCB

Term	Abbreviation
Free height	FRHGT
Free-machining steel	FMS
Freezing point	FP
Frequency	FREQ
Frequency modulation	FM
Friction	FRICT
Front	FR
Full hard	FH
Full indicator reading	FIR.
Furnish	FURN
Gage	GA
Gallon	GAL
Gallons per hour	GPH
Gallons per minute	GPM
Gallons per second	GPS
Galvanize	GALV
Galvanized iron	GALVI
Galvannealed	GALVND
Galvanometer	GALVNM
Gaseous oxygen	GOX
Gasket	GSKT
Gate valve	GTV
Gear	GR
General	GENL
Generator	GEN
Glove valve	GLV
Government	GOVT
Government furnished equipment	GFE
Governor	GOV
Grade	GR
Grain	GR
Gram	G
Gram calorie	GCAL
Graphite	GPH
Gravity	G
Gray	GRA, GY
Grease	GRS
Green	GRN, G
Grind	GND
Grommet	GROM
Gross ton	GT
Ground	GND
Ground support equipment	GSE

Term	Abbreviation
Ground-to-air	G-A
Ground-to-ground	G-G
Guided missile	GM
Gyroscope	GYRO
Half-hard	½ H
Half-round	½ RD
Handbook	HDBK
Handle	HDL
Handling	HDLG
Hand wheel	HD WHL
Hard	HD
Hard chromium	HD CR
Hard-drawn	HD DRN
Harden	HDN
Harden & grind	H&G
Hardness	HDNS
Hardware	HDW
Hazardous	HAZ
Head	HD
Heading	HDG
Headless	HDLS
Headquarters	HQ
Heat	HT
Heater	HTR
Heat resisting	HT RES
Heat treat	HT TR
Heavy	HVY
Height	HGT
Henry	H
Hertz	Hz
Hexagon	HEX
Hexagonal head	HEX HD
High carbon	HC
High-carbon steel	HCS
High-carbon steel, heat-treated	HCSHT
High frequency	HF
High pressure	HP
High speed	HS
High-speed steel	HSS
High-tensile steel	HTS
High-velocity aircraft rocket	HVAR
High voltage	HV

20-7

Term	Abbreviation
Horizontal	HORIZ
Horizontal center line	HCL
Horizontal location of center of gravity	HCG
Horizontal reference line	HRL
Horsepower	HP
Horsepower-hour	HP HR
Hot-press	HP
Hot-rolled steel	HRS
Hour	HR
Housing	HSG
Hydraulic	HYDR
Hydrogen	H
Hydrostatic	HYDRST
Identification	IDENT
Ignition	IGN
Illuminate	ILLUM
Illustrate	ILLUS
Impeller	IMPLR
Impregnate	IMPRG
Improved plow steel	IMP PS
Impulse	IMPLS
Inboard	INBD
Inch	IN.
Inches per second	IPS
Inch-pound (work)	IN LB
Inclined	INCLN
Include	INCL
Inclusive	INCL
Incoming	INCM
Incorporated	INC
Increase	INCR
Index list	IL
Indicate	IND
Indicated air speed	IAS
Indicated horse power	IHP
Indicator	IND
Inductance-capacitance	LC
Inductance-capacitance resistance	LCR
Inductance coil	L
Industrial	INDL
Information	INFO
Initial velocity	IV

Term	Abbreviation
Inside diameter	ID
Inside mold line	IML
Inside radius	IR
Inspect	INSP
Install, installation	INSTL
Instantaneous	INST
Instruction	INSTR
Instrument	INSTR
Instrumentation	INSTM
Interchangeable	INTCHG
Intercommunication	INTERCOM
Interconnection	INTCON
Interface control drawing	ICD
Interference	INTRF
Interior	INTR
Interlock	INTLK
Intermediate frequency	IF
Internal	INTL
Internal pipe thread	IPT
International annealed copper standard	IACS
International Organization for Standardization	ISO
International Pipe Standard	IPS
International Standard Thread (metric)	IST
Iron (abbreviate only in conjunction with other materials	I
Iron pipe	IP
Iron pipe size	IPS
Irregular	IRREG
January	JAN
Job order	JO
Joggle	JOG.
Joint	JT
Joule	J
Journal	JNL
July	JUL
Junction	JCT
Junction box	JB
June	JUN
Kelvin	K
Keyseat	KST

Term	Abbreviation
Keyway	KWY
Kilocycle	KC
Kilocycles per second	KC/S
Kilogram	KG
Kilograms per cubic meter	KG/CU M
Kilograms per second	KGPS
Kilohertz	KHZ
Kilohm	K
Kilometer	KM
Kilometers per second	KM/S
Kilovolt	KV
Kilovolt ampere hour	KVAH
Kilowatt	KW
Kilowatt hour	KWH
Kinetic energy	KE
Laboratory	LAB
Lacquer	LAQ
Laminate	LAM
Landing	LDG
Large	LGE
Lateral	LATL
Latitude	LAT.
Leading edge	LE
Left	L
Left hand	LH
Length	LG
Length over-all	LOA
Letter	LTR
Lifting eye	LE
Light	LT
Lightening	LTG
Lightening hole	LTGH
Limit	LIM
Limited	LTD
Limit switch	LIM SW
Linear	LIN
Liquid	LIQ
Liquid Oxygen	LOX
Liquid rocket engine	LRE
Liter	L
Loading	LDG
Locate	LCT
Locked	LKD
Locked closed	LKD C
Locked open	LKD O
Locking	LKG
Locknut	LKNT
Lockscrew	LKSCR
Lock washer	LK WASH.
Lockwire	LKWR
Logarithm	LOG.
Logarithm (natural)	LN
Longeron	LONGN
Logic	LGC
Long	L
Longitude	LONG.
Longitudinal	LONG.
Low-alloy steel	LAS
Low carbon	LC
Lower	LWR
Low frequency	LF
Low pressure	LP
Lubricant	LUBT
Lubricate	LUB
Lumen	LM
Luminous	LUM
Lux	LX
Machine	MACH
Machine screw	MSCR
Machine steel	MST
Magnaflux	M
Magnesium	MAG
Magnet	MAG
Magnetic amplifier	MAGAMP
Magneto	MGN
Maintenance	MAINT
Major	MAJ
Male and female	M&F
Malleable	MAL
Malleable iron	MI
Manifold	MANF
Manual	MNL
Manufacture	MFR
Manufactured	MFD
Manufacturing	MFG
March	MAR
Mark	MK
Master	MA

Term	Abbreviation
Master change record	MCR
Material	MATL
Mathematical, mathematics	MATH
Maximum	MAX
Maximum material condition	MMC
Maximum working pressure	MWP
Mean effective pressure	MEP
Measure	MEAS
Mechanical, mechanism	MECH
Medium	MDM
Mega (prefix)	M
Megacycle (See Megahertz)	
Megahertz	MHz
Megawatt	MW
Megohm	MEGO
Melting point	MP
Memorandum	MEMO
Metal	MET
Meter	MTR
Microampere	UA
Microampere (peak)	Ua
Microfarad	UF
Microhenry	UH
Mircohm	UOHM
Micrometer	MIC
Microsecond	USEC
Microphone	MIC
Microvolt	UV
Mile	MI
Miles per gallon	MPG
Miler per hour	MPH
Military	MIL
Military standard	MS
Milli (10^{-3}) (prefix)	m
Milliampere	mA
Milligram	mG
Millihenry	mH
Millimeter	mM
Millisecond	mS
Millivolt	mV
Milliwatt	mW
Minimum	MIN
Minor	MIN
Minute	MIN
Miscellaneous	MISC
Missile	MSL
Mixture	MXT
Model (for use in military nomenclatures)	M
Model (for general use)	MOD
Modification	MOD
Modify	MOD
Molded	MLD
Molding	MLDG
Mold line	ML
Molecular weight	MOL WT
Month	MO
Mount	MT
Mounted	MTD
Mounting	MTG
Mounting center	MTGC
Multiple	MULT
Music wire	MUW
Namely	VIZ
Nameplate	NPL
National	NATL
National Aeronautics and Space Administration	NASA
National Aerospace Standards	NAS
National course (thread)	NC
National Electric Code	NEC
National extra fine (thread)	NEF
National fine (thread)	NF
National form (thread)	N
National gas outlet (thread)	NGO
National pipe	NP
National special (thread)	NS
National stock number	NSN
National taper pipe (thread)	NPT
Nautical	NAUT
Naval	NAV
Naval Bronze	NAV BRZ
Near Face	NF
Near side	NS
Neck	NK
Negative	NEG
Neutral	NEUT

Term	Abbreviation
New British Standard (imperial wire gage)	NBS
Newton	N
Next assembly	NA
Nickel	NKL
Nickel steel	NS
Nipple	NIP.
Nitrogen	N
No change	NC
Nomenclature	NOMEN
Nominal	NOM
Noncorrosive metal	NCM
Nondestructive testing	NDT
Normal	NORM.
Normalize	NORM.
Normally closed	NC
Normally open	NO.
North	N
Not applicable	NA
Not to scale	NTS
November	NOV
Nozzle	NOZ
Number	NO.
Numerical control	NC
Nut place	NTPL
Nylon	NYL
Obsolete	OBS
October	OCT
Ohm (for use only on diagrams)	Ω
Ohmmeter	OHM.
On center	OC
Opening	OPNG
Opposite	OPP
Optional	OPTL
Orange	ORN, O
Ordnance	ORD
Origin	ORIG
Original	ORIG
Oscillator	OSC
Oscilloscope	SCOPE.
Ounce	OZ
Ounce-inch (work)	OZ-IN.
Outboard	OUTBD
Outside diameter	OD
Outside face	OF.
Outside mold line	OML
Outside radius	OR.
Oval head	OVH
Over-all	OA
Overhead	OVHD
Oxide	OXD
Oxygen	OXY
Package	PKG
Packing	PKG
Page	P
Pair	PR
Panel	PNL
Pan head	PNH
Panic bolt	PANB
Paragraph	PARA
Parallel	PRL
Parkway	PKWY
Part	PT
Parting line (castings)	PL
Parts list	PL
Part number	PN
Pascal	PA
Passivate	PSVT
Patent	PAT.
Pattern	PATT
Per	/
Percent	PCT
Perforate	PERF
Permanent	PERM
Perpendicular	PERP
Phase	PH
Phenolic	PHEN
Phillips head	PHH
Phosphate	PHOS
Phosphor bronze	PH BRNZ
Phosphorous	P
Photograph	PHOTO
Physical	PHYS
Pico (prefix)	p
Pico farad	pF
Pico henry	pH
Picture	PIX

Term	Abbreviation
Piece	PC
Pint	PT
Pitch	P
Pitch circle	PC
Pitch diameter	PD
Plate	PL
Plate (electron tube)	P
Platinum	PLAT
Plus or minus	PORM
Plywood	PLYWD
Pneumatic	PNEU
Point	PT
Point of tangency	PT
Pole	P
Polyphase	PLYPH
Portable	PORT.
Position	POSN
Positive	POS
Potential	POT.
Potentiometer	POT.
Pound	LB
Pound-foot (torque)	LB FT
Pound-inch (torque)	LB IN.
Pounds per square inch	PSI
Pounds per square inch absolute	PSIA
Pounds per square inch gage	PSIG
Power	PWR
Power amplifier	PA
Power factor	PF
Power supply	PWR SPLY
Power take-off	PTO
Preamplifier	PREAMP
Prefrbricated	PREFAB
Preferred	PFD
Preliminary	PRELIM
Prepare	PREP
Pressed	PRSD
Pressure	PRESS.
Pressure reducing valve	PRV
Primary	PRI
Printed circuit	PC
Printed circuit board	PCB
Printed wiring	PW
Printed wiring board	PWB
Product, production	PROD.
Project	PROJ
Proposed	PRPSD
Protective	PROT
Pulley	PUL
Purchase Order	PO
Purple	PRP
Push-pull	PP
Pyrometer	PYROM
Quadrant	QDRNT
Qualified products list	QPL
Quality	QUAL
Quality assurance	QA
Quality control	QC
Quantity	QTY
Quart	QT
Quarter	QTR
Quarter-hard	¼ H
Quarter-phase	¼ PH
Quarter-round	¼ RD
Quick-acting	QA
Quick disconnect	QDISC
Radius	RAD
Ratchet	RCHT
Rate	RT
Reactive volt-ampere	VAR
Received	RCVD
Receiver	RCVR
Receptacle	RCPT
Reciprocating	RECIP
Rectangle	RECT
Rectifier	RECT
Red	RED, R
Reducer	RDCR
Reference	REF
Reference designation	REF DES
Reference line	REFL
Regardless of feature size	RFS
Regular	RGLR
Reinforce	REINF
Release	RLSE
Relief	RLF
Relocated	RELOC

Term	Abbreviation
Removable	REM
Remove	RMV
Replace	REPL
Reproduce, reproduction	REPRO
Request	REQ
Required	REQD
Requirement	REQT
Research	RES
Research and Development	R&D
Reservoir	RSVR
Resistance	RES
Resistor	RES
Retainer	RTNR
Retractable	RETR
Return	RTN
Reverse	RVS
Revise	REV
Revision	REV
Revolution	REV
Revolutions per minute	RPM
Revolutions per second	RPS
Rheostat	RHEO
Rhodium	RHOD
Right	R
Right hand	RH
Rivet	RVT
Rocker	RKR
Rocket	RKT
Rocket assist take-off	RATO
Rockwell hardness	RH
Rolled	RLD
Root mean square	RMS
Root-sum square	RSS
Rotary	RTRY
Rotate	ROT.
Rough	RGH
Round	RND
Roundhead	RDH
Rubber	RBR
Safety	SAF
Sand blast	SD BL
Saturate	SAT.
Schedule	SCHED
Schematic	SCHEM
Screen	SCRN
Screw	SCR
Seamless	SMLS
Seamless steel tubing	SSTU
Secant	SEC
Second	SEC
Section	SECT.
Segment	SEG
Senior	SR
Separate	SEP
September	SEPT
Sequence	SEQ
Serial	SER
Serrate	SERR
Service	SERV
Servo	SVO
Servomechanism	SERVO.
Set screw	SSCR
Shaft horsepower	SHP
Shank	SHK
Sheet	SH
Shop order	SO
Short wave	SW
Shoulder	SHLDR
Siemens	S
Signal Corps	SIGC
Silver	SIL
Silver braze or brazing	SB
Silver solder	SILS
Similar	SIM
Sine	SIN.
Single conductor	1/C
Single contact	SC
Single pole	SP
Single pole, double throw	SPDT
Single pole, single throw	SPST
Single throw	ST
Sketch	SK
Slate	SLT, S
Sleeve	SLV
Slide	SL
Slip joint	SJ
Slope	SLP
Slotted	SLTD

Term	Abbreviation
Socket	SKT
Socket head	SCH
Solenoid	SOL.
Solid height	SOL HGT
South	S
Spanner	SPNR
Spare, spare part	SP
Special	SPCL
Special equipment	SE
Specific	SP
Specification	SPEC
Specification control drawing	SCD
Specific gravity	SP GR
Specific heat	SP HT
Specimen	SPEC
Speed light	SP LT
Spherical	SPHER
Spindle	SPDL
Split ring	SR
Spot face	SF
Spot weld	SW
Spring	SPR
Square	SQ
Square centimeter	SQ CM
Square foot	SQ FT
Square head	SQH
Square inch	SQ IN.
Square kilometer	SQ KM
Square root	SQRT
Square root of mean square	RMS
Stabilizer	STAB.
Stainless steel	SST
Stairway	STWY
Standard	STD
Standpipe	SP
Station	STA
Steel	STL
Stiffener	STIF
Stock	STK
Stove bolt	SB
Straight	STR
Straight shank	SS
Street	ST

Term	Abbreviation
Strength	STR
Stringer	STGR
Structure	STRUCT
Structural	STRL
Subassembly	SUBASSY
Subject	SUBJ
Substitute	SUBST
Suction	SUCT
Sulfur	S
Superseded	SUPSD
Supplement	SUPPL
Supply	SPLY
Surface	SURF.
Switch	SW
Switchboard	SWBD
Swivel	SWVL
Symbol	SYM
Symmetrical	SYMM
Synchronize, Synchronizer, Synchronizing	SYNC
Synchronous	SYN
Synthetic	SYNTH
System	SYS
Tabulate	TAB.
Tachometer	TACH
Tangent	TAN.
Tapping	TPG
Technical	TECH
Technical manual	TM
Technical order	TO
Teeth	T
Telephone	TEL
Television	TV
Temperature	TEMP
Tempered	TMPD
Template	TEMPL
Temporary construction hole	TCH
Tensile strength	TS
Tension	TNSN
Tentative	TNTV
Terminal	TERM.
Tesla	T
Tetrachloride	TET
That is	IE

20-14

Term	Abbreviation
Theoretical	THEOR
Thermal	THRM
Thermocouple	TC
Thermometer	THERM.
Thermostat	THERMO
Thick	THK
Thread	THD
Threads per inch	TPI
Three-conductor	3/C
Three-phase	3PH
Three-pole	3P
Three-way	3 WAY
Through	THRU
Title block	T/B
Tobin Bronze	TOB BRZ
Tolerance	TOL
Ton	T
Tongue and groove	T&G
Tool steel	TS
Torque	TRQ
Total indicator reading	TIR
Transducer	XDCR
Transfer	XFR
Transformer	XFMR
Transistor	XSTR
Transmitter	XMTR
Transportation	TRANSP
Treatment	TRTMT
Triple pole	3P
Triple throw	3T
True position	TP
True position tolerance	TPTOL
Trunnion	TRUN
Truss head	TRH
Tubing	TBG
Tungsten	TUNG
Turbine	TURB
Turbine generator (driver)	TURBO GEN
Turbo alternator	TURBO ALT
Turbo-jet propulsion	TJP
Turnbuckle	TRNBKL
Turned	TRND
Two-conductor	2/C
Two-phase	2PH

Term	Abbreviation
Two-way	2 WAY
Typical	TYP
Ultimate	ULT
Ultra-high frequency	UHF
Ultraviolet	UV
Underwater-to-air missle	UAM
Underwater-to-surface missile	USM
Unfinished	UNFIN
Unified	UN
Unified coarse (thread)	UNC
Unified extra fine (thread)	UNEF
Unified fine (thread)	UNF
Unified special (thread)	UNS
United States	US
United States gage	USG
Universal	UNIV
Used with	U/W
Vacuum	VAC
Vacuum tube	VT
Vacuum tube voltmeter	VTVM
Valve	V
Vanadium	V
Variable	VAR
Velocity	VEL
Ventilate	VENT.
Ventilator	VENT.
Vernier	VERN
Versed sine	VERS
Versus	VS
Vertical	VERT.
Vertical center line	VCL
Vertical location of the center of gravity	VGC
Vertical reference line	VRL
Very low frequency	VLF
Vibrate, vibration	VIB
Violet	VIO.
Viscosity	VISC
Volt	V
Volt-ampere reactive	VAR
Voltage regulator	VR
Volt ampere	VA
Voltmeter	VM
Volt ohm milliammeter	VOM

Term	Abbreviation
Volume	VOL
Vulcanize	VULC
Washer	WSHR
Water line	WL
Watt	W
Watt-hour	WHR
Watt-hour meter	WHM
Wattmeter	WM
Weber	Wb
Week	WK
Weight	WT
West	W
Wheel	WHL
White	WHT, W
Width	WD
Withdrawn	W/D
Without	W/O
Woodruff	WDF
Working circle	WC
Working point	WP
Working pressure	WPR
Wrought brass	W BRS
Wrought iron	WI
Wrought steel	WS
Yard	YD
Year	YR
Yellow	YEL, W
Yield point (psi)	YP
Yield strength (psi)	YS
Zone	Z

ABBREVIATIONS OF SELECTED INDUSTRY ASSOCIATIONS

Abrasive Grain Association .... AGA
Aeronautical Radio, Inc .... ARINC
Aerospace Industries Assn .... AIA
Air-conditioning & Refr Inst .... ARI
*Air Diffusion Council .... ADC
*Air Transport Association .... ATA
*Alloy Casting Institute .... ACI
*Aluminum association .... AA
*Amer Assn of State Hwy &
 Transportation Officials ..... AASHTO
*Amer Assn of Textile
 Chemists & Colorists .... AATCC
*Amer Bureau of Shipping .... ABS
 American Concrete Institute .... ACI
*Amer Conference of Governmental
 Industrial Hygienists .... ACGIH
 American Dry Milk Institute ...... ADMI
*American Frozen Food Institute ... AFFI
 American Gage Design
 Committee .... AGDC
 American Gas Association .... AGA
 American Gear Mfrs
 Association................. AGMA
 American Home Laundry
 Mfrs Assn .... AHLMA
*American Inst of
 Chemical Engs .... AIChE
 American Inst of
 Steel Constr .... AISC
 American Inst of
 Timber Constr .... AITC
*American Insurance
 Association .... AIA
 American Iron & Steel Inst .... AISI
*American Natl Standards
 Inst .... ANSI
*American Paper Institute .... API
*American Petroleum Institute ...... API
 Amer Power Net Assn (Now: NTA)   -
 (Now: NTA)
 American Railway
 Engrg Assn .... AREA
*Amer Soc of Agricultural
 Engs...................... ASEA
*American Soc of Civil Engs ...... ASCE

*American Society for NDI .... ASNT
*Amer Soc for Quality Control .... ASQC
 Amer Soc for Testing & Matls .... ASTM
 Amer Soc of Heating, Refr &
 Air-conditioning Engrs ...... ASHRAE
 Amer Soc of Mechanical Engs ... ASME
 American Stds Assn (Now: ANSI)   -
*American Water Works Assn ... AWWA
 American Welding Society .... AWS
 Anti-Friction Bearing
 Mfrs Assn .... AFBMA
 Assoc General Contr
 of America .... AGCA
 Assn of Amer Feed
 Control Ofcl............... AAFCO
*Assn of American Railroads ...... AAR
 Assn of Iron & Steel
 Engineers .... AISE
 Biological Stain Commission ...... BSC
*Boating Industry Association ...... BIA
 Braided Trimming Mfrs Assn
 (Now: NTA)   -
 California Redwood Association ... CRA
 Cemented Carbide
 Producers Assn .... CCPA
*Composite Can & Tube
 Institute .... CCTI
*Compressed Gas Association ..... CGA
 Copper Development
 Association .... CDA
*Cooling Tower Institute .... CTI
 Crayon & Water Color
 Crafts Inst .... CWCCI
*Diesel Engine Mfrs Association .. DEMA
*Edison Electric Institute .... EEI
 Elastic Braid Mfrs Assn (Now: NTA) -
 Elastic Fabric Mfrs Inst (Now: NTA) -
 Electrical Apparatus
 Service Assn .... EASA
 Electronic Industries Association ... EIA

20-17

ABBREVIATIONS OF SELECTED INDUSTRY ASSOCIATIONS

*Engineers Joint Council .... EJC

Galvanized Ware Mfrs Council .. GWMC
*Gas Processors Association ...... GPA
*Graphic Arts Technical
  Foundation .... GATF
Gummed Industries Association ... GIA

Hangar & Indl Door
  Technical Council .... HIDTC
Heat Exchange Institute .... HEI
Hydraulic Institute .... HI

*Illuminating Engineers Society ..... IES
*Industrial Fasteners Institute .... IFI
*Inst of Electrical &
  Electronic Engs .... IEEE
*Institute of Printed Circuits .... IPC
Instrument Society of America ..... ISA
Insulated Power Cable
  Engineers Assn .... IPCEA
*International Air Transport
  Association .... IATA
*Intl Conference of Building
  Officials .... ICBO
*International Standards
  Organization .... ISO

*Joint Electronic Device
  Engg Council .... JEDEC
*Joint Industrial Council .... JIC

Library Binding Institute .... LBI
Lithograhic Technical Fdn
  (Now: GATF)   -

Magnetic Materials
  Producers Assn .... MMPA
*Mfrs Standardization Soc of
  the Valves & Fitting Industry ..... MSS
*Mfg Chemists Association .... MCA
Maple Flooring
  Manufacturers Assn .... MFMA
*Material Handling Institute .... MHI

Metal Lath Association .... MLA
*Metal Powder Industry
  Federation .... MPIF
Monorail Manufacturers
  Association .... MMA
*Narrow Fabrics Institute .... NFI
*Natl Assn of Arch Metal Mfrs .. NAAMM
National Assn of Broadcasters .... NAB
National Assn of Chain
  Manufacturers .... NACM
*Natl Assn of Corrosion
  Engineers .... NACE
*Natl Assn of Relay
  Manufacturers .... NARM
Natl Cottonseed Products Assn .. NCPA
*Natl Council on Radiation Protec-
  tion and Measurement .... NCRP
Natl Electrical Mfrs
  Association .... NEMA
*Natl Elevator Industries, Inc. .... NEII
Natl Elevator Mfg Ind (Now: NEII)   -
Natl Fibre Can & Tube Assn (Now: CCTI)
*Ntl Fire Protection Association ... NFPA
*National Microfilm Association ... NMA
Natl Printing Ink
  Research Inst .... NPIRI
Natl Sanitation Foundation .... NSF
Natural Gas Processors Assn
  (Now: GPA)   -
*Northern Textile Association ...... NTA

Optical Manufacturers
  Association .... OMA
Outboard Industry Assn (Now: BIA)   -

Paper bag Institute, Inc .... PBI
*Pipe Fabrication Institute .... PFI
*Portland Cement Association ..... PCA
*Pulp Mfrs Research League ..... PMRL

*Radio Technical Commission
  for Aeronautics .... RTCA
Recording Industry Assn
  of Amer .... RIA

ABBREVIATIONS OF SELECTED INDUSTRY ASSOCIATIONS

Resistance Welder Mfrs
   Association .... RWMA

Rubber Manufacturers
   Association .... RMA

*Scientific Apparatus Mfrs Assn .. SAMA
Shoe Lace Mfrs Assn (Now: NFI)   -
*Sheet Metal & Air-conditioning
   Contractors Natl Assn ...... SMACNA
Soap & Detergent Council .... SDA
Society of Automotive Engineers ... SAE
*Soc of Naval Arch &
   Marine Engs .... SNAME
Southern Pine Inspection Bureau .. SPIB
Specialty Paper & Board Affiliates
   (Now: Amer Paper Inst)   -
*Steel Door Institute .... SDI
*Steel Structures Painting
   Council..................... SSPC
Steel Window Institute .... SWI
Sulphite Pulp Mfrs
   Research League (Now: PMRL)   -
Sump Pump Manufacturers
   Assn .... SPMA

Tech Assn of the Pulp &
   Paper Ind................... TAPPI
Tire & Rim Association.......... TRA
*Tubular Exchanger Mfrs Assn .... TEMA

*Underwriters Laboratories .... UL
USA Standard Inst (Now: ANSI)   -

Note: Entries without asterisks (*) are those listed in DOD-4120.3-M and MIL-STD-12. Those with an asterisk are additional associations frequently encountered in design reference, including some which supersede former listings.

NOTES

Section 21 — TOOL DRAFTING PROCEDURE

21.1 SCOPE — This tool Drafting Procedures section provides standardized regulations for the preparation and interpretation of Company tooling drawings. This Section is concerned primarily with format and procedures peculiar to tooling drawings.

21.2 APPLICABLE DOCUMENTS — NONE

21.3 DEFINITIONS

21.3.1 TOOLING DRAWINGS — Drawings of special equipment for testing, handling, checking, or the fabrication of parts, subassemblies, and assisting in the process of manufacturing these items. They may include drawings of jigs, dies, fixtures, molds, patterns, tapes, gages, or other special equipment or manufacturing aids.

21.3.2 TOOL NUMBER — A part or identifying number assigned to a tool.

21.4 RESPONSIBILITIES OF THE TOOL DESIGNER

21.4.1 The tool designer is responsible for the preparation of tool designs that will adequately meet the requirements of the tool order. The tool drawing must present a true picture of the tool design, be clear, concise, complete, accurate, and subject to only one interpretation.

21.4.2 The tool drawing should indicate the function of the tool and the part number for which the tool was designed.

21.4.3 The designer shall present layouts, dimensional calculations, catalogs, design memoranda, part prints and reference prints to the tool drawing checker.

21.4.4 Final correction of the drawing, after checking, should be made by the original tool designer, so that he may avoid making the same or similar errors in the future. Final correction of the drawing shall normally take precedence over other work.

21.4.5 A drawing changed after it has been submitted to check, or prior to release, shall be resubmitted to the checker for approval.

21.5 RESPONSIBILITIES OF THE TOOL DESIGN CHECKER

21.5.1 The checker is responsible for verifying the functional feasibility of the tool, dimensional accuracy of tool design drawings and assuring conformance to the standard drafting procedures. The checker is the last person to sign off the drawing except when the Supervisor of the Tool Design Group is required to sign off prior to release. The checking procedure is as stated in Section 2.

21.6 DRAWING NUMBERING SYSTEM

21.6.1 TOOL NUMBER ASSIGNMENT — Once assigned, a tool number is peculiar to that application, do not reassign it.

21.6.1.1 The tool number identifies one tooling drawing with one title, regardless of the number of sheets on which it is drawn.

21.6.1.2 The tool number also identifies the physical tooling manufactured in accordance with the tooling drawing.

21.7 ASSIGNMENT OF DETAIL NUMBERS

21.7.1 Assign or cite a part number for each part of a tool, including commercial parts. The detail number can identify:
  (1) An individual part of a tool or assembly.
  (2) An assembly of parts which is a portion of the tool.
  (3) An assembly of parts which is a portion of another assembly.
  (4) Shown and opposite parts and assemblies (See 21.7.3)
  (5) Through tabulations—each of a series of similar parts of a tool.
  (6) Other tools with other tool numbers, if these tools become components of the tool shown on the drawing.

21.7.2 When a tool or other item is shown on a drawing for reference purposes (Not a part of the tool), draw it with a phantom outline, and appropriately identified.
  Example:
   XXXXXXX SPINNING FIXTURE ⟶

Standard construction balls, pins, and rolls that are required for fabrication and inspection of the tool shall be drawn in phantom, and appropriately identified.
  Example:
   .500 DIA CONSTRUCTION BALL ⟶

Construction balls, rolls, and pins are assigned detail numbers as tool parts only if they are used in the function of the tool.

21.7.3 Shown and opposite parts shall be assigned detail numbers that run in sequence whenever practical. The opposite shall be understood to be a mirror opposite of the shown component. Differences must be clearly indicated by views or notes.

21.7.4 When a tool component is detailed in a place other than where called out, the title should consist of a detail number and the word "DETAIL" or "ASSY" as applicable.

21.7.5 DETAIL NUMBER CHANGES

21.7.5.1 Changes to a tool component after formal drawing release shall be as follows:
  (1) When it is practical to rework the original component, the original detail number is retained.
  (2) When it is not practical to rework the original component, the original detail number is made obsolete, except when the change is minor, as when a bolt size, type, or length is changed.

21.7.5.2 Obsolete details or assemblies are deleted in the Parts List and "lined-out" on the field of the drawing. Once a detail number has been deleted, it is not used again on the same drawing.

21.8 DRAWING FORMAT

21.8.1 DRAWING ARRANGEMENT — The title block, general notes, drawing numbers, etc. are located on the drawing as shown in Section 5.

21.8.2 CLASSIFICATION AND ESPIONAGE ACT STAMPS — These notes shall not be added to drawings unless required by the cognizant requestor or the security department. When required, the appropriate security classification shall be added to the drawing as directed by the security representative. The security classification must be added prior to any reproduction or duplication of the drawing.

21.8.3 MARGINS — The margins are made as shown in Section 5. The "J" size drawing has a 4 inch protective flap at the right hand end of the drawing along with a 1 inch margin on top, bottom, and left end of the drawing.

21.8.4 DATA BLOCKS

21.8.4.1 Blocks are located as shown in Section 5. Blocks within the title block which do not pertain to individual designs are to be crossed out. Parts list format may be varied to suit specific requirements.

21.9 DRAWING TITLES

21.9.1 TITLE ASSIGNMENT

21.9.1.1 Titles on all tool design drawings shall consist of a basic name, a modifier, and a modifying phrase.

21.9.1.2 The basic name (noun or noun phrase) shall describe the tool, or usage of the tool.
> Example:
> FIXTURE, JIG, DIE

21.9.1.3 The method of manufacture of the tool shall not be used in the basic name.
> Example:
> WELDED, MACHINED, CAST

21.9.1.4 The singular form of the noun or noun phrase is used as the basic name except where the only form of the noun is plural.
> Example:
> BELLOWS, TONGS, SCISSORS

or, where the drawing contains multiple single items or tabulations.
> Example:
> SPACERS, BOLTS, WASHERS

21.9.1.5 The modifier may be a single word or a qualifying phrase. The modifier shall serve to narrow the area of concept established by the basic name.

21.9.1.6 The modifying phrase shall be separated from the preceding modifier by a comma. This phrase further identifies the function or application (what it is used for), or location (where it is used), of the item.
> Example:
> FIXTURE, WELD, SUB-ASSY
> TEMPLATE, CONTOUR
> FORWARD NOSE

21.10 DRAWING NOTES

21.10.1 Local and general note composition and methods of placing this data on drawings shall be as shown in Section 7.

21.11 PARTS LIST (See Section 5)

21.11.1 GENERAL REQUIREMENTS — The Parts List shall list all components and assemblies that are assigned detail numbers on the field of the drawing. Standard tool drawing lettering of .156 inch in height shall be used in the Parts List.

21.11.2 ASSEMBLY NUMBER COLUMN — The detail number on each assembly as defined here shall appear in the Part or Identifying Number Column of the Parts List.

21.11.3 PART OR IDENTIFYING NUMBER COLUMN — The detail number of each tool component shall appear in the Part or Identifying Number Column of the Parts List.

21.11.4 QUANTITY REQUIRED COLUMN — Entries are made here to indicate the total number of each detail required to fabricate one tool as described by the tool drawing.

21.11.4.1 ASSEMBLY — An assembly is composed of two or more details joined as a unit:
> (a) that is intended by the designer to be fabricated independently from the rest of the tool.
> (b) that will be attached, removed, or handled separately from the rest of the tool in the normal process of use.

21.11.5 NOMENCLATURE/DESCRIPTION COLUMN — A complete description of each detail number shall be listed in this column for ordering purposes.

21.11.5.1 Any reference to supplier's part number shall appear in the "Part or Identifying Column" with the supplier's "Code Ident No." listed in the adjacent column.

21.11.5.2 When a supplier does not have a code ident. no., a delta " ⚠ " referring to the General Notes which will list manufacturer's name and address.

21.12 WELDING PROCEDURES (See Section 15)

21.13 THREADS (See Section 16)

21.14 DRAWING CHANGE PROCEDURE

21–4

21.14.1 GENERAL PROCEDURES — The change procedures and standard forms described in Section 18 will be used for initiating changes to tooling and recording changes to tool drawings.

21.14.1.2 When changes do not affect interchangeability of the tool, and it is capable of performing the function as before, the change shall be incorporated on the drawing as a letter revision.

21.14.1.3 When modifications are made to the extent that the original capabilities of a tool are destroyed, the modified tool should be re-identified with a new tool number, and an appropriate drawing prepared.

21.14.1.4 When identical tooling is in use simultaneously at more than one facility, and modifications are necessary to fit conditions peculiar to one facility that are not applicable or desirable at the others, a reidentification with new tool numbers is necessary.

NOTES

Section 22 — SHEET METAL DRAWINGS

22.1   SCOPE — This section establishes drafting and dimensioning practices pertinent to the preparation of drawings for parts to be fabricated from sheet metal. Additional information is presented to assist the draftsman in establishing proper dimensions for certain features of formed sheet metal parts.

22.2   APPLICABLE DOCUMENTS

DOD-STD-100	Engineering Drawing Practices
MIL-STD-403	Preparation for and Installation of Rivets and Screws, Rocket and Missile Structure
NAS 523	Code, Rivet
ANSI Y14.5	Dimensioning and Tolerancing for Engineering Drawings

22.3   DEFINITIONS

22.3.1   BEND ALLOWANCE — The length of material around a bend from bend line to bend line. (See Figures 1 and 2)

22.3.2   BEND ANGLE — The angle through which sheet metal is bent. It is measured from the flat through the bend to the finished angle after bending and is not to be confused with the included angle between the flange and adjacent leg. (See Figure 2)

22.3.3   BEND RADIUS — The minimum radius required to bend the material to prevent cracking or requiring additional work when forming the flanges. See Figures 1 through 4)

22.3.4   BEND RELIEF — For optimum forming, interfering material at flange extremities is removed to a point behind the bend line to prevent cracking of the material when forming the flanges. See Figures 6 through 8

22.3.5   BEND LINE — The line of tangency where a bend changes to a flat surface. There are two bend lines for each bend. (See Figure 1)

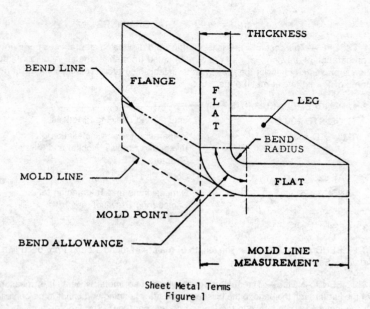

Sheet Metal Terms
Figure 1

22.3.6 BEVEL, CLOSED — The condition of a flange bent more than 90° from its flat condition.

22.3.7 BEVEL, OPEN — The condition of a flange bent less than 90° from its flat condition.

22.3.8 BLANK — A flat sheet metal shape approximately of the correct size to make a finished part. The part is usually trimmed to size after forming.

22.3.9 CENTER LINE OF BEND — A radial line from the center of the bend radius which bisects the included angle between bend lines. (See Figure 3)

22.3.10 DEVELOPED LENGTH — The length of a flat part which can be bent to make a part depicted on a drawing. This length is always shorter than the sum of mold line dimensions on the part. (See Figure 5)

22.3.11   DIMPLING — Stretching a relatively small shallow indentation into sheet metal. Stretching metal into a conical flange for use of a countersunk rivet head or screw. Dimpling is stronger than countersinking and should be performed on thin panels.

22.3.12   FORM BLOCK LINE — The inside mold line of a part.

22.3.13   FLAT PATTERN — A flat layout of a formed sheet metal part which can be bent to make the finished part without trimming after forming.

22.3.14   JOGGLE — An offset in the face of a part which has an adjacent flange. (See Figures 9 and 10)

22.3.15   MOLD LINE — The line of intersection of two flat surfaces of a formed sheet metal part. The term alone usually refers to the outside mold line. See form block line also. (See Figure 1)

22.3.16   SET BACK — The amount of deduction in length when a flat pattern is developed across a bend. It represents the saving in material by going around a bend radius rather than around a square corner.

22.3.17   TOOL HOLES OR PIN HOLES — Holes without a functional purpose in the end product, used for aligning a part in the proper position on a die or form block, or for other fabrication or tooling purposes.

22.3.18   MEDIAN LINE — A neutral axis through a bend where there is no stretching or compressing, located approximately 44% of the material thickness from the inside surface of the bend. (See Figure 2)

22.4   GENERAL — The fabrication of sheet metal parts involves methods which are quite different from those used for fabricating machined parts and its is necessary that the draftsman understand these differences in order to properly prepare drawings. Sheet metal parts are frequently cut out of flat stock using developed patterns (templates) and then formed to finished parts. Other parts may be made from flat blanks and deep drawn or formed into complex contours by dies in a punch press or hydropress, or by drop hammer dies. These parts are usually trimmed after forming to produce the finished part. These are only some of the methods which force the establishment of certain special procedures for sheet metal drawings.

22.5 CHARACTERISTICS OF SHEET METAL BENDS — When metal is bent, there is usually a local thinning or thickening of the material. This happens because the material is compressed on the inside of the bend and stretched on the outside, causing some displacement and plastic deformation of the material. It has been determined that there is a line through bend, where no stretching or compression takes place. This line, called the median line, is located approximately 44% of the material thickness from the inside surface of the bend. The location of the median line forms the basis for bend allowance calculations. (See Figure 2)

22.5.1 BEND FORMULAS — The following paragraphs should aid in the preparation of flat pattern development for undimensioned drawings (see paragraph 22.15) and to provide information for calculating dimensions for bend reliefs. Flat development of sheet metal parts is not permitted on engineering drawings except undimensioned drawings.

22.5.1.1 BEND ALLOWANCE — The following empirical formula has been developed to determine the distance around a bend for all bend radii one inch or less: (.0078T + .0174R) times the number of degrees of bend where T = material thickness and R = inside bend radius. (See Figure 2)

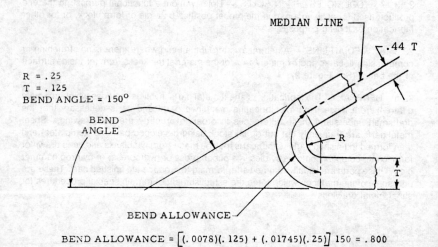

$$\text{BEND ALLOWANCE} = \left[(.0078)(.125) + (.01745)(.25)\right] 150 = .800$$

Bend Allowance
Figure 2

22.5.1.2 DISTANCE FROM BEND LINE TO MOLD LINE — This distance can be calculated by trigonometry using the following formulas: (See Figure 3)

$$X = (R + T)X\left(\tan\frac{a}{2}\right) \qquad X = \frac{(R + T)}{\cot\frac{a}{2}}$$

Where:
- X = distance from bend line to mold line
- R = inside bend radius
- T = material thickness
- a = angle in degrees

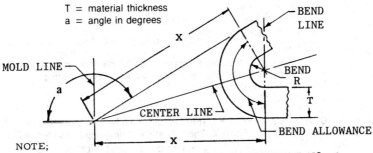

NOTE;
FOR 90° FLANGES, X = R+T SINCE THE TANGENT OF 45° = 1.

EXAMPLE:

If R = .25
T = .125
a = 150°

$X = (R + T)X\left(\tan\frac{a}{2}\right)$

X = 1.400

Bend Line - Mold Line Calculation
Figure 3

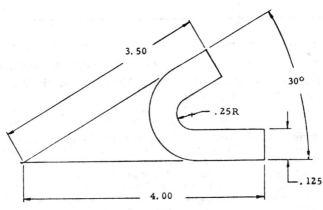

Bend Dimensioning
Figure 4

22.5.1.3 SET BACK — This is the deducation made in the length of a flat pattern development corresponding to the amount of material saved by bending around a radius instead of a sharp corner. It is equal to twice the distance from the bend line to the mold line minus the bend allowance. The formula is written as follows:

"K" = 2X — BA

Where:
K = Set back
X = Distance from bend line to mold line
BA = Bend allowance

Example:
Using the preceding examples, the "X" distance was calculated to be .1.400 and the bend allowance was .800; therefore,
K = 2 (1.400) — .800
K = 2.000

22.5.1.4 DEVELOPING A FLAT PATTERN — Using the bend in the previous example; the part is dimensioned as in Figure 4. The flat pattern is developed as shown in Figure 5.

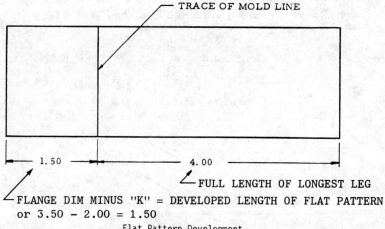

FLANGE DIM MINUS "K" = DEVELOPED LENGTH OF FLAT PATTERN
or 3.50 - 2.00 = 1.50

Flat Pattern Development
Figure 5

22.6 BEND RELIEF CUTOUTS — Whenever sheet metal bends intersect one another, it is necessary to remove material from the intersection area to prevent interference and buckling. Material must be removed at least .03 behind the intersection of bend lines. (See Figure 6.) There are many possible configurations for bend reliefs, two of which are illustrated in Figure 6. The usual practice is to radius the inside of a relief cut out, but this is not essential for the function of a relief cut out; a sharp corner will do as well if it removes material .03 beyond the intersectino of bend lines.

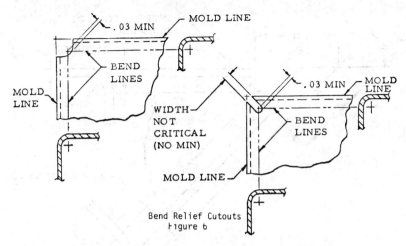

Bend Relief Cutouts
Figure 6

22.6.1 DIMENSIONING RELIEF CUTOUTS — The distance from the mold line to the bend line can be calculated by using the formula in Paragraph 22.5.1.2. This distance can be used in various ways to establish dimensions for the relief cutout, depending on the shape of the cutout. Two examples are shown in Figures 7 and 8 to illustrate the possibilities.

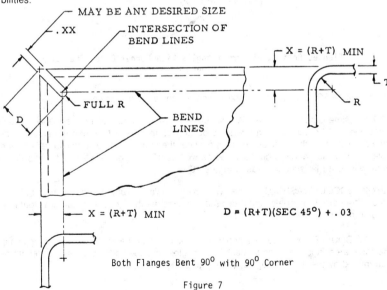

Both Flanges Bent 90° with 90° Corner
Figure 7

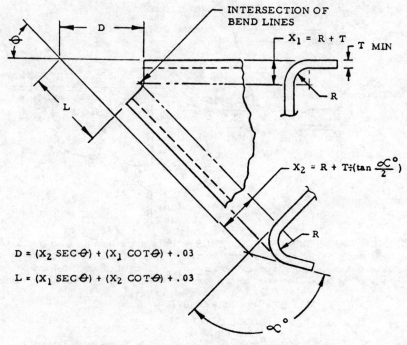

Flanges Bent at Different Angles With Corner Other Than 90°
Figure 8

22.7 JOGGLES — Joggles are shown and dimensioned as shown in Figure 9.

22.7.1 Dimensions "F" and "W" should have a tolerance of not less than ±.03. Dimension "D" should have a tolerance of not less than ±.010. When the angle in the face of the part is other than 90° to the mold line, the angle must be specified. It is preferable to allow the flange to be displaced as well as the face of the part, but if this is not permissible, it may be shown as in Figure 10.

22.7.2 JOGGLE WASHOUT — The "W" dimension varies with different materials, being greater for harder materials, and is dependent on the depth of joggle. Joggle washout may be determined by using Table I or II.

22.7.3 DOUBLE JOGGLES — When both the face of a part and the flange are joggled in the same area, it is preferred to make both joggles coincide and use a washout length as determined by the deepest joggle. (See Figure 11)

Example 1:

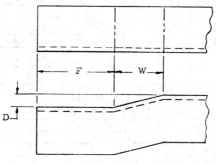

Example 2:

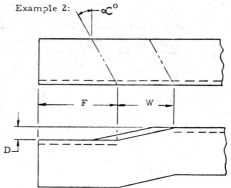

Joggles
Figure 9

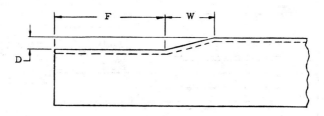

Joggle Without Flange Displacement
Figure 10

22-9

Table I
Standard Joggle Washout Factors for Steels

THICKNESS	1010 (B=4)		1020 1095 (B=4)		4130 * (B=4)				Corrosion-Resistant (B=4)			
					Normalized		Annealed		Annealed		1/2 Hard	
	A	C	A	C	A	C	A	C	A	C	A	C
.016			.099	.157					.009	.157	.009	.157
.018	.009	.161			.017	.286			.009	.161	.017	.286
.020			.010	.165					.010	.165	.017	.290
.025	.010	.175	.018	.300	.018	.300			.010	.175	.018	.300
.031			.018	.312					.011	.187	.018	.312
.032	.011	.189										
.035	.019	.320	.019	.320	.034	.570			.019	.320	.026	.445
.042	.020	.334	.020	.334	.034	.584			.020	.334	.027	.459
.050	.021	.350	.028	.475	.035	.600			.021	.350	.035	.600
.063									.022	.376	.037	.626
.065	.022	.380	.030	.505	.037	.630			.022	.380	.037	.630
.078	.031	.531	.039	.656	.046	.781			.031	.531	.046	.781
.083			.039	.666	.047	.791						
.094									.033	.563	.055	.938
.095	.033	.565	.055	.940			.033	.565				
.109	.042	.718	.057	.968			.042	.718				
.125	.044	.750	.059	1.000			.044	.750	.044	.750	.074	1.250
.156	.063	1.062	.077	1.312			.055	.937				
.188	.066	1.125	.096	1.625			.066	1.125	.066	1.125		
.250	.088	1.500	.118	2.000			.088	1.500	.088	1.500		

If joggle depth is greater than "A".   If joggle depth is less than "A".

Washout = B x Depth   Washout = $\sqrt{\text{Depth} (C - \text{Depth})}$

*Factors for 4137, 4337, and 8630 steels
are the same as for 4130 steel.

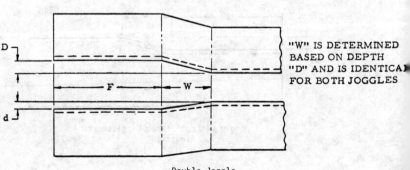

Double Joggle
Figure 11

Table II

STANDARD JOGGLE WASHOUT FACTORS FOR ALUMINUM ALLOYS

THICKNESS	2024-0 (B=3)		2024-T4 (B=4)		5052-0 6061-0 (B=3)		6061-T4 (B=4)		6061-T6 (B=4)		7075-0 (B=4)		7075-T6 (B=8)	
	A	C	A	C	A	C	A	C	A	C	A	C	A	C
.012	.015	.149	.016	.274	.009	.152	.009	.152	.009	.152				
.016	.016	.157	.017	.282	.009	.160	.009	.160	.009	.160	.017	.282	.008	.532
.020	.016	.165	.017	.290	.010	.168	.010	.168	.010	.168	.017	.290	.008	.540
.025	.030	.300	.019	.300	.010	.178	.010	.178	.018	.300	.018	.300	.008	.550
.032	.031	.314	.026	.439	.011	.192	.011	.192	.018	.314	.018	.314	.009	.564
.040	.033	.330	.027	.455	.019	.330	.019	.330	.027	.455	.019	.330	.013	.830
.051	.035	.352	.035	.602	.021	.352	.021	.352	.028	.477	.028	.477	.017	1.102
.064	.050	.503	.044	.753	.022	.378	.030	.503	.037	.628	.030	.503	.021	1.378
.072	.064	.644	.060	1.019	.023	.394	.038	.644	.045	.769	.045	.769	.025	1.644
.081	.066	.662	.068	1.162	.032	.537	.046	.737	.054	.912	.054	.912	.030	1.912
.091	.068	.682	.077	1.307	.033	.557	.055	.932	.062	1.057	.070	1.182	.034	2.182
.102	.083	.892	.093	1.579	.041	.704	.063	1.079	.071	1.204	.086	1.454	.038	2.454
.125	.100	1.000	.118	2.000	.052	.875	.074	1.250	.081	1.375	.096	1.625	.050	3.250
.156	.131	1.312	.151	2.562	.062	1.062	.084	1.437	.099	1.687	.129	2.188	.062	4.062
.188	.175	1.750	.221	3.750	.074	1.251	.110	1.876	.132	2.251	.155	2.625	.075	4.875
.250	.237	2.375	.323	5.500	.096	1.625	.176	3.000	.176	3.000	.206	3.500	.100	6.500

22.8 BEADS — Beads are raised or depressed areas in sheet metal parts, usually for the purpose of providing rigidity. It should be recognized that beads require stretching the material and normally are formed into parts with the use of dies. The plan view shows an outline of the bead at the mold line in phantom lines. It is necessary to show sections or breakouts to dimension the bead depth and the bead radii. When there is a straight section between two bend radii, the angle must be dimensioned. Figure 12 illustrates dimensioning for a bead.

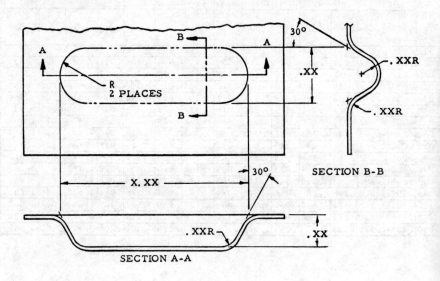

Beads
Figure 12

22.9 LIGHTENING HOLES — Lightening holes are cutouts in sheet metal parts usually for the purpose of reducing weight. They may be plain round holes or of other shapes and are dimensioned in the same manner as other cutouts. If a bead is incorporated around the periphery, the bead is dimensioned as explained in Paragraph 22.8 and the cutout is dimensioned to finished dimensions after forming the bead. It is not economical to specify hole size tolerances closer than $\pm.06$ on holes with beaded edges and even larger tolerances should be specified if the hole is not circular. It should be recognized that cutouts with beaded edges require the use of dies.

22.10   DIMPLING AND COUNTERSINKING

22.10.1   Dimpling and countersinking are the two methods used to install flat head fasteners in a flush condition. Thin materials are dimpled and thick materials are countersunk. The determination whether to dimple or countersink can be made by consulting tables in MIL-STD-403 which show the minimum and maximum material thickness applicable to dimpling and the minimum for countersinking. There is some overlap between the maximum for dimpling and the minimum for countersinking and the method used in this overlap area should be an engineering decision. All installations of flush head screws or rivets for rocket and missile structure should be prepared for 100° fasteners in accordance with the requirements of MIL-STD-403.

22.10.2   DRAWING CALLOUTS

22.10.2.1   When dimpling two or more sheets or a combination of dimpling the top sheet and countersinking the lower sheet is done, the dimple in the top sheet must nest perfectly in the dimple or countersink in the lower sheet. To achieve this condition, the sheets are matchdrilled with a pilot drill, dimpled and then the hole is drilled to finished size. For this reason, dimples are called out only on assembly drawings where all dimpled parts are shown in their assembled position. The finished hole size, only, is specified. Finished hole sizes for dimpled parts are shown in MIL-STD-403. Dimples are not dimensioned but are called out as follows:

   (1) DIMPLE FOR .093 DIA RIVET — ITEMS 1 AND 3 PER MIL-STD-403.
   (2) DIMPLE FOR .250 DIA SCREW — ITEMS 1 AND 3 PER MIL-STD-403.
   (3) DIMPLE FOR .250 DIA SCREW — ITEM 1 CSK 100°0' + 0°30' X .513/.523 DIA — ITEM 3 PER MIL-STD-403.

22.10.2.2   Dimples are shown the same as countersinks in the plan view. In section views, dimples are shown as a depression in the metal, matching the contour of the fastener head.

22.10.2.3   Countersinks are shown and dimensioned as shown in Section 10. Dimensions for countersinks alone and countersinks used in combination with dimples are shown in MIL-STD-403.

22.11   DIMENSIONING AND TOLERANCING SHEET METAL PARTS

22.11.1   DIMENSIONING AND TOLERANCING PARTS WITH STRAIGHT BENDS — Parts with straight bends which can be formed on a brake are dimensioned to outside mold lines and to inside bend radii. Dimensioning to inside mold lines is done only when the tolerance on the metal thickness cannot be allowed to affect the tolerance on the inside dimensions. When this condition exists, the part usually requires die forming to hold close tolerances. Parts to be formed on a brake should have tolerances of not less than ±.03 for bend radii and between bends. Tolerances larger than ±.03 are preferred. Smaller tolerances are possible for a dimension from a bend to the edge of the part, but trimming after forming is usually necessary when this is required. Figure 13 shows acceptable methods for dimensioning brake formed parts.

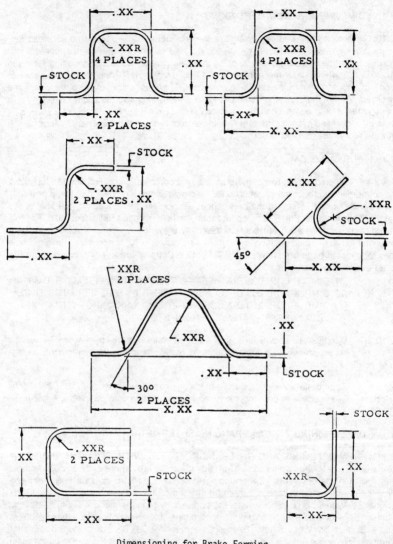

Dimensioning for Brake Forming
Figure 13

22.11.2 DIMENSIONING AND TOLERANCING DRAWN PARTS OR PARTS WITH CURVED BENDS — Parts which require stretching and/or compressing the material as in deep drawn parts or those with bends which are not straight are formed by means of dies or form blocks. Tolerances should be no less than ±.03 and larger tolerances are preferred. Dimensions may be given to either inside or outside mold lines depending on design requirements, but all dimensions are given to the same side of the material. Bend radii are given to the inside of the bend. Acceptable methods of dimensioning drawn parts are shown in Figure 14.

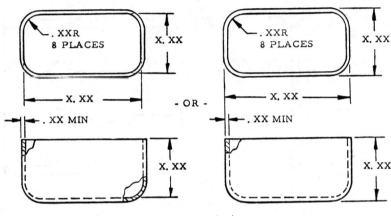

Drawn Part Dimensioning
Figure 14

22.11.3 DIMENSIONING SPUN PARTS — Parts formed by spinning are dimensioned to the inside surface so that dimensions given can be used directly to make the spinning block. Small tolerances can be held with increased cost, but tolerances should be as large as the design will permit. Thickness tolerance dpends on the severity of forming. It is preferred to specify thickness as the minimum acceptable limit.

22.11.3.1 CHAIN DIMENSIONING — When it is not convenient to show dimension lines for each dimension, it is permissible to use chain dimensions. In this method a group of dimensions from a common datum are placed in a single line with single arrowheads for each dimension except the first dimension, and the longest dimension which are on separate lines with arrowheads at each end of the lines. (See Figure 15)

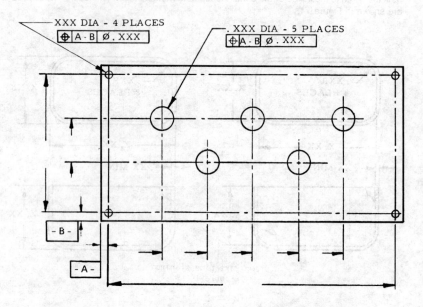

NOTE: All DIMENSIONS LOCATING TRUE POSITION ARE BASIC

Chain Dimensioning
Figure 15

22.11.3.2 TABULAR DIMENSIONING — When there are many holes in a part and it is not practical to use normal dimensioning methods, tabular dimensioning may be used. Datums are identified and arrowheads are used to indicate the "X" and "Y" directions. Each hole is identified with a letter and numeral. The same letter with a different numeral is assigned to all holes of identical size. A tabulation shown the hole size, number of holes, "X" and "Y" dimensions from datums and the positional tolerance for each hole. (See Figure 16)

22-16

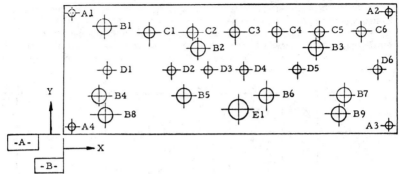

HOLE SYMBOL	HOLE DIAMETER	BASIC LOCATION X →	Y ↑	TRUE POSITION CONTROL SYMBOL
A1	.121	.250	4.750	⌖ |A|B| ⌀.020
A2	↑	10.750	4.750	↑
A3	↓	10.750	.250	
A4	.121	.250	.250	⌖ |A|B| ⌀.020
B1	.510	.750	4.500	⌖ |A|B| ⌀.125
B2	↑	4.125	3.250	↑
B3		9.000	3.250	
B4		.625	2.062	
B5		3.750	2.062	
B6		5.500	2.062	
B7		9.875	2.062	
B8	↓	.750	.875	↓
B9	.510	9.750	.875	⌖ |A|B| ⌀.125
C1	.281	3.000	4.250	⌖ |A|B| ⌀.062
C2	↑	4.500	4.250	↑
C3		5.625	4.250	
C4		6.875	4.250	
C5	↓	9.125	4.250	↓
C6	.281	9.750	4.250	⌖ |A|B| ⌀.062
D1	.221	.875	2.500	⌖ |A|B| ⌀.030
D2	↑	2.375	2.500	↑
D3		4.000	2.500	
D4		6.250	2.500	
D5	↓	8.000	2.500	↓
D6	.221	10.250	2.500	⌖ |A|B| ⌀.030
E1	.750	6.000	1.000	⌖ |A|B| ⌀.125

Tabular Dimensioning
Figure 16

22-17

22.11.3.3 PATTERN OF HOLES — When a group of holes must mate with another part, it is preferable to locate one hole in the pattern from established datums and using this hole center as an implied auxiliary datum, locate other holes in the pattern from this auxiliary datum. If overcrowded conditions exist in the area where this occurs or the pattern is repeated in other areas, remove a detail of the area and show dimensions from the auxiliary datums on the detail. (See Figure 17)

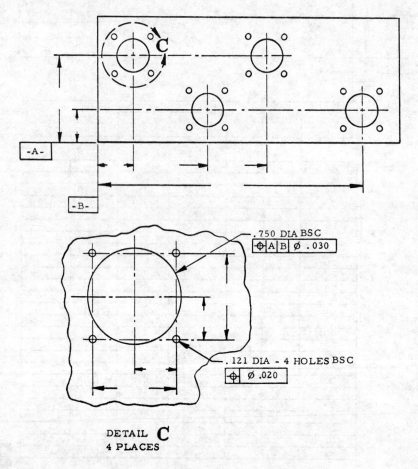

Hole Pattern Dimensioning
Figure 17

22.12 LIMITATIONS IN FORMING SHEET METAL

22.12.1 MINIMUM FLANGE WIDTHS FOR POWER BRAKE — Because of physical characteristics of power brakes, there are limitations on the minimum flange width that can be formed. The minimum flange widths depend on the material thickness and bend radius specified and are listed in Table III.

22.12.2 MINIMUM DISTANCE BETWEEN BENDS FOR POWER BRAKE — The minimum distance between bends formed by a power brake may be determined by adding the minimum flange width from Table III to the bend radius plus the material thickness.

Table III
Minimum Flange Widths Formed on Power Brake

Gage	PARTS UP TO 48.0 LONG Bend Radii												
	.03	.06	.09	.12	.16	.19	.22	.25	.28	.31	.38	.44	.50
							W						
.016	.080	.141	.203	.266	.328	.391	.453	.516	.578	.641			
.020	.091	.145	.207	.270	.332	.395	.457	.520	.582	.645	.770		
.025	.107	.150	.212	.275	.337	.400	.462	.525	.587	.651	.775	.900	1.025
.032	.128	.158	.219	.282	.344	.407	.469	.532	.594	.657	.782	.907	1.032
.040	.152	.182	.227	.290	.352	.415	.477	.540	.602	.665	.790	.915	1.040
.051		.215	.246	.301	.363	.426	.488	.551	.613	.676	.801	.926	1.051
.064			.285	.316	.376	.439	.501	.564	.626	.689	.814	.939	1.064
.072			.309	.340	.384	.447	.509	.572	.634	.697	.822	.947	1.072
.081				.367	.398	.456	.518	.581	.643	.706	.831	.957	1.081
.091				.397	.428	.466	.528	.591	.653	.716	.841	.966	1.091
.102					.462	.493	.539	.602	.664	.727	.852	.977	1.102
.125						.562	.593	.625	.687	.750	.875	1.000	1.125
.156							.717	.749	.781	.906	1.031	1.156
.188									.876	.938	1.063	1.188
.250											1.125	1.188	1.250

Minimum Distance Between Bends (L) Formed on Power Brake

Example:
T (Thickness) = .051
R (Bend Radius) = .25
W (Minimum) (from Table III) = .55
L (Minimum)
= W + R + T
= .55 + .25 + .05
= .85

22.12.3 MINIMUM BEND RADIUS — Refer to Tables IV through VIII for the minimum bend radius to be used for any specific material and material thickness. Whenever possible, use larger radii than the minimum values shown. Use the same bend radius for all bends in a part to avoid additional setup time for fabrication.

22.12.4 TOOLING OR PIN HOLES — When tooling holes or pin holes are required the drawing shall contain one of the following notes:
TOOLING HOLES PERMISSIBLE, LOCATION OPTIONAL.
(or)
TOOLING HOLES PERMISSIBLE THIS SURFACE ONLY.

22.13 UNDIMENSIONED DRAWINGS — An undimensioned drawing shows, to precise full scale, on environmental stable material, a part or assembly of parts.

22.13.1 USES AND ADVANTAGES — This type of drawing may be used for any kind of part, but has been most successfully applied to assembly drawings containing a large percentage of sheet metal and extruded parts which are joined with rivets or screws. The most significant advantages gained by using undimensioned drawings are:
1. Holes for joining parts are automatically coordinated for all parts being joined.
2. The drawing provides all template information so that no intermediate step is necessary to lay out templates.
3. Clearances, edge distances and interferences show more clearly, reducing the number of changes after release.

22.13.2 DRAWING REQUIREMENTS — The assembly is shown with all detail parts correctly located and drawn in the finished form to an accuracy of $\pm.005$ inch. The flanges of all sheet metal parts are then unfolded in flat pattern, using phantom lines, to the same accuracy. If parts are to be formed on form blocks, the form block line (FBL) is shown. All information required to make the necessary fabrication templates is shown, including location of rivet, screw or tooling holes, direction and angle of bends, bend radii and cutouts. (See Figure 18)

22.13.2.1 HOLES — Rivet holes are shown as intersecting center lines and coded in accordance with NAS 523. Screw holes are shown as circles with crossing center lines and are called out with a leader line. Tool holes are shown as crossing center lines with the letters "TH" in the upper left hand quadrant and the diameter in decimals in the lower right hand quadrant.

22.13.2.2 FORMING INFORMATION — In a flat pattern drawing, the direction of bend (up or down), the width of the flange, the bend angle and the bend radius are shown in the flange flat pattern. The mold line is coded, and, if a form block line (FBL) is shown and is identified. (See Figure 18)

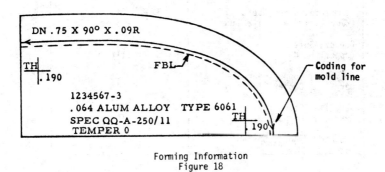

Forming Information
Figure 18

22.13.2.3 CUTOUTS — Cutouts are drawn to an accuracy of ±.005 inch, with the note "CUTOUT IN (Part No.)" placed in the cutout area. If necessary for clarity, arrowheads are shown with leaders from the note to points on the periphery of the cutout.

Table IV
Recommended Minimum Bend Radii - Brass, Copper and Bronze

Material	Minimum Bend Radius t = Material Thickness
Commercial Brass, Soft	2t
Commercial Brass, 1/2 Hard	3.5t
Copper, Soft	2t
Copper, Hard	11t
Beryllium Copper, Annealed	2t
Beryllium Copper, 1/4 Hard	4t
Beryllium Copper, 1/2 Hard	8t
Beryllium Copper, Hard	15t
Phosphor Bronze, Spring	25t

Table V
Approximate Bend Radii ① for 90-Degree Cold Bend - Aluminum Alloys

Alloy	Temper	1/64 in.	1/32 in.	1/16 in.	1/8 in.	3/16 in.	1/4 in.	3/8 in.	1/2 in.
1100	-0	0	0	0	0	0-1t	0	0	1t-2t
	-H12	0	0	0	0	0-1t	0-1t	0-1t	1t-3t
	-H14	0	0	0	0	0-1t	0-1t	0-1t	2t-3t
	-H16	0	0	0-1t	1/2t-1-1/3t	1t-2t	1-1/2t-3t	2-1/2t-3-1/2t	3t-4t
	-H18	0-1t	1/2t-1-1/2t	1t-2t	1-1/2t-3t	2t-4t	2t-4t	3t-5t	3t-6t
Alclad 2014	-0	0	0	0	0	0-1t	0-1t	1-1/2t-3t	3t-5t
	-T3	1t-2t	1-1/2t-3t	2t-4t	3t-5t	4t-6t	4t-6t	5t-7t	5-1/2t-8t
	-T4	1t2t-3t	2t-4t	3t-5t	4t-6t	4t-6t	5t-7t	5-1/2t-8t	
	-T6	2t-4t	3t-5t	3t-5t	4t-6t	5t-7t	6t-10t	7t-10t	8t-11t
2024	-0 [2]	0	0	0	0	0-1t	0-1t	1-1/2t-3t	3t-5t
	-T3 [2,3]	1-1/2t-3t	2t-4t	3t-5t	4t-6t	4t-6t	5t-7t	6t-8t	6t-9t
	-T36 [2]	2t-4t	3t-5t	4t-6t	5t-7t	5t-7t	6t-10t	7t-10t	8t-11t
	-T4	1-1/2t-3t	2t-4t	3t-5t	4t-6t	4t-6t	5t-7t	6t-8t	6t-9t
	-2t-5t	4-1/2t-6t	5t-7t	6-1/2t-8t	7t-9t	8t-10t	9t-11t	9t-12t	
	-T86	4t-5-1/2t	5t-7t	6t-8t	7t-10t	8t-11t	9t-11t	10t-13t	10t-13t
3003	-0	0	0	0	0	0	0	0	1t-2t
	-H12	0	0	0	0	0-1t	0-1t	0-1t	1t-3t
	-H14	0	0	0	0-1t	0-1t	1/2t-1-1/2t	1t-2-1/2t	1-1/2t-3t
	-H16	0-1t	0-1t	1/2t-1-1/2t	1t-2t	1-1/2t-3t	2t-4t	2-1/2t-4t	3t-5t
	-H18	1/2t-1-1/2t	1t-2t	1-1/2t-3t	2t-4t	3t-5t	4t-6t	4t-7t	5t-8t
3004	-0	0	0	0	0	0	0-1t	1/2t-1-1/2t	1t-2t
	-H32	0	0	0	0-1t	0-1t	0-1t	1t-2t	1-1/2t-2-1/2t
	-H34	0	0	0-1t	1/2t-1-1/2	1t-2t	1-1/2t-3t	2t-3t	2-1/2t-3-1/2t
	-H36	0-1t	1/2t-1-1/2t	1t-2t	1-1/2t-3t	2t-45	2t-4t	2-1/2t-5t	3t-5-1/2t
	-H38	1/2t-1-1/2t	1t-2t	1-1/2t-3t	2t-4t	3t-5t	4t-6t	4t-7t	5t-8t

Table V (continued)
Approximate Bend Radii[1] for 90-Degree Cold Bend

| Alloy | Temper | \multicolumn{8}{c}{Radii for Various Thicknesses Expressed in Terms of Thickness "t"} | | | | | | | |
		1/64 in.	1/32 in.	1/16 in.	1/8 in.	3/16 in.	1/4 in.	3/8 in.	1/2 in.
5005	-O -H12 -H14 -H16 -H18 -H32 -H34 -H36 -H38	0 0 0-1t 1/2t-1-1/2t 0 0 0-1t 1/2t-1-1/2t	0 0 0-1t 1t-2t 0 0 0-1t 1t-2t	0 0 1/2t-1-1/2t 1-1/2t-3t 0 0 1/2t-1-1/2t 1-1/2t-3t	0 0-1t 1t-2t 2t-4t 0 0-1t 1t-2t 2t-4t	0 0-1t 1-1/2t-3t 3t-5t 0-1t 1-1/2t-3t 3t-5t	0-1t 1/2t-1-1/2t 2t-4t 4t-6t 0-1t 1/2t-1-1/2t 2t-4t 4t-6t	0 0-1t 1t-2-1/2t 2-1/2t-4t 4t-7t 0-1t 1t-2-1/2t 2-1/2t-4t 4t-7t	1t-2t 1t-3t 1-1/2t-3t 3t-5t 5t-8t 1-1/2t-3t 3t-5t 5t-8t
5050	-O -H32 -H34 -H36 -H38	0 0 0-1t 1/2t-1-1/2t	0 0 0-1t 1t-2t	0 0 1/2t-1-1/2t 1-1/2t-3t	0 0-1t 1t-2t 2t-4t	0 0-1t 1-1/2t-1/2t 1-1/2t-3t 3t-5t	0 1/2t-1-1/2t 1t-2t 2t-4t 4t-6t	2t-3-1/2t 2t-4t 4t-7t	2-1/2t-4t 3t-4t 3t-5t 5t-8t
5052	-O -H32 -H34 -H36 -H38	0 0 0-1t 1t-2t	0 0 0-1t 1t-2t	0 0 1/2t-1-1/2t 1-1/2t-3t	0 0-1t 1t-2t 2t-4t	0 0-1t 1t-2t 2-1/2t-4t 3t-5t	0-1t 1/2t-1-1/2t 1-1/2t-3t 2t-4t 4t-6t	0-1t 1/2t-1-1/2t 1t-2t 2t-3t 2-1/2t-5t 4t-7t	1t-2t 1-1/2t-2-1/2t 2-1/2t-3-1/2t 3t-5-1/2t 5t-8t
5083	-O			0-1/2t	0-1t	0-1t	1/2t-1-1/2t	1-1/2t-2t	1-1/2t-2-1/2t
5086	-O -H32 -H34 -H36 -H112	0 0-1/2t 0-1t 0-1/2t	0 0-1t 0-1-1/2t 0-1/2t	1/2t-1-1/2t 1t-1-1/2t 1/2t-1t	0-1t 1t-2t 2t-3-1/2t 1-1/2t-3-1/2t	0-1t 1-1/2t-2t 2t-3t 2-1/2t-4t 1t-1-1/2t	1/2t-2t 1-1/2t-2-1/2t 2t-3t 3t-4-1/2t 1t-2t	1/2t-1-1/2t 2t-2-1/2t 2t-3-1/2t 3t-5t 1t-2t	1/2t-1-1/2t 2-1/2t-3t 3t-4t 3-1/2t-5-1/2t 1-1/2t-2-1/2t
5154	-O -H32 -H34 -H36 -H38 -H112	0 0-1t 0-1t 1t-2t	0 0-1t 1/2t-1-1/2t 1-1/2t-3t	0 0-1t 1/2t-1-1/2t 2t-4t	0 0-1t 1t-2t 2t-4t 3t-5t	0-1t 1t-2t 2t-3t 4t-6t	1/2t-1-1/2t 1-1/2t-2t 2t-4t 4t-6t 1/2t-3t	1/2t-1-1/2t 2t-4t 2-1/2t-4-1/2t 2-1/2t-5t 5t-8t 2-1/2t-4t	1t-2t 2-1/2t-4t 3t-6t 5t-8t 3t-5t
5456	-O -H321 -H323 -H343			1t-2t 1t-2t	0-1t 2t-3t 2t-3t 1-1/2t-3t	1/2t-1t 3t-4t 2t-4t 1-1/2t-2t-3-1/2t 2t-4t	1/2t-2t 3t-4t 2t-4t 2-1/2t-4-1/2t	1/2t-1-1/2t 3t-4t	1/2t-2t 3t-4t
5457	-O -H38	0 1/2t-1-1/2t	0 1t-2t	0 1-1/2t-3t	0 2t-4t	3t-5t	4t-6t	0 4t-7t	1t-2t 5t-8t
6061	-O -T4 [2] -T6 [2]	0 0-1/2t 0-1t	0 0-1t 1/2t-1-1/2t	0 1/2t-1-1/2t 1t-2t	0 1/2t-1-1/2t 1-1/2t-3t	0 1-1/2t-3t 2t-3t	0-1t 2t-4t 3t-4t	1/2t-2t 2-1/2t-4t 3-1/2t-5-1/2t	1t-2-1/2t 3t-5t 4t-6t
7075	-O -T6 [2]	0 2t-4t	0 3t-5t	0 4t-7t	1/2t-1-1/2t 5t-7t	1t-2t 6t-10t	1-1/2t-3t 7t-11t	2-1/2t-4t 7t-11t	3t-5t 7t-12t
7079	-O -T6	0 2t-4t	0 3t-5t	0 4t-6t	1/2t-1-1/2t 5t-7t	1t-2t 5t-7t	1-1/2t-3t 6t-10t	2-1/2t-4t 7t-11t	3t-5t 7t-12t
7178	-O -T6 [2]	0 2t-4t	0 3t-5t	0 4t-6t	1/2t-1-1/2t 5t-7t	1/2t-1-1/2t 5t-7t	1-1/2t-3t 6t-10t	2-1/2t-4t 7t-11t	3t-5t 7t-12t

[1] Minimum permissible radius over which sheet may be bent varies with nature of forming operation, type of forming equipment, and design and condition of tools. Minimum working radius for a given material or hardest alloy and temper for a given radius can be ascertained only by actual trial under contemplated conditions of fabrication.

[2] Alclad sheet can be bent over slightly smaller radii than the corresponding tempers of the uncoated alloy.

[3] Immediately after quenching, this alloy can be formed over appreciably smaller radii.

Table VI
Recommended Minimum Bend Radii Corrosion Resistant Alloys

T	19-9DI. & 17-7PH 18-8, Annealed Stabilized and Unstabilized	18-8, 1/4 H	18-8, 1/2 H	18-8, Hard
.010	.03	.03	.03	.09
.012	.03	.03	.03	.09
.016	.03	.03	.06	.09
.018	.03	.03	.06	.09
.020	.03	.03	.06	.12
.022	.03	.06	.06	.12
.025	.03	.06	.06	.16
.028	.03	.06	.09	.16
.032	.03	.06	.09	.16
.036	.03	.09	.09	.19
.040	.06	.09	.09	.22
.045	.06	.09	.12	.25
.050	.06	.09	.12	.25
.063	.06	.12	.16	.31
.071	.09	.16	.19	.38
.080	.09	.16	.19	.41
.090	.09	.16	.22	.47
.100	.12	.19	.25	.50
.112	.12	.22	.28	.56
.125	.16	.25	.31	.62
.160	.16	.31	.38	.78
.188	.19	.38	.44	.94
.250	.25	.50	.59	1.25

Table VII
Recommended Minimum Bend Radii - Steel and Steel Alloys

T	SAE 950 1010 1020 1025	Alloys 4130 Cond N
.020	.06	.06
.025	.06	.06
.032	.06	.09
.036	.06	.09
.040	.06	.12
.050	.09	.12
.063	.12	.16
.080	.16	.22
.090	.19	.25
.112	.22	.28
.125	.25	.31
.160	.31	.41
.188	.38	.47
.250	.50	.62
.375	.75	.94

Table VIII
Recommended Minimum Bend Radii - Magnesium Alloys at 70°F (Room Temperature)

T	FS1-O AMS 52S-O	FS1-H24 AMS 52S-H24
.012	.06	.12
.016	.09	.16
.020	.09	.19
.025	.12	.25
.032	.16	.31
.040	.19	.41
.050	.25	.50
.063	.31	.66
.071	.38	.75
.080	.41	.81
.090	.47	.91
.100	.50	1.03
.125	.62	1.25
.160	.78	1.56
.188	.94	1.88
.250	1.25	2.50
.375	1.88	3.75
.500	2.50	5.00

NOTES

Section 23 — PACKAGING DRAWINGS

23.1 SCOPE — This Packaging Design Drafting Procedures Section provides a standard for the preparation and interpretation of Company packaging drawings. This section is concerned with drawing format and procedure peculiar to packaging drawings. Packaging Drawing Procedures normally are the same as those of engineering drawings. Drawing procedures not covered here may be found in other sections of the Manual.

23.2 APPLICABLE DOCUMENTS

MIL-STD-129	Marking for Shipment and Storage
MIL-STD-1189	Standard Symbology for Marking Items of Supply, Unit Prices, Outer Containers, and Selected Documentation.

23.3 DEFINITIONS (For Complete Definitions, See: FED-STD-75.)

23.3.1 A DETAILED PACKAGING INSTRUCTION — A procedure or drawing applicable to all phases of the preservation, packaging and packing operation; ranging from the initial cleaning to the final shipping container.

23.3.2 PRESERVATION AND PACKAGING — The application or use of adequate protective measures to prevent deterioration including, as applicable, the use of appropriate cleaning and drying methods, preservatives, protective wrappings, cushioning, interior containers and complete identification marking, up to but not including the exterior shipping container, except when the unit container is also the shipping container.

23.3.4 PRESERVATION AND PACKAGING LEVELS — Preservation and packaging which will afford adequate protection against corrosion, deterioration and physical damage for:

LEVEL "A" — Shipment, handling, indeterminate storage and world-wide redistribution.

LEVEL "B" — Multiple reshipment, handling and storage conditions for periods normally not exceeding one year.

LEVEL "C" — Shipment from supply source to the first receiving activity for immediate use (supplier to user).

23.3.5 THE UNIT PACKAGE — The first tie, wrap, or container, applied to a single item or a group of items of a single part number, preserved or unpreserved, which involves a complete or identifiable package suitable for packing inside a shipping container; or items not held by a tie, wrap, or container, preserved or unpreserved, that are individually identified and placed directly into an intermediate package or exterior shipping container.

23.3.6 THE INTERMEDIATE PACKAGE is an interior container which contains two or more unit packages of identical items.

23.3.7 PACKING — The insertion of the unit or intermediate packages or unpackaged material into a shipping container and the application of required bracing, blocking, cushioning, and waterproof barriers. Packing is complete with the final closure, strapping, and marking of the shipping containers.

23.4 GENERAL INFORMATION

23.4.1 RESPONSIBILITIES OF DRAFTING AND DRAFTING MANAGEMENT — (See Section 2)

23.4.2 RESPONSIBILITIES OF THE CHECKER — (See Section 2)

23.4.3 RESPONSIBILITIES OF ENGINEERING AND ENGINEERING MANAGEMENT — (See Section 2)

23.4.4 The packaging drawing should reference the part number of the item to be packaged in the general notes, and the part may be outlined in phantom showing its position in the package.

23.5 DRAWING APPLICATION

23.5.1 DRAWING NOTES — General notes either apply to the package as a whole or are those which would become unnecessarily repetitive if placed at each point of application.

23.5.2 DELTA GENERAL NOTES — Used instead of Local Notes in limited space area, such as Title Block, Parts List, or on the field of the drawing to prevent repetition.

23.5.3 General notes are located in the upper left hand corner of the drawing. (See Section 5) They are numbered consecutively starting with "1"

23.5.4 Once a note has been assigned a number, the note may be revised but must not be replaced by a note of different application. (See Section 18)

23.5.5 On multisheet drawings, the general notes start on Sheet 1. (See Section 5)

23.5.6 The following are recommended examples of general notes used on packaging drawings as applicable.

    △ IDENTIFY PER _____ (use applicable dash number) WITH PART NUMBER.

    △ STENCIL IN ½" MINIMUM CHARACTERS ICC- (Insert applicable ICC number and gross weight to Interstate Commerce Commission approved containers only. Example: ICC-23F35).

    △ STENCIL IN ½" MINIMUM CHARACTERS BA- (Insert applicable BA number on Bureau of Explosives approved containers only).

    △ ESTIMATED NET WEIGHT OF PART x lbs.

    △ SKIDS TO BE ATTACHED BY NAILING THROUGH THE BASE INTO THE SKID A MINIMUM OF 1/3 THE SKID THICKNESS WITH TWO STAGGERED ROWS 2" APART AND 6" MAXIMUM BETWEEN NAILS.

    △ ITEM TO BE SUPPLIED AND INSTALLED BY THE COMPANY.

    △ On all main assembly drawings and major subassemblies, use:
    PACKAGING, PACKING, AND MARKING FOR OUT-OF-PLANT SHIPPING PER _____.

    △ BAR CODING SYMBOLOGY MARKINGS SHALL BE AS SPECIFIED IN MIL- STD-1189.

23.5.6 The proprietary and security classification notes, as specified in Section 5, will be placed on the drawing only when directed by the responsible engineer.

23.6 PARTS LIST

23.6.1 GENERAL INFORMATION

23.6.1.1 The Parts List for all packaging design drawings shall conform to DRM Section 9, except as modified herein.

23.6.1.2 Stock Size — Lumber callouts in the Parts List shall be called out in the nominal sizes.

Example: 2 x 4; 1 x 6

Comparison of Actual and Nominal Dimensions for Surfaced Lumber

Thickness (Inches)		Width (Inches)	
Nominal	Actual	Nominal	Actual
1	25/32	3	2-5/8
1-1/4	1- 1/16	4	3-1/2
1-1/2	1- 5/16	5	4-1/2
1-3/4	1- 7/16	6	5-1/2
2	1- 5/8	7	6-1/2
2-1/2	2- 1/8	8	7-1/4
4	3- 5/8	9	8-1/4
		10	9-1/4
		11	10-1/4
		12	11-1/4

23.6.1.3 Whenever possible, consider specifying commercial standard lengths and widths.

23.6.1.4 When special sizes of lumber are required, they shall be called out in the size required.

Example: 3/4 x 2-3/8

23.7 GENERAL DIMENSIONING PRACTICES FOR PACKAGING

23.7.1 Conditions not specifically covered should conform to the DRM, Section 10.

23.7.2 Dimensions shown on packaging drawings are usually expressed in fractions.

23.7.3 Tolerance on dimensions is usually $\pm 1/8$ inch.

23.7.4 The bilateral or unilateral tolerance system as defined in Section 10 may be used, but only one system shall appear on a single drawing.

23.7.5 The primary dimensions that indicate the size of containers are always given as the inside dimensions, in inches, in the following sequence:
   LENGTH X WIDTH X DEPTH for Square or Rectangular Containers.
   DIAMETER X LENGTH, depth or height for Round Containers.

23.8 PACKAGING IDENTIFICATION AND MARKING

23.8.1 Identification markings not covered will be found in Section 14.

23.8.2 MARKINGS

23.8.2.1 Routing marking requirements for shipment or storage (MIL-STD-129) are subject to change and should not be shown on packaging drawings. Special markings are either detailed or listed in the general notes.

23.8.2.2 Containers over 10 feet in length or those which are unbalanced shall be identified by stenciling or printing in 1 inch letters, the words "CENTER OF BALANCE" immediately above or along a 1 inch wide by 3 inch long vertical line locating the center of balance on both sides.

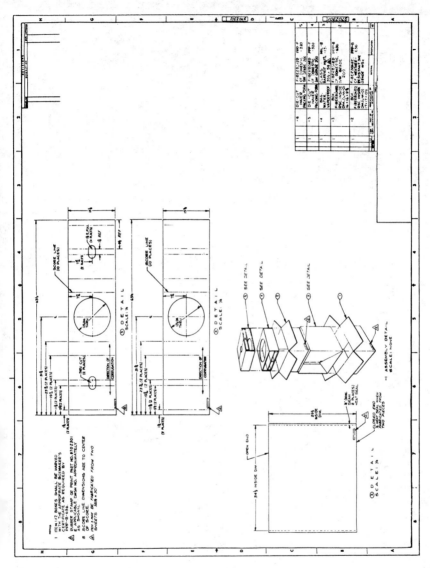

Packaging Drawing
Figure 1

NOTES

Section 24 — FACILITY DRAWINGS

24.1 SCOPE — This Section provides a standard for the preparation and interpretation of Facility Drawings. It is concerned with drawing types, format and procedure normally associated with the architectural and construction industry.

24.1.1 This section covers only those drawing types, format and practices that are either not included or are excepted in other sections of this manual.

24.1.2 The practices recommended in this section are concerned primarily with facility drawings that show designs of systems which are or will be interconnected to hardware. During the interconnected phase, the facility system or systems can be interpreted to be an integral part of the hardware. Then the facility system configuration may be subject to the same change and control as required of the hardware.

24.1.3 The general requirements and practices included are those necessary to prepare drawings required for "ON SITE" facility construction. The additional requirements necessary for shop detail and "OFF SITE" manufactured equipment are contained in other sections of this manual.

24.2 APPLICABLE DOCUMENTS

MIL-STD-15-2	Electrical Wiring Equipment Symbols for Ships Plans
DOD-STD-100	Engineering Drawing Practices
IEEE-91 (ANSI Y32.14)	Graphic Symbols for Logic Diagrams
TM-5-581B	Construction Drafting
NAVFAC-DM6	Construction Drafting
ANSI C37.20	**Switch Gear Assy, including Metal Enclosed Bus**
ANSI Y32.4	Graphic Symbols for Plumbing
ANSI Y32.9	Graphic Electrical Wiring Symbols for Architectural and Electrical Layout Drawings

24.3 DEFINITIONS

24.3.1 INTERCONNECTED FACILITY SYSTEMS DRAWINGS

24.3.1. DIAGRAMMATIC DRAWING — A diagrammatic drawing delineates features and relationship of items forming an assembly or system by means of symbol and lines. A diagrammatic drawing is a graphic explanation of the manner by which an installation, assembly or system (e.g., mechanical, electrical, electronic, hydraulic, pneumatic) performs its intended function. American National Standards Institute Standards ANSI Y14.15, ANSI Y32.2, ANSI C37.20, MIL-STD-15, Part 2 and ANSI Y32.9 provide directions for use of symbology relative to diagrammatic drawings.

24.3.1.1.1 SCHEMATIC DIAGRAM — A schematic or elementary diagram shows, by means of graphical symbols, the electrical connections and functions of a specific circuit arrangement. A schematic diagram facilitates tracing the circuit and its functions without regard to the actual physical size, shape, or location of the component devices or part. (See Section 17)

24.3.1.1.2   CONNECTION DIAGRAM — A connection or wiring diagram shows the electrical connections of an installation or of its component devices or parts. It may cover internal or external connections, or both, and contains such detail as is needed to make or trace connections that are involved. A connection diagram usually shows general physical arrangement of the component devices or parts. (See Section 17) A connection diagram may be prepared in the form of a "running (wire) list." (See Figure 5)

24.3.1.1.3   INTERCONNECTION DIAGRAM — An interconnection diagram is a form of connection or wiring diagram which shows only external connections between units, sets, groups, and systems.

24.3.1.1.4   SINGLE LINE DIAGRAM — A single line or one line diagram shows, by means of single lines and graphic symbols, the course of an electric circuit, or system of circuits, and the component devices or parts used therein. (See Section 17)

24.3.1.1.5   LOGIC DIAGRAM — A logic diagram shows, by means of graphic symbols, the sequence and function of logic circuitry. (See Figure 2) Logic diagrams shall be prepared per requirements of IEEE-91 (ANSI Y32.14)

24.3.1.1.6   MECHANICAL SCHEMATIC DIAGRAM — A mechanical schematic diagram illustrates the operational sequence or arrangement of mechanical devices. (See Sect 4)

24.3.1.1.7   PIPING DIAGRAM — A piping (hydraulic, pneumatic, or fluid) diagram depicts the interconnection of components by piping, tubing or hose, and, when desired, sequential flow of fluids in the system. (See Figure 3) Sufficient detail shall be shown to explain (a) the arrangement of the piping, valves, etc., or (b) operational sequence. Symbolic line representation may be used to distinguish functions of various parts. When the objective is to show arrangement, the following characteristics may be shown: routing of fluids, physical locations and arrangements of components, pipe diameters, types and sizes of fittings, flow, pressure, volume, etc.

24.3.2   HARDWARE is any separable assembly interconnected to a facility for purpose of test, evaluation or checkout.

24.3.3   KIT DRAWING (See Section 4)

24.3.4   PLANT EQUIPMENT DRAWINGS (See Section 4)

24.3.5   CONSTRUCTION DRAWING — A construction drawing delineates the design of buildings, structures, or related construction, ashore or afloat, individually or in groups, and is normally associated with the architectural-construction-civil engineering operations. Construction drawings establish all the interrelated elements of an architectural-civil engineering design, including pertinent services, equipment, utilities, and other engineering details. Maps, except those accompanying or used in conjunction with construction drawings, sketches, presentation drawings, perspectives, and renderings are not considered to be construction drawings.

24.3.5.1 ERECTION DRAWING — An erection drawing shows procedures and operation sequence for erection or assembly of individual items or assemblies of items. An erection drawing shall show the location of each part in the structure, identification markings, types of fastenings required, approximate weight of heavy structural members, controlling dimensions, and any other information which will contribute to erection of the structure.

24.3.5.2 PLAN DRAWING — A plan drawing depicts a horizontal projection of a structure, showing the payout of the foundation, floor, deck, roof, or utility system. (See Figure 4) As applicable, a plan drawing shall show shapes, sizes and materials of the foundation, its relation to the superstructure and its elevation with reference to a fixed datum plane, location of walls, partitions, bulkheads, stanchions, companionways, openings, columns, stairs, shapes and sizes of roofs, parapet walls, drainage, skylights, ventillators, etc. A plan drawing shall specify materials of construction and shall show the arrangement of structural framing. As applicable, the location of equipment or furniture may be indicated. Also a plan drawing for services may depict individual layouts for heating, plumbing, air conditioning, electrical or other utility systems.

24.3.5.3 PLOT (PLAT) PLAN DRAWING — A plot (plat) drawing depicts areas on which structures are clearly indicated with detailed information regarding their relationship to other structures, existing and proposed utilities, topography, boundary lines, roads, walks, fences, the property lines and locations, contours and profiles, shrubbery, sewer and waterlines, building lines, location of structures to be constructed, existing structures, finished grades and other pertinent data.

24.3.5.4 VICINITY PLAN DRAWING — A vicinity plan drawing (or vicinity map used with construction drawings) delineates the relationship of a site to features of the surrounding area, such as towns, bodies of water, railroads, highways, etc.

24.3.6 SPECIAL PURPOSE DRAWINGS — Special purpose drawings are other than end product drawings used to supplement end product requirements. These kinds of engineering drawings may be required for management control, logistic purposes, configuration management, manufacturing aids, and other functions unique to a Government Design Activity.

24.3.6.1 BOOK-FORM DRAWINGS — A book-form drawing is an assemblage of related data disclosing by means of pictorial delineations or technical tabulations, or combinations thereof, the engineering requirements of an item, a family of items, or a system. A book-form drawing is used for special purpose application in which it is expeditious to provide a document consisting of numerous small sheets, suitable for binding into book form. (See Figure 5) A book-form drawing shall preferably be prepared on ''A'' size drawing formats. Other standard size formats may be used provided the final original document size sheets are reduced to 11-inch height and can be folded to 8.5-inch width, with resultant legibility maintained. Book-form drawings shall not be prepared to circumvent the requirements for furnishing the types of drawings normally required for the delineation of an item or system.

24.3.6.2 RUNNING (WIRE) LIST — A running (wire) list is a book-form drawing consisting of tabular data and instructions required to establish wiring connections within or between units of an equipment, or between equipments, sets, or assemblies of a system. A running (wire) list is a form of interconnection diagram. (See Figure 5) Normally, the principal sections of a running (wire) list shall be as follows:
    a. Title sheet.
    b. Revision status of sheets tabulation.
    c. Table of contents.
    d. Referenced documents.
    e. Illustrations. The figures necessary to supplement the tabulations shall be provided.
    f. List of units. Units shall be listed by unit or reference designation and nomenclature.
    g. Summaries of cabling information by units.

24.3.7 FACILITIES DRAWINGS

24.3.7.1 Facilities drawings define the design of buildings, structures, sites, or related construction either individually or in groups. They shall establish all interrelated elements in engineering features of the design, including pertinent services, equipment, and utilities. The general requirements and practices are those necessary to prepare the drawings required for "on site" facility construction. The additional requirements necessary for shop detail and "off site" manufactured equipment are not included.

24.3.7.2 FACILITY DRAWING TYPES

24.3.7.2.1 CIVIL DRAWINGS — Graphic and symbolic representations of existing and/or planned surface features of a region showing the necessary construction required to develop a site. Natural and man-made features or objects (such as hills, valleys, streams, swamps, buildings and structures, power transmission lines, railroads, etc.), indicating their geometric configuration and physical relationship to other structures and boundary lines, are shown. Certain important imaginary lines (such as state, community, and property boundaries, zoning boundaries, building setbacks, coordinate grid system, etc.) are also indicated for record and reference purposes. Included in the general planning and layout of construction required to develop a site are drawings depicting structure location, grading, roads and paving, underground piping, yard structures, etc.

24.3.7.2.2 STRUCTURAL DRAWINGS — Engineering drawings that delineate such items as structural steel framing for buildings, towers, and other structures and the construction details for bridges, barges, buildings, and many other facilities components. These drawings establish the basis for the construction of the structural components of facilities. The delineation of structural drawings, by the use of symbols, dimensions, specifications, schedules, lists, reference codes, etc., describe the size and placement of beams, reinforcing steel, concrete, rivets, bolts, plates, welds, columns, etc.
Structural drawings are of three types:
    a. Structural concrete.
    b. Structural steel.
    c. Structural shop drawings.

24.3.7.2.3 ARCHITECTURAL DRAWINGS — The architectural requirements for buildings and other structures, including magnitude, appearance, interior and exterior materials, location for construction details of walls, partitions, foundations, floors, roofs, doors, windows, etc., and location and/or details of equipment such as lockers, shelves, tables, etc. For structures that are basically structural concrete, structural steel, or a combination of these, the architectural drawings become key or composite drawings. These drawings depict "go together" of all components, plus all other nonstructural details such as wall and roof materials and application, stair and handrail details, window and louver installation, suspended or acoustical ceiling details, built-in counters, cabinets, etc., and all other miscellaneous steel and iron work.

24.3.7.2.4 MECHANICAL DRAWING delineates piping to convey solids, liquids, or gases, the construction details for mechanical devices and air-conditioning installations, and the construction details for tanks, fire protection systems, etc. These drawings establish the requirements for construction and/or planning of interrelated elements of the facility design including pertinent services, equipment, and other features required to insure the performance of the mechanical equipment. These drawings incorporate dimensions, symbols, reference to codes, conventions, schedules, diagrams, etc. in describing the size and routing of pipes, the kind of material to be used, equipment criteria, duct sizes and shapes, amount of flow and the temperature of material in pipes and ducts, valve types and location, floor and wall penetrations, tank construction, and other facets of mechanical design.

24.3.7.2.5 ELECTRICAL DRAWINGS — Graphic representations of facilities electrical design requirements. These drawings shall be provided when essential for planning, procurement, construction, evaluation, recording, repair, maintenance, and use of the particular facilities.

24.3.8 DRAWING PACKAGE/SET — All drawings necessary to define the design of a facility or a system. It must contain all of those drawings prepared to delineate the work of a single contractor or a subcontractor.

24.3.9 CONSTRUCTION SPECIFICATIONS — Engineering requirements intended for primary use in construction which give clear, accurate, technical requirements for items, materials, utilities, and services including the procedure by which it will be determined that the design requirements of the facility have been met. The construction specification contains the scope of work; references applicable codes, construction practices, and installation requirements; recommends guides for materials, certain construction testing, and general design requirements that can be more adequately described in the specifications rather than in separate callouts on numerous drawings.

24.3.9.1 The construction specification and the set of drawings comprise the design package. Duplication of information between drawings and specifications shall be avoided.

24.3.9.2 In the case of conflict between the construction drawings and construction specifications, the specifications govern.

24.4 REVISIONS (CHANGES)

24.4.1 Revisions shall be made by erasure, crossing out, addition of information, or redrawing.

24.4.2 CROSSING OUT — When the crossing-out method is used, a series of parallel lines shall be placed on the face of the drawing (preferably at a 45° angle) in a manner not to obscure the information thus changed. The superseding data, if any, or reference to its location shall be placed near the portion crossed out and shall be indicated by a revision letter.

24.4.3 REVISION LOCATIONS — To be identified by one of the following methods:
  a. Revision symbols on field of drawing. (See Figure 6)
  b. Description in the revision block. (See Figure 6)
  c. Revision authorization document referenced on drawing. (See Figure 7)
  d. Zone in the revision authorization block. (See Figure 6)
  e. Combinations of a, b, c, and d above, when required for clarity.

24.4.3.1 EXCEPTION OF USE — When revision symbols are used and many changes are involved in one area such that separate revision symbols would crowd the drawing, a single revision symbol may be used to identify the changes provided sufficient supplementary information is included within the referenced revision authorization document.

24.4.4 DEVIATION OF USE OF REVISION SYMBOL — Revision symbols shall not be used on book-form drawings, master art-work or undimensioned drawings, and schematic or wiring diagrams where the use of such symbols may conflict with other symbols as used on these types of drawings.

24.5 IDENTIFICATION MARKING

24.5.1 See Section 14 for marking and identification of assemblies, end items, parts and plant equipment.

24.5.2 Systems may be identified for functional operation by assigning line numbers or schematic diagrams.

24.5.2.1

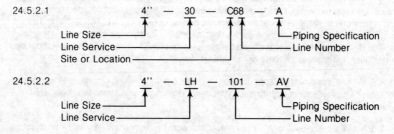

24.5.2.2

24.5.2.3 MISCELLANEOUS LINE IDENTIFICATION PROCEDURES

24.5.2.3.1 Stub connections on manifold and headers carry the manifold or header line number.

24.5.2.3.2 Parallel suction lines to two pumps, from common header, may carry the same line number.

24.5.2.3.3 Parallel discharge lines from two pumps to a common header may carry the same line number.

24.5.2.3.4 Distribution headers retain same line number through size changes. Branches off main header are assigned different line numbers than header.

24.5.2.3.5 When material specification change occurs in a pipe line, a different line number shall be assigned.

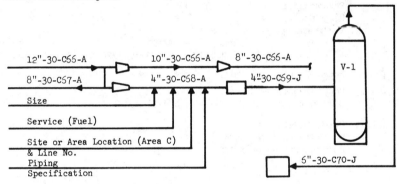

Example of System Identification

24.6 DRAWING SCALES

24.6.1 TYPES OF DRAWING SCALES — Use one of the following scale methods for selecting a proper drawing scale.

24.6.1.1 FRACTIONAL METHOD — The fractional method expresses in the form of a common fraction, the ratio of the size of the object as drawn to its true size.

24.6.1.2 EQUATION METHOD — The equation method expresses, in the form of an equation, the relatinoship of the size of the object as drawn to its true dimension.

24.6.2 TYPICAL SCALES — The following typical scales are recommended for various types of drawings.

24.6.2.1 CIVIL
Area Plot Plan: 1" = 200'
Grading, Plot Plan; Paving and Utilities Plan: 1" = 20' or 1" = 10"

24.6.2.2 STRUCTURAL/ARCHITECTURAL
Buildings and structure: 1/8" or 1/4" = 1'0"
Foundation details: 1/2" or 3/4" = 1'0"
Steel structure details: 3/4" or 1" = 1'0"

24.6.2.3 MECHANICAL
Piping: 3/8" = 1'0"
Vessels: Scale to suit sheet size
Plumbing: 1/8" or 1/4" = 1'0"
Isometrics: No Scale
Mechanical Specialties: Full Size 1/1
    Enlarged 2/1, 4/1, 10/1
    Reduced 1/2, 1/4, 1/10, 1/20 up to 1/100

24.6.2.4 ELECTRICAL
Wiring Diagram: No Scale
Building and Structures: 1/8", 1/4" or 1/2" = 1'0"

24.6.2.5 INSTRUMENTATION
Wiring Diagrams: No Scale
Building and Structures: 1/8" or 1/4" = 1'0"

24.6.3 BAR SCALES (Graphs)

24.6.3.1 Bar scales (graphs) shall be used on all drawings with the exception of schematics, sketches, wiring diagrams and tabulations, piping and shop drawings.

24.6.3.2 Place the Bar Scale a minimum of 2-1/2" to the left of the title block, and 1" above and parallel to the border line.

24.6.3.3 If more than one scale is used on any drawing, show only the maximum and minimum scales (Bar Graphs) used. Indicate other scale at the respective views. (See Figures 8 and 9)

24.7   DRAWING FORMAT AND SIZES

24.7.1   For drawing sizes and format see Section 5.

24.7.1.1   All drawings in bid/construction packages shall use one size (normally "F" size).

24.7.1.2   All drawings in a bid/construction package shall be single sheet and individually numbered.

24.7.1.3   The index sheet in a bid/construction package shall list all applicable drawings, reference drawings, and standard drawings.

24.7.1.4   The package of drawings shall be numbered consecutively by sheet number in respect to the total bid/construction package.

24.7.1.5   "Reference Only" Drawings are not considered as contract drawings and do not contain a sheet number of the package.

24.7.2   When more than one engineering discipline is involved such as civil, mechanical, electrical, etc., secondary top drawings should be made to show features peculiar to the discipline. Secondary top drawings and subordinate drawings must show the drawing type in the title.

24.7.3   SAMPLE FORMAT BLOCKS

24.7.3.1   TITLE BLOCK (See Section 5)

24.7.3.2   PARTS LIST

24.7.3.2.1   On drawings for shop detail, off-site manufactured equipment, or fabrication drawings, see Section 9.

24.7.3.2.2   Parts lists are not normally used on construction drawings.

24.7.3.3   REFERENCE DRAWING BLOCK — This block is for the entry of reference drawing numbers and titles. The following examples indicate the intended use of this block: (See Figure 10)
   a. Package index of all drawings.
   b. Floor plan drawings shall reference their related plot plans, steel framing plans and details.
   c. Installation drawings shall reference their related civil, structural, or plot plan drawings.
   d. Electro/Mechanical Assembly drawings shall reference their related schematic and wiring diagrams.
   e. Bay group drawings shall refernce their related interconnection diagrams, block diagrams, system schematics or tabulations.
   f. Schematic diagram drawings shall reference their related wiring diagrams, block diagrams or tabulations.

24.7.3.4 DRAWING NOTES (See Section 7)

24.7.3.5 TABULATION BLOCKS (Schedules) — (See Figures 4 and 10)

24.7.3.6 REVISION BLOCKS (Changes) — (See Figures 6 and 7)

24.7.3.7 LEGEND BLOCKS (See Figure 10)

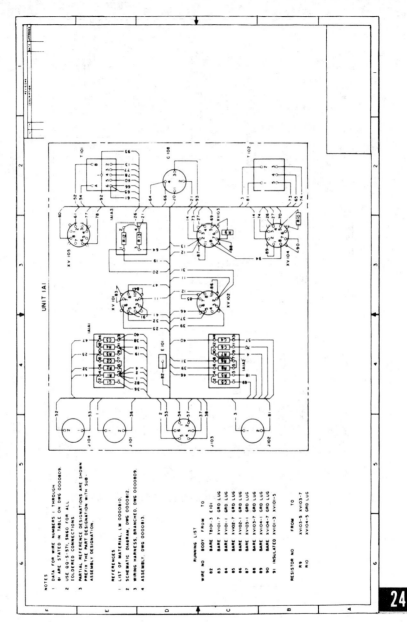

Connection Diagram
Figure 1
24-11

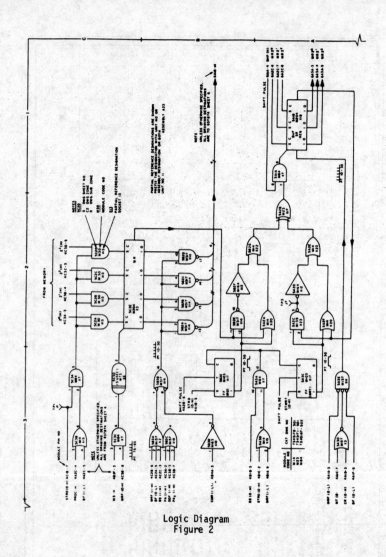

Logic Diagram
Figure 2

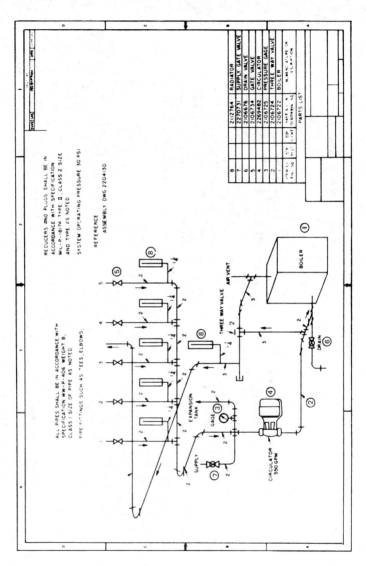

Piping Diagram
Figure 3

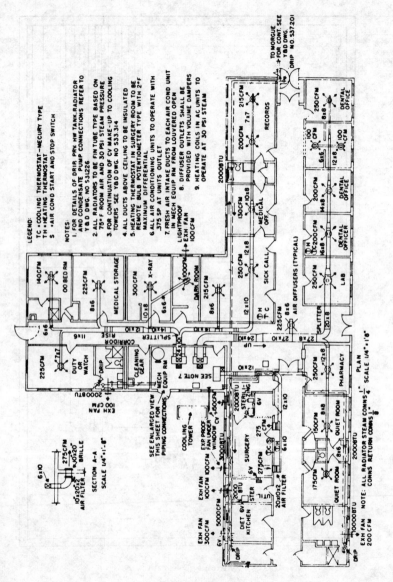

Plan Drawing
Figure 4

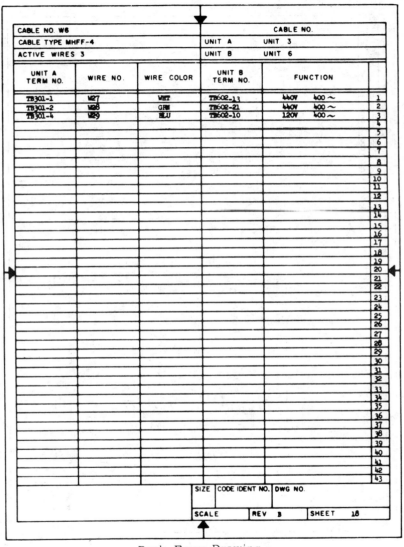

Book-Form Drawing
Cable Running Sheet
Figure 5

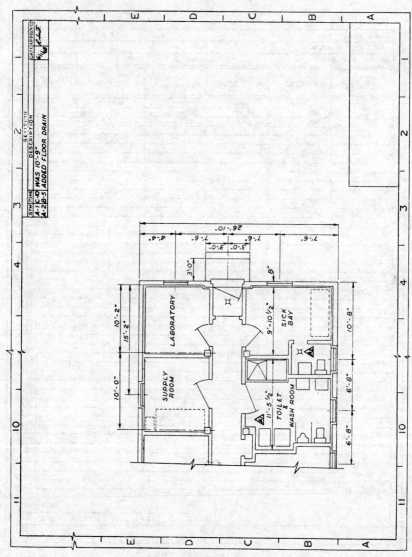

Revision Symbols on Field of Drawing
Figure 6

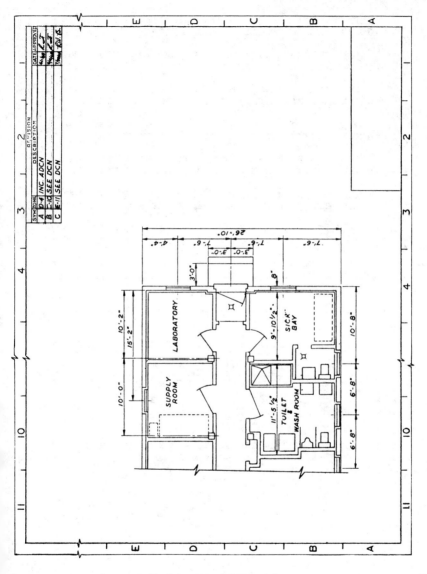

Revision Block Changes
Figure 7

Graphic Scale
Architectural Type
Figure 8

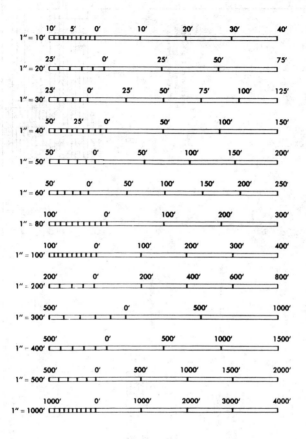

Graphic Scale
Engineering Type
Figure 9

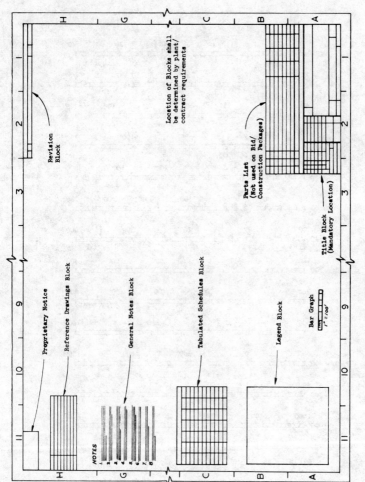

Location of Information Blocks on Drawing Vellum
Figure 10

Section 25 — DEFINITIONS

ACCEPTANCE: The act of an authorized representative of the Government by which the Government assumes for itself, or as an agent of another ownership of existing and identified supplies tendered, or approves specific services rendered, as partial or complete performance of the contract on the part of the contractor.

ACCESSORY: An item used in conjunction with or to supplement an assembly, unit or set, contributing to the effectiveness thereof without extending or varying the basic function of the assembly, unit, or set. An accessory may be used for testing, adjusting or calibrating purposes. (Examples: Recording camera for radar set, emergency power supply.)

ACTUAL SIZE: (See: Size, Actual).

ALTERED ITEM   An item taken from existing stock and altered (reworked) to meet new design requirements. (E.g., standard variable resistor with a shortened, flatted shaft.)

ALLOWANCE: The intentional difference between the MMC limits of size of mating parts; the minimum clearance (positive allowance) or maximum interference (negative allowance) between such parts. Also see "Fit".

ANGULAR DIMENSION: (See: Dimension, Angular).

ANGULARITY: The condition of a surface or axis at a specified angle from a datum plane or axis.

ANGULARITY TOLERANCE: (See: Tolerance, Augularity).

ASSEMBLY: A number of parts or subassemblies (or any combination thereof) joined together to perform a specific function.   (Examples: Power shovel-front, fan assembly, audio frequency amplifier.)

   NOTE: The distinction between an assembly and a subassembly is determined by the individual application. An assembly in one instance may be a subassembly in another where it forms a portion of an assembly.

ASSOCIATED LIST: A tabulation of engineering information pertaining to an item depicted on an engineering drawing or on a set of engineering drawings.

ATTACHMENT: An item used for physical connection to an assembly, unit, or set, contributing to the effectiveness thereof by extending or varying the basic function of the assembly, unit, or set. (Examples: Hoisting attachment on a truck, milling attachment for a lathe.) (compare "accessory")

AUXILIARY DIMENSION: (British term from BS308. Equivalent to "Reference Dimension").

AVERAGE DIAMETER: (See: Diameter, Average).

AXIS: A straight line (center line, ₵) about which a feature of revolution revolves; or about which opposite-hand features are symmetrical (original definition).

   NOTE: In numerical control nomenclature, three mutually perpendicular axes form the basis of a cartesian coordinate system. (See: ANSI-Y 14.5, appendix A).

BASE LINE: A configuration identification document (or a set of such documents) formally designated and fixed at a specific time during the life cycle of a configuration items. Base lines, plus approved changes from those base lines, constitute the current configuration identification. For configuration management there are three base lines, as follows:

   (a) Functional base line. The initial approved functional configuration identification.

   (b) Allocated Base line. The initial approved allocated configuration identification.

   (c) Product Base line. The initial approved or conditionally approved product configuration identification.

BASIC DIMENSION: (See: Dimension, Basic)

BASIC SIZE: (See: Size, Basic)

BILATERAL TOLERANCE: (See: Tolerance, Bilateral)

BULK MATERIALS: Those necessary constituents of an assembly or part such as oil, wax, solder, cement, ink, damping fluid, grease, powered graphite, flux, welding rod, thread, twine and chain for which the quantity required is not readily predetermineable or if knowing the quantity, the physical nature of the material is such that it is not adaptable to dipiction on a drawing; or which can be cut to finished size by the use of such hand or bench tools as shears, pliers, knives, etc., without any further machining operations and the configuration is such that it can be fully described in writing without the necessity of pictorial presentation. In addition, high usuage, low-cost items and hardware generally available; such as hinges, locks, light bulbs, fan belts, clamps, rivets, terminals sleeving, wire, nuts, bolts, screws and washers, etc. are considered bulk materials providing such materials are normally available in commercial channels and are normally procured in bulk quantities.

BURR: A featherlike cross section developed along the cut edge of a piece of material.

CENTER LINE: (See: Symmetry)

CENTER PLANE: (See: Datum)

CHECKER DRAWING: A person authorized by the design activity to validate engineering documentation.

CHECKING DRAWING: (See: "Validation").

CIRCULAR RUNOUT: (See: Runout, Circular)

CIRCULARITY: (See: Roundness)

CLEARANCE: The difference between actual sizes of mating features (e.g., the diametral space between hole and shaft), when that difference is positive (adapted from ANS1-SR11). Also see: Allowance, Fit.

    MAXIMUM CLEARANCE: In a clearance or transition fit, the difference between the maximum size of a female feature and the minimum size of a male feature. (i.e., a least material condition (LMC) fit). (adapted from: ANS1-SR11.)

    MINIMUM CLEARANCE: In a clearance fit, the difference between the minimum size of a female feature and the maximum size of a male feature (i.e., a maximum material condition (MMC) fit.) (adapted from: ANS1-SR11.)

CLEARANCE FIT: (See: Fit, Clearance)

COAXIALITY: (See: Concentricity)

CODE IDENTIFICATION NUMBER: A five digit number, assigned to each design activity, used in conjunction with a part or identify number in a parts list. This number (also referred to as a Federal Supply Code for Manufacturers (FSCM) is assigned by cataloging handbooks H4-1 and H4-2.

COMMERCIAL ITEM: A supply or service which (a) regularly is used for other than Government purposes and (b) is sold or traded in the course of conducting normal business operations. A service, per se, normally is not subject to delineation on engineering drawing.

COMPONENT: (See "part" or "unit".)

COMPANY FACILITY: The company facility that controls the drawing after release.

CONCENTRICITY: The condition of features of revolution or symmetry which have a common axis. (E.G., two or more concentric features such as circles, cylinders cones, spheres, squares, prisms, etc.) (Adapted from: ANS1-Y145, MIL-STD-8)

    NOTE: ANS1-Y14.5-1973 defines the term "coaxiality", but uses symtsology for "concentricity (coaxiality)."

CONCENTRICITY TOLERANCE: (See: Tolerance, Concentricity)

CONCEPTUAL: At an early stage of concept, probably not reflecting finalized configuration.

CONTRACT: An agreement or order for the procurement of supplies or services; an award or notice of award; a contract of a fixed-price, cost, cost-plus-a-fixed-fee, or incentive type; a contract providing for the issurance of job orders, task orders, or task letters thereunder; a letter contract or purchase order; a supplemental agreement with respect to any of the foregoing.

CONTRACTING OFFICER: Any person who, in accordance with Departmental procedures, is currently authorized to enter into and administer contracts and make determinations and findings with respect therto, or with any part of such authority. The term also includes the authorized representative of the contracting officer acting within the limits of his authority.

CONTRACTOR: Any individual, partnership, public or private corporation, association, institution, or other entity which is a party of the contract.

COORDINATE DIMENSIONING: (See: Dimensioning, Coordinate)

DASH NUMBER (PART SUFFIX): A number suffixed to a drawing number to identify individual parts or assemblies depicted and controlled by the drawing.

DATUM: Points, lines, planes or cylinders, and other geometric shapes assumed to be exact for purposes of computation, from which the location or geometric relationship (form) of features of a part may be established.

DESIGN ACTIVITY: An activity having responsibility for the design of an item; may be a Government activity or a contractor, vendor or others.

DESIGN AGENT: An activity contracted to or tasked to develop details of a design for which the design activity retains responsibility.

DEVELOPMENT: The systematic use of scientific knowledge which is directed toward the production of, or improvements in, useful products to meet specific performance requirements, but exclusive of manufacturing and production engineering.

DIAMETER, AVERAGE: An Average diameter on a nonrigid part is necessary to ensure the actual diameter of the feature can be restrained to the desired shape at assembly. An average diameter is the average of several diametral measurements across a circular or cylindrical feature. Normally, a sufficient number of measurements, usually no less than four, are taken to assure the establishment of an average diameter. If practicable, an average diameter may be determined by a peripheral tape measurement.

DIMENSION, ANGULAR: Angular dimensions are expressed in degrees, minutes and seconds. These are expressed by symbols; for degrees $°$, for minutes $'$, and for seconds $''$. Where degrees are indicated alone, the numerical value shall be followed by the symbol $°$. Where only minutes or seconds are specified, the number of minutes or seconds shall be preceded by $0°$ or $0°0'$, as applicable. Where desired, the angle may be given in degrees and decimal parts of a degree and the tolerance in decimal parts of a degree.

DIMENSION, BASIC: Basic Dimension. A numerical value used to describe the theoretically exact size, shape or location of a feature or datum target. It is the basis from which permissible variations are established by tolerances on other dimensions, in notes or by feature control symbols.

DIMENSION, COORDINATE: Rectangular coordinate dimensioning is where all dimensions are measured from two or three mutually perpendicular datum planes.

DOCUMENT: A specification, drawing, sketch, list, standard, pamphlet, report, or other information, relating to the design, procurement, manufacture, test or inspection of an item.

DOCUMENT IDENTIFICATION NUMBER: Consists of numbers or combinations of letters, numbers, and dashes. This number is assigned to a document, in addition to the title, for identification purposes.

DRAWING FORMAT: The standardized form, usually preprinted, upon which various constant information (design activity identification, standard tolerance block, etc.) is provided; together with spaces for variable information (drawing number, title, etc.)

DRAWING NUMBER: Consists of letters, numbers or combination of letters and numbers, which may or may not be separated by dashes. The number is assigned to a particular drawing for identification (file retrieval) purposes by the design activity.

DRAWING TYPE: Name applied to a drawing, descriptive of its context and end use. (See: section 4).

DUPLICATE ORIGINAL: A replica of an original engineering drawing made by a photo-duplicating technique, or a combination of a photo-duplicating technique and drafting on a medium (vellum, plastic base material, etc.) suitable for reproducing other reproducible and non-reproducible drawings.

END PRODUCT: An item in its final deliverable condition.

ENGINEERING DATA: Drawings, associated lists, accomanying documents, manufacturer specifications, and standards, or other information relating to the design, manufacture, procurement, test, or inspection of items or services.

ENGINEERING DEFINITION: A description expressed in engineering terms in sufficient detail to enable meeting the requirements of design, development, engineering, production, procurement or logistic support.

ENGINEERING DOCUMENT RELEASE: The process of transferring custody of an engineering document, or change thereto, from the preparing activity to a control activity which is responsible for its reproduction, distribution, storage, and the maintenance of change history records.

ENGINEERING DRAWING: An engineering document which discloses, by means of pictorial and/or textual presentations, the form and function of an item.

EXCHANGEABILITY OF ITEMS:

(a) INTERCHANGEABLE ITEM: One, which (1) possesses such functional and physical characteristics as to be equivalent in performance, reliability, and maintainability, to another item of similar or identical purposes; and (2) is capable of being exchanged for the other item (a) without selection for fit or performance, and (b) without alteration of the items themselves or of adjoining items, except for adjustment.

(b.) REPLACEMENT ITEM: One which is interchangeable with another item, but which differs physically from the original item in that the installation of the replacement item requires operations such as drilling, reaming, cutting, filing, shimming, etc., in addition to the normal application and methods of attachment.

(c.) SUBSTITUTE ITEM: One which possesses such functional and physical characteristics as to be capable of being exchanged for another only under specified conditions or in particular applications and without alteration of the items themselves or of adjoining items.

FEDERAL SUPPLY CODE FOR MANUFACTURERS (FSCM): Five digit code applicable to all activities which have produced or are producing items used by the Federal Government; also applies to Government activities which control design, or are responsible for the development of certain specifications, drawings, or standards which control the design of items. These codes are assigned in conformance with Cataloging Handbook H4, Federal Supply Code for Manufacturers, name to Code. Organizations which neither manufacture nor control design (such as dealers, agents, or vendors of items produced by others) are not included in H4.

FIND NUMBER: The item number from a parts list, used in the field of an assembly drawing to locate the item and to cross-reference to the item on the parts list. On an electronic assembly, a reference designation (per ANS1-Y32.16) may be used as a "find number".

FIT, CLEARANCE: The general term used to signify range of tightness or looseness which results from application of a specific combination of allowances and tolerances in mating parts.

FORMULATION: A mixture such as an explosive, filler, propellant, pyrotechnic, etc. Each formulation is discretely identified. Formulations are not to be construed as "bulk materials".

GOVERNMENT PROCUREMENT QUALITY ASSURANCE (PQA): The function by which the Government determines whether a contractor has fulfilled his contract obligations pertaining to quality and quantity. This function is related to and generally preceeds the act of acceptance.

GROUP: A collection of units, assemblies, or subassemblies which is not capable of performing a complete operational function. A group may be a subdivision of a set or may be designed to be added to extend the function or the utility of the set. (Example: Antenna group.)

INSEPARABLE: Incapable of being disassembled without destroying the intended function of the item.

INDIVIDUAL: The program manager or his designee.

INTERCHANGEABLE ITEM: (See: Exchangeability of items).

ITEM: A non-specific term used to denote any product, including system, material, part, subassemblie, set, accessorie, etc.

ITEM LEVELS: Item levels (as defined elsewhere from the simplest division to the more complex are as follows:

- Part
- Subassembly
- Assembly
- Unit
- Group
- Set
- Subsystem
- System

LIMITED PRODUCTION: Manufacture under model-shop conditions, as opposed to mass production under factory (production line) conditions.

LEVEL, DRAWING: (See Section 2)

LEVEL, EQUIPMENT: (See "item level").

MANUFACTURER: "Manufacturer" is a person or firm (a) who owns, operates or maintains a factory or establishment that produces on the premises the materials, supplies, articles, or equipment required under the contract or of the general character described by the specifications, standards, and publications: or (b) who, if newly entering into a manufacturing activity of the type described above, has made all necessary prior agreements for manufacture space, equipment, and personnel, to perform the manufacturing operations required for contract performance.

MATCHED PARTS: "Matched parts" are those parts, such as special application parts, which are machine matched, or otherwise mated, and for which replacement as a matched set or pair is essential.

MAY: Optional or non-mandatory.

MODELS (DEVELOPMENT AND PRODUCTION):

(a.) EXPLORATORY DEVELOPMENT: An item (preliminary parts or circuits) used for experimentation or tests to investigate or evaluate the feasibility and practicality of a concept, device, circuits, or system in breadboard or rough experimental form, without regard to the eventual overall fit or final form.

(b.) ADVANCED DEVELOPMENT: An item used for experimentation or tests to (a) demonstrate the technical feasibility of a design, (b) determine its ability to meet existing performance requirements, (c) secure engineering data for use in further development and, where appropriate, (d) establish the technical requirements for contract definition. Dependent upon the complexity of the equipment and the technological factors involved, it may be necessary to produce several successive models, to achieve additional objectives. The final advanced development model approaches the required form factor and employs standard parts (or nonstandard parts approved by the agency concerned). Serious consideration is given to military requirements such as reliability, maintainability, human factors and environmental conditions.

(c.) ENGINEERING DEVELOPMENT (SERVICE TEST): An item used in tests to determine tactial suitability for military use in real or simulated environments for which the item was designed. It closely approximates an initial production design, has the required form, employs standard parts (or nonstandard parts approved by the agency concerned) and meets the standard military requirements such as reliability, maintainability, human factors, environmental conditions, etc.

(d.) PREPRODUCTION (PROTOTYPE): An item suitable for complete evaluation of form, fit, and performance. It is in final form in all respects, employs standard parts (or nonstandard parts approved by the agency concerned), and is completely representative of final equipment.

(e.) PRODUCTION: An item in its final form of final production design made by production tools, jigs, fixtures and methods. It employs standard parts (or nonstandard parts approved by the agency concerned).

NON-PART DRAWING: An engineering drawing that provides requirements, procedures, instruction, etc., applicable to an item, when it is not convenient to include this information on the applicable part drawing. Examples include test requirements drawing, wiring diagram drawing, index drawing, etc.

ORIGINAL DRAWING: An original of a drawing is the drawing or "Marked" copy thereof on which is kept the revision record recognized as official by the design activity.

PART: One piece, or two or more pieces joined together which are not normally subject to disassembly without destruction or impairment of designed use. (Examples: Outer front wheel bearing of 3/4 ton truck, electron tube, composition resistor, screw, gear, mica capacitor, audio transformer, milling cutter.) Note: the term "component" should not be used to mean "part".

PART DRAWING: An engineering drawing that defines an item and assigns a part or control number to identify its configuration.

PART NUMBER: A number (or combination of numbers and letters) assigned to uniquely identify a specific item. The part number shall be or shall include the design activity drawing number.

PARTS LIST FORMAT: When parts lists are prepared integral with the drawing they shall include, as a minimum, columns shown in ANSI Y14.1. Column entries shall follow rules established for appropriate columns as provided in Section 9.

PART OR IDENTIFYING NUMBER (PIN): The part number or other identifying numbers which is entered in a parts list to identify a constituent item (part, assembly, material, process etc.) of an assembly.

PRODUCTION: The process of converting raw materials by fabrication and assembly into required material. It includes functions of production-scheduling, inspection, quality control, and related processes.

QUALITY ASSURANCE: A planned and systematic pattern of all actions necessary to assure that an item conforms to established technical requirements.

REFERENCED DOCUMENTS: A document which is cited on a drawing or list.

RELEASE: The transfer of custody of an engineering document, or change thereto, from the preparing activity to a control activity which is responsible for its reproduction, distribution, storage, and maintenance of change history records.

REPAIRABLE: Capable of being restored to original condition. An item whose parts are welded, encapsulated (or otherwise permanently joined) is usually non-repairable.

REPLACEMENT DRAWING: A new original drawing substituted for the previous original drawing of the same drawing numbers.

REPLACEMENT ITEM: See: "Exchangeability of Items".

REVISION: Any change to an original drawing after that drawing has been released for use.

REVISION AUTHORIZATION: A document such as a "Notice of Revision", "Engineering Change Notice" or "Revision Directive" which describes the revision in detail and is issued by the activity having the authority to revise the drawing.

REVISION SYMBOL: A letter (which maybe accompanied by a suffix number), used to identify particular revisions on the face of the drawing or in a revision description block. (Also see "suffix number").

ROUNDNESS: Roundness is a condition of a surface of revolution where:

(a) For a cylinder or cone, all points of the surface intersected by any plane perpendicular to a common axis are equidistant from that axis.

(b) For a sphere, all points of the surface intersected by any plane passing through a common center are equidistant from that center.

RUNOUT, CIRCULAR: Runout is a composite tolerance used to control the functional relationship of one or more features of a part of a datum axis. The types of features controlled by runout tolerances include those surfaces constructed around a datum axis and those constructed at right angles to a datum axis.

SECURITY CLASSIFICATION AND NOTATION: The applicable security classification, espionage, and downgrading notations shall be shown on engineering drawings requiring security classification in accordance with the DOD Industrial Security Manual for Safeguarding Classified Information, DOD 5220.22-M

SELECTED ITEM: An item, taken from existing stock and selected for refined (tighter) tolerances or performance without alteration. (E.g., a + 2% tolerance item selected from a + 5% standard line.)

SEPARABLE: Capable of being disassembled, without destruction of its intended use.

SET: A unit or units and necessary assemblies, subassemblies and parts connected together or used in association to perform an operational function. (Example: Radio receiving set, sound measuring set, radar homing set, which include parts, assemblies and units such as cables, microphone and measuring instruments.) ("Set" is also used to denote a collection of related items such as "tool set", "drawing set," or a "set" of tires.)

SHALL: Mandatory and binding

SIZE, ACTUAL: The measured size.

SIZE, BASIC: (See: Dimension, Basic)

SPECIFICATION: A clear and accurate description of the technical requirements for a material, a product or a service, including the procedure by which it can be determined that the requirements have been met.

STAKING: A method of securing screws, nuts, or bolts when a lock washer or locking wire is impracticable, may consist of staking varnish, glyptal, or upsetting metal with a punch. Staking will be used only when called for by print or visual aids.

STANDARD: A document which establishes engineering and technical limitations and application for items, materials, processes, methods, designs, drafting room, and other engineering practices. A standard may be issued by a design activity, a government agency, or by an industry association.

STANDARD, SHEET FORM: A standard prepared in format similar to MS sheet or MIL-STD unit page.

SUBASSEMBLY: Two or more parts which form a portion of an assembly or a unit replaceable as a whole, but having a part or parts which are individually replaceable. (Examples: Gun mount stand, window recoil mechanism, floating piston, telephone dial, IF strip, mounting board with mounted parts, power shovel dipper stick.)

SUBCONTRACTOR: A vendor who supplies items per the drawings/specifications of a higher-level design activity.

SUBSTITUTE ITEM: (See: "Exchangeability of items").

SUBSYSTEM: A combination of sets, groups, etc., which performs an operational function within a system and is a major subdivision of the system. (Examples: Data processing subsystem, guidance subsystem.)

SUFFIX NUMBER: Included in the revision symbol to distinguish between different changes located by the same revision letter; also to cross-reference such changes on the field of the drawing to the description thereof entered in the revision block.

SUPPLIER: (See: "vendor".)

SYMMETRICALLY OPPOSITE PARTS: Those parts which are mirror images of each other.

SYMMETRY: Symmetry is a condition in which a feature (or features) is symmetrically disposed about the center plane of a datum feature.

SYSTEM:
  (a.) GENERAL: A composite of equipment, skills, and techniques capable of performing or supporting an operational role, or both. A complete system includes all equipment, related facilities, material, software, services, and personnel required for its operation and support to the degree that it can be considered self sufficient in its intended operational environment. (Example: Dew Line.)

  (b.) ELECTRICAL-ELECTRONIC: A combination of two or more sets, which may be physically separated when in operation, and such other assemblies, subassemblies and parts necessary to perform an operational function or functions. (Examples: AEW electronic system, antiaircraft defense system, telephone carrier system, GCA electronic system, fire control system including the tracking radar, computer, and gun mount.)

TOLERANCE: The total amount by which a specific dimension is permitted to vary. The tolerance is the difference between the maximum and minimum limits.

TOLERANCE, ANGULARITY: Angularity is the condition of a surface or axis at a specified angle (other than 90°) from a datum plane or axis. An angularity tolerance specifies one of the following:

   (a) A zone defined by two parallel planes at the specified basic angle from a datum plane (or axis) within which the surface of the considered feature must lie.

   (b) A tolerance zone defined by two parallel planes at the specified basic angle from a datum plane (or axis) within which the axis of the considered feature must lie.

TOLERANCE, BILATERAL: A tolerance in which variation is permitted in both directions from the specified dimension.

TOLERANCE, CONCENTRICITY: Concentricity is the condition where the axes of all cross-sectional elements of a feature's surface of revolution are common to the axis of a datum feature.

TOLERANCE, UNILATERAL: A tolerance in which variation is permitted in one direction from the specified dimension.

UNIT: An assembly or any combination of parts, subassemblies and assemblies mounted together, normally capable of independent operation in a variety of situations. (Examples: Hydraulic jack, electric motor, electronic power supply, internal combustion engine, electric generator, radio receiver.) This term replaces the term "component." Note: The size of an item is a consideration in some cases. An electric motor for a clock may be considered as a part inasmuch as it is not normally subject to disassembly.

VALIDATION: The process by which the preparing activity for a document determines that the document reflects accurate and current requirements, including reference to current documents that are clearly and specifically applicable to the document being validated.

VENDOR: A design activity, manufacturer, wholesaler, or agent other than the prime contractor, from whom are acquired items used in the performance of the contract.

APPENDIX 1
Section A1
INTERNATIONAL SYSTEM OF UNITS (SI)
METRIC APPLICATION

A1.1 SCOPE — The scope of this APPENDIX is to present the information required by a draftsman or engineer to enable him to create a metric drawing, or to convert an existing inch drawing to a metric drawing.

A1.2 APPLICABLE DOCUMENTS.

ISO 1000	SI Units and Recommendations for the Use of their Multiples and of certain other units
NBS MISC PUB 286	Units of Weights and Measures
NBS-SP330	The International System of Units (SI)
NASA-SP7012	The International System of Units Physical Constants and Conversion Factors
ASTM-E380 (ANSI Z 210.1)	Metric Practice Guide

A1.3 DEFINITIONS — (U.S. BUREAU OF STANDARDS)

A1.3.1 UNIT OF LENGTH — **The meter**

A1.3.1.1 The **meter** is the length equal to 1 650 763.73 wavelengths in vacuum of the radiation corresponding to the transition between the levels $2p_{10}$ and $5d_5$ of the krypton - 86 atom.

A1.3.2 UNIT OF MASS — The kilogram

A1.3.2.1 The unit of mass is the mass of the international prototype of the kilogram: a platinum-iridium alloy cylinder, kept by the International Bureau of Weights and Measures near Paris, France.

A1.3.3 UNIT OF TIME — The second

A1.3.3.1 The second is the duration of 9 192 631 770 periods of the radiation corresponding to the transition between the two hyperfine levels of the ground state of the cesium 133 atom.

A1.3.4 UNIT OF ELECTRIC CURRENT— The ampere

A1.3.4.1 The ampere is that constant current which, if maintained in two straight parallel conductors of infinite length, of negligible circular cross section, and placed one metre apart in a vacuum, would produce between these conductors a force equal to 2×10^{-7} newton per **meter** of length.

A1.3.5 UNIT OF THERMODYNAMIC TEMPERATURE — The kelvin

A1.3.5.1 The kelvin (the SI unit of thermodynamic temperature) is the fraction 1/273.16 of the thermodynamic temperature of the triple point of water.

A1.3.6 UNIT OF AMOUNT OF SUBSTANCE — The mole

A1.3.6.1 The mole is the amount of substance of a system which contains as many elementary entities as there are atoms in 0.012 kilogram of carbon 12. The elementary entities must be specified and may be atoms, molecules, ions, electrons, other particles, or specified groups of such particles.

A1.3.7 UNIT OF LUMINOUS INTENSITY — The candela

A1.3.7.1 The candela is the luminous intensity, in the perpendicular direction, of a surface of 1/600 000 square **meter** of a blackbody at the temperature of freezing platinum under a pressure of 101 325 newtons per square **meter**.

A1.3.8 UNIT OF PLANE ANGLE — The radian

A1.3.8.1 The radian is the plane angle between two radii of a circle which cut off on the circumference an arc equal in length to the radius.

A1.3.9 UNIT OF SOLID ANGLE — The steradian

A1.3.9.1 The steradian is the solid angle which, having its vertex in the center of a sphere, cuts off an area of the surface of the sphere equal to that of a square with sides of length equal to the radius of the sphere.

A1.4 SI-THE SYSTEM of INTERNATIONAL UNITS — SI was created, and is controlled by the General Conference of Weights and Measures—an international treaty organization. SI was established to provide a system of well defined, coordinated units for science and industry, which would be the international language of measurement.

A1.5 SI UNITS — There are three classes of units in the international system:
1.) BASE UNITS
2.) SUPPLEMENTARY UNITS
3.) DERIVED UNITS

A1.5.1 BASE UNITS — The seven base units are:
1.) The **meter** — the unit of length
2.) The kilogram — the unit of mass
3.) The second — the unit of time
4.) The ampere — the unit of electric current
5.) The kelvin — the unit of thermodynamic temperature
6.) The mole — the unit of amount of substance
7.) The candela — the unit of luminous intensity

The seven base units are defined in paragraphs A1.3 through A1.3.7.

A1.5.2 SUPPLEMENTARY UNITS — The two supplementary units are:
1.) The radian
2.) The steradian

The two supplementary units are defined in paragraphs A1.3.8 and A1.3.9 and illustrated in Figures 1 and 2.

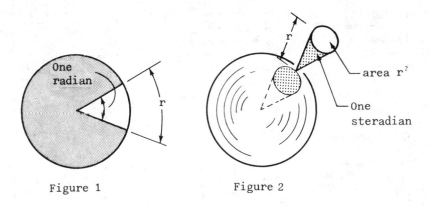

Figure 1   Figure 2

A1.5.3 DERIVED UNITS — Derived units are formed from base and supplemental units by multiplying, dividing, or raising the unit (s) to a positive or negative power. SEE TABLE 1.

TABLE 1

QUANTITY	UNIT	SYMBOL	DERIVATION
Area	square meter	m^2	
Volume	cubic meter	m^3	
Frequency	hertz	Hz	s^{-1}
Density	kilogram per cubic meter	kg/m^3	
Velocity	meters per second	m/s	
Angular velocity	radians per second	rad/s	
Acceleration	meters per second squared	m/s^2	
Angular acceleration	radians per second squared	rad/s^2	
Force	newton	N	$kg \cdot m/s^2$
Pressure	pascal	N/m^2	
Kinematic viscosity	square meter per second	m^2/s	
Dynamic viscosity	pascal second	$Pa \cdot s$	
Work, energy	joule	J	$N \cdot m$
Quantity of heat	joule	J	$N \cdot m$
Power	watt	W	J/s
Electric charge	coulomb	C	$A \cdot s$
Voltage	volt	V	W/A
Electric field strength	volt per meter	V/m	
Electric resistance	ohm	Ω	V/A
Electric capacitance	farad	F	$A \cdot s/V$
Magnetic flux	weber	Wb	$V \cdot s$
Inductance	henry	H	$V \cdot s/A$
Magnetic flux density	tesla	T	Wb/m^2
Magnetic field strength	ampere per meter	A/m	
Magnetomotive force	ampere	A	
Flux of light	lumen	lm	$cd \cdot sr$
Luminance	candela per square meter	cd/m^2	
Illumination	lux	lx	lm/m^2

A1.6 NON-SI UNITS WHICH MAY BE USED WITH SI — There are certain useful units of measurements which are not SI units, but are recognized, and may be used with SI. These units are:
1.) degree (of angle)
2.) degree Celsius (temperature)
3.) decibel
4.) minute
5.) hour
6.) day
7.) month
8.) year
9.) **Liter**

A1.7 OBSOLETE UNITS AND PREFIXES WHICH ARE TO BE AVOIDED —
1.) Erg
2.) Dyne
3.) Poise
4.) Stokes
5.) Gauss
6.) Oersted
7.) Maxwell
8.) Stilb
9.) Phot
10.) Micron
11.) Myria
12.) Fermi
13.) Metric Carat
14.) Torr
15.) kilogram-force
16.) Calorie
17.) Stere

A1.7.1 The word, "centrigrade" (relative to temperature) has been obsolete since 1948. The proper term is, "degree Celsius".

A1.8 SI PREFIXES

A1.8.1 SI unit-multiples and unit-submultiples are designated by adding a prefix to the unit's name.

A1.8.2 The prefix, and its multiplying value are shown in Table 2.

TABLE 2

PREFIX NAME	PREFIX VALUE	SCIENTIFIC NOTATION	
exa	one million million million times	1 000 000 000 000 000 000	10^{18}
peta	one thousand million million times	1 000 000 000 000 000	10^{15}
tera	one million million times	1 000 000 000 000	10^{12}
giga	one thousand million times	1 000 000 000	10^{9}
mega	one million times	1 000 000	10^{6}
kilo	one thousand times	1 000	10^{3}
hecto *	one hundred times	100	10^{2}
deka *	ten times	10	10^{1}
unit name	one time	1	
deci *	one tenth of	0.1	10^{-1}
centi *	one hundredth of	0.01	10^{-2}
milli	one thousandth of	0.001	10^{-3}
micro	one millionth of	0.000 001	10^{-6}
nano	one thousandth millionth of	0.000 000 001	10^{-9}
pico	one millionth millionth of	0.000 000 000 001	10^{-12}
femto	one thousandth millionth millionth of	0.000 000 000 000 001	10^{-15}
atto	one millionth millionth millionth of	0.000 000 000 000 000 001	10^{-18}

* Avoid using these multiples and submultiples whenever possible. Prefixes representing steps of 1000 are recommended.

A1.9 SI SYMBOLS — All branches of knowledge use symbols. Consequently certain letters may have one meaning when written as a capital letter and a different meaning when written in lower case. Consequently, care must be taken to use the correct symbol, properly! SEE TABLES 3, 4, 5 and 6.

A1.9.1 Tables 3 and 4 show symbols which are written in capital letters:

TABLE 3

UNIT NAME	PROPERTY	INDIVIDUAL HONORED	SYMBOL
ampere	electric current	Andre Ampere	A
degree Celsius	Celsius temperature	Anders Celsius	°C
coulomb	quantity of electricity	Chas. de Coulomb	C
farad	electric capacitance	Michael Faraday	F
henry	inductance	Joseph Henry	H
hertz	frequency	Heinrich Hertz	Hz
joule	energy	James Joule	J
kelvin	absolute temperature	Wm. (Lord) Kelvin	K
newton	force	(Sir) Isaac Newton	N
ohm	electrical resistance	George Simon Ohm	Ω
pascal	pressure, stress	Blaise Pascal	Pa
siemens	electrical conductance	Werner von Siemens	S
tesla	magnetic flux density	Nikola Tesla	T
volt	electromotive force	Alessandro (Count) Volta	V
watt	power	James Watt	W
weber	magnetic flux	Wilhelm Weber	Wb

A1.9.1.1 Although a person's name is begun with a capital letter, with the exception of Celsius, that same name is written with lower case letters when the name is used as a unit's name.

$$\begin{array}{ccc} \text{James Watt} & \text{10 watts} & \text{10 W} \\ \text{Andre Ampere} & \text{50 amperes} & \text{50 A} \end{array}$$

- The person's name is a proper noun.
- The unit's name is a common noun.
- But the symbol is written as a capital when it is derived from a proper name.

A1.9.1.2 The unit, degree Celsius is a compound symbol: °C. Celsius (both name and symbol) require a capital "C".

A1.9.1.3 When writing the multi-letter symbols Hz, Pa, and Wb, capitalize the first letter only.

A1.9.1.4 There are six prefix symbols, which are written with capital letters. See table 4.

TABLE 4

PREFIX	SYMBOL
exa	E
peta	P
tera	T
giga	G
mega	M
liter	L

A1.9.2 Table 5 and table 6 present symbols for units and prefixes which are written in lower case letters.

TABLE 5

UNIT NAME	SYMBOL
metre	m
kilogram	kg
second	s
mole	mol
candela	cd
radian	rad
steradian	sr
minute	min
hour	h
day	d

TABLE 6

PREFIX NAME	SYMBOL
kilo	k
hecto	h
deka	da
deci	d
centi	c
milli	m
micro	µ
nano	n
pico	p
femto	f
atto	a

Note: liter was changed to a capitol "L" by IEEE-STD-260-1977

A1.9.3 Prefix and unit combinations, and unit and unit combinations shall be indicated by combining the appropriate symbols. Proper use of capital and lower case letters shall be observed in order to preserve the integrity of the symbols.
Examples:
  megahertz   MHz
  kilohertz   kHz
  kilometer   km
Other examples of combined symbols can be seen in Table 1.

A1.9.4 Symbols are not abbreviations. Consequently they are not followed by a period, or a plural "S."

A1.10 SI DRAFTING SYMBOLS

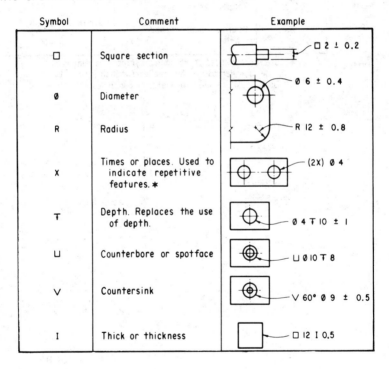

Symbol	Comment	Example
□	Square section	□ 2 ± 0.2
∅	Diameter	∅ 6 ± 0.4
R	Radius	R 12 ± 0.8
X	Times or places. Used to indicate repetitive features. *	(2X) ∅ 4
⊤	Depth. Replaces the use of depth.	∅ 4 ⊤ 10 ± 1
⊔	Counterbore or spotface	⊔ ∅ 10 ⊤ 8
V	Countersink	V 60° ∅ 9 ± 0.5
I	Thick or thickness	□ 12 I 0.5

A1.10 SI DRAFTING SYMBOLS (continued)

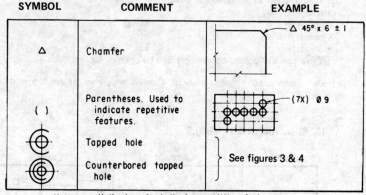

* Use a capitalized X to indicate repetitive features;
use a lowercase x for the dimensional joiner "by".
Do not use to indicate material quantities.

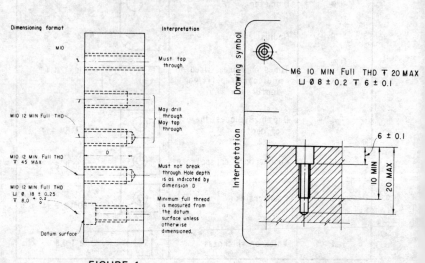

FIGURE 1
Examples of various tapped hole
designations and the interpretation

FIGURE 2
The dimensioning and interpretation
of a counterbored, tapped hole

APPENDIX
Section A2 — CONVERSION OF UNITS
(ALSO SEE SECTION 10.11 FOR CONVERSION TABLES)

A2.1 SCOPE — This section portrays the methods of converting the quantities of various systems of American measurement to the International System of Units (SI) and from the SI units to American units.

A2.2 APPLICABLE DOCUMENTS

ISO 1000	SI Units and Recommendations for the Use of their Multiples and of Certain Other Units.
NBS-SP330	The International System of Units (SI)
ASTM E380 (ANSI Z 210.1)	Metric Practice Guide

A2.3 DEFINITIONS — Not Applicable

A2.4 CONVERSION OF LINEAR UNITS —

A2.4.1 One inch equals 25.4 **millimeters**, exactly. Therefore, to convert any inch dimension to a **millimeter** dimension multiply the inch dimension by 25.4

A2.4.2 To convert a **millimeter** dimension to an inch dimension exactly, divide the **millimeter** dimension by 25.4.

A2.5 CONVERSION OF EXPLICIT TOLERANCES — Two methods of tolerance conversion are presented:

A2.5.1 Method A rounds to values nearest to the limits of the tolerance. Proceed as follows:
  1.) For each dimension, calculate the maximum and minimum limits.
  2.) Convert the corresponding inch values into **millimeter** values, as directed in paragraph A2.4.1
  3.) The use of the conversion factor (1" = 25.4 mm) generally produces converted values containing more decimal places than are required for the desired degree of accuracy. Therefore, the converted valves must be rounded to a practical degree of accuracy, as shown in Table 1.

A2.5.2 Method B rounds each limit toward the interior of the tolerance zone. Proceed as in Method A, steps (1) and (2). Then round each converted value toward the interior of the tolerance zone.

TOTAL TOLERENCE IN INCHES	ROUND THE MILLIMETER CONVERSION TO
More than .4	Whole millimeter
.4 to .04	1 decimal place
.04 to .004	2 decimal places
.004 to .0004	3 decimal places
.0004 to .00004	4 decimal places

TABLE 1

A2.6 CONVERSION OF TEMPERATURE UNITS

A2.6.1 The SI unit of temperature is the kelvin. (Ref. paragraph A1.3.5).

A2.6.2 Kelvin temperature is also called "absolute temperature" and "thermodynamic temperature." No conversion is required between values bearing these designations.

A2.6.3 Kelvin units and Celsius degrees are equal temperature increments, but the kelvin temperature scale and the Celsius temperature scale differ in the location of their zeros. See table 2.

A2.6.3.1 Subtract 273.15 from kelvin temperature (from "absolute temperature" or "thermodynamic temperature") to get Celsius temperature.

A2.6.3.2 Add 273.15 to Celsius temperature to get kelvin temperature. See table 2.

	Thermodynamic Temperature	Celsius Temperature
Water boils	373.15 K	100 °C
Body temperature	310.15 K	37 °C
Triple point of water	273.16 K	0.01 °C
Celsius Zero	273.15 K	0.00 °C
Absolute Zero	0.00 K	-273.15 °C

TABLE 2

NOTE:

1.) As laboratory techniques improved it was observed that the triple point of pure water was not 0.00 °C but was actually 0.01 °C. Therefore, you will see absolute zero as being 273.15 °C below Celsius zero, and as being 273.16 °C below the triple point of pure water.

2.) There must be a space placed between the temperature value and the symbol, °C as noted below:

$$100 \ °C$$

A2.6.4 The kelvin and the Rankine temperature scales share the same zero point (absolute zero) but the size of their degrees differ. The kelvin unit is 1.8 times larger than the Rankine degree.

A2.6.4.1 To change Rankine temperature to kelvin temperature, divide the Rankine temperature by 1.8. The quotient is kelvin units.

Rankine to kelvin

$$\text{kelvin} = \frac{\text{Rankine}}{1.8}$$

A2.6.4.2 To change Rankine temperature to Celsius temperature, divide the Rankine temperature by 1.8. From this quotient, subtract 273.15 to get degrees Celsius.

Rankine to Celsius

$$\text{Celsius} = \frac{\text{Rankine}}{1.8} - 273.15$$

A2.6.5 Although it is proper to say, "degree Celsius," it is incorrect to speak of kelvin degrees. Kelvin temperature is referred to as, "zero kelvin, or 240 kelvin, or 380.95 kelvin," etc. To change degrees Celsius to kelvin temperature, add 273.15 to degrees Celsius.

Celsius to kelvin

$$\text{kelvin} = \text{Celsius} + 273.15$$

A2.6.6 The kelvin and Fahrenheit temperature scales differ in their zero points and in the size of their degrees. The kelvin unit is 1.8 times larger than the Fahrenheit degree, and their zeros are 255.37 kelvin apart.

A2.6.7 To change Fahrenheit temperature to kelvin, add 459.67 to the Fahrenheit temperature and divide that sum by 1.8. The quotient is in kelvin units.

Fahrenheit to kelvin

$$\text{kelvin} = \frac{(\text{Fahrenheit} + 459.67)}{}$$

A2.6.8 To change Fahrenheit temperature to degrees Celsius, subtract 32 from the Fahrenheit temperature, and divide that difference by 1.8. The quotient is in degrees Celsius.

Fahrenheit to Celsius

$$\text{Celsius} = \frac{\text{Fahrenheit} - 32}{1.8}$$

A2.6.8.1 Chart No. 2 equates values of Fahrenheit and Celsius scales.

RELATIVE TEMPERATURES
CHART NO. 1

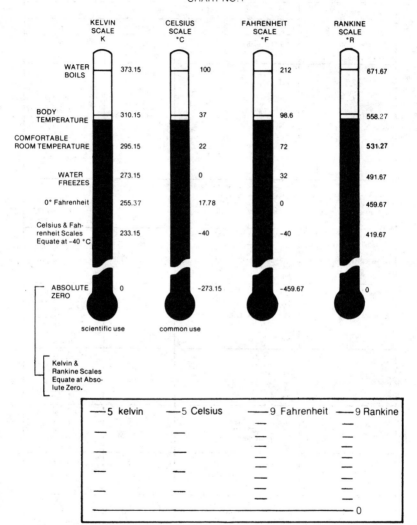

Relative Size of Degrees

A2.7 FREQUENTLY USED CONVERSIONS — A partial list of measurements more frequently used are listed in Table 3.

MEASUREMENT	TO CONVERT		
	FROM	TO	MULTIPLY BY
Acceleration meters/second^2 (m/s^2)	foot/second^2 (ft/s^2) m/s^2	m/s^2 ft/s^2	0.304 8 3.280 8
Area meter^2 (m^2)	acre $foot^2 (ft^2)$ hectare (ha) $inch^2$ (in^2) $mile^2$ acre ha m^2 m^2	m^2 m^2 m^2 m^2 m^2 ha acre ft^2 in^2	4 046.86 0.092 903 10 000. 0.000 645 16 2 589 988. 0.404 7 2.471 10.763 92 1 550.003
Bending Moment or Torque newton meter (Nm)	kilogram-force-meter(kgfm) ounce-force-inch (ozfin) pound-force-inch (lb fin) pound-force-foot (lb fft) Nm	Nm Nm Nm Nm lb fft	9.806 650 0.007 062 1.129 848 1.355 818 0.737 562
Energy (work) joule (J)	British thermal unit (Btu) calorie (cal) foot-pound-force (lb f) kilowatt-hour (kWhr) watt-hour (Whr) watt-second (Ws)	J J J J J J	1 055.056 4.186 8 1.355 818 3 600 000. 3 600. 1.000
Force newton (N)	dyne kilogram-force (kgf) kilopond ounce-force (ozf) pound force (lb f) N N	N N N N N kgf lb f	0.000 01 9.806 650 9.806 650 0.278 014 4.448 222 0.101 972 0.224 809
Length metre (m)	foot (ft) inch (in) mile (U.S. statute) (mi) m mm km	m m m ft in mi	0.304 8 0.025 400 1 609.344 3.280 84 0.039 37 0.621 371
Mass kilogram (kg)	ounce-mass (ozm) pound-mass (lb m) ton (long, 2240 lb m) ton (short, 2000 lb m) kg	kg kg kg kg lb m	0.028 349 52 0.453 592 4 1 016.047 907.184 7 2.204 6
Mass/Volume or Density kilogram/meter^3 (kg/m^3)	pound-mass/$foot^3$ (lb m/ft^3) pound-mass/$inch^3$ (lb m/in^3) kg/m^3	kg/m^3 kg/m^3 lb m/ft^3	16.018 46 27 679.90 0.062 428
Power watt (W)	Btu per hour (Btu/hr) foot-pound-force/second * horsepower (550 ftlbf/s) W W	W W W Btu/hr ftlbf/s	0.293 071 1.355 82 745.699 9 3.412 14 0.737 56

* (ftlbf/s)

TABLE 3

A2

MEASUREMENT	TO CONVERT		
	FROM	TO	MULTIPLY BY
Pressure or Stress Force/Area=Pascal (Pa)	kilogram-force/centimeter^2 (kgf/cm^2)	Pa	98 066.50
	kilogram-force/meter^2 (kgf/m^2)	Pa	9.806 650
	kilogram-force/millimeter^2 (kgf/mm^2)	Pa	9 806 650.
	pound-force/foot^2 (lb f/ft^2)	Pa	47.880 26
	pound-force/inch^2 (psi)	Pa	6 894.757
	MPa	kgf/mm^2	0.101 972
	MPa	psi	145.037 745
Temperature Kelvin (K)	degree Celsius (°C)	K	$tK = tC+273.15$
	degree Fahrenheit (°F)	K	$tK = (tF+459.67)/1.8$
	degree Fahrenheit (°F)	°C	$tC = (tF-32)/1.8$
	K	°C	$tC = tK-273.15$
	°C	°F	$tF = (1.8tC+32)$
	K	°F	$tF = 1.8(tK-273.15)+32$
Velocity or speed metres/second (m/s)	foot/second (ft/s)	m/s	0.304 8
	kilometer/hour (km/h)	m/s	0.277 778
	mile/hour (mph)	m/s	0.447 040
	mile/hour (mph)	km/h	1.609 3
	m/s	ft/s	3.280 8
	kph	mph	0.621 4
Volume metre^3 (m^3)	foot^3 (ft^3)	m^3	0.028 316 8
	gallon (U.S. liquid) (gal)	m^3	0.003 785 412
	inch^3 (in^3)	m^3	0.000 016 387
	liter (L)	m^3	0.001 000
	yard^3 (yd^3)	m^3	0.764 555
	liter (L)	qt	1.056 688
	m^3	yd^3	1.307 951
	quart (U.S. liquid) (qt)	1	0.946 353
	gal	1	3.785 412
Volume/Time metre^3/second (m^3/s)	foot^3/minute (ft^3/min)	m^3/s	0.000 47
	foot^3/second (ft^3/s)	m^3/s	0.028 317
	m^3/s	ft^3/s	35.314 475

TABLE 3 (Cont.)

WEIGHT CONVERSION TABLE - Pounds Into Kilos

lb	kg	lb	kg	lb	kg	lb	kg	lb	kg	lb	kg	lb	kg	lb	kg	lb	kg
1	.5	101	46.0	201	91.5	301	137.0	401	182.0	501	227.5	601	273.0	701	318.0	801	363.5
2	1.0	102	46.5	202	92.0	302	137.0	402	182.5	502	228.0	602	273.5	702	318.5	802	364.0
3	1.5	103	47.0	203	92.5	303	137.5	403	183.0	503	228.0	603	274.0	703	319.0	803	364.5
4	2.0	104	47.5	204	93.0	304	138.0	404	183.5	504	229.0	604	274.0	704	319.5	804	365.0
5	2.5	105	48.0	205	93.5	305	138.5	405	184.0	505	229.5	605	274.5	705	320.0	805	365.5
6	3.0	106	48.0	206	93.5	306	139.0	406	184.5	506	230.0	606	275.0	706	320.5	806	366.0
7	3.5	107	48.5	207	94.0	307	139.5	407	185.0	507	230.0	607	275.5	707	321.0	807	366.5
8	4.0	108	49.0	208	94.5	308	140.0	408	185.5	508	230.5	608	276.0	708	321.5	808	367.0
9	4.5	109	49.5	209	95.0	309	140.5	409	186.0	509	231.0	609	276.5	709	322.0	809	367.0
10	5.0	110	50.0	210	95.5	310	141.0	410	186.0	510	231.5	610	277.0	710	322.5	810	367.5
11	5.0	111	50.5	211	96.0	311	141.5	411	186.5	511	232.0	611	277.5	711	323.0	811	368.0
12	5.5	112	51.0	212	96.5	312	142.0	412	187.0	512	232.5	612	278.0	712	323.0	812	368.5
13	6.0	113	51.5	213	97.0	313	142.0	413	187.5	513	233.0	613	278.0	713	323.5	813	369.0
14	6.5	114	52.0	214	97.5	314	142.5	414	188.0	514	233.5	614	278.5	714	324.0	814	369.5
15	7.0	115	52.5	215	97.5	315	143.0	415	188.5	515	234.0	615	279.0	715	324.5	815	370.0
16	7.5	116	53.0	216	98.0	316	143.5	416	189.0	516	234.0	616	279.5	716	325.0	816	370.5
17	7.5	117	53.0	217	98.5	317	144.0	417	189.5	517	234.5	617	280.0	717	325.5	817	371.0
18	8.0	118	53.5	218	99.0	318	144.5	418	190.0	518	235.0	618	280.5	718	326.0	818	371.0
19	8.5	119	54.0	219	99.5	319	145.0	419	190.0	519	235.5	619	281.0	719	326.5	819	371.5
20	9.0	120	54.5	220	100.0	320	145.5	420	191.0	520	236.0	620	281.5	720	327.0	820	372.0
21	9.5	121	55.0	221	100.5	321	146.0	421	191.5	521	236.5	621	282.0	721	327.5	821	372.5
22	10.0	122	55.5	222	101.0	322	146.0	422	191.5	522	237.0	622	282.5	722	327.5	822	373.0
23	10.5	123	56.0	223	101.5	323	146.5	423	192.0	523	237.5	623	283.0	723	328.0	823	373.5
24	11.0	124	56.5	224	101.5	324	147.0	424	192.5	524	238.0	624	283.5	724	328.5	824	374.0
25	11.5	125	57.0	225	102.0	325	147.5	425	193.0	525	238.0	625	283.5	725	329.0	825	374.5
26	12.0	126	57.0	226	102.5	326	148.0	426	193.5	526	238.5	626	284.0	726	329.5	826	375.0
27	12.0	127	57.5	227	103.0	327	148.5	427	194.0	527	239.0	627	284.5	727	330.0	827	375.5
28	12.5	128	58.0	228	103.5	328	149.0	428	194.5	528	239.5	628	285.0	728	330.5	828	376.0
29	13.0	129	58.5	229	104.0	329	149.5	429	195.0	529	240.0	629	285.5	729	331.0	829	376.0
30	13.5	130	59.0	230	104.5	330	150.0	430	195.5	530	240.5	630	286.0	730	331.5	830	376.5
31	14.0	131	59.5	231	105.0	331	150.5	431	196.0	531	241.0	631	286.5	731	332.0	831	377.0
32	14.5	132	60.0	232	105.5	332	151.0	432	196.0	532	241.5	632	287.0	732	332.5	832	377.5
33	15.0	133	60.5	233	105.5	333	151.0	433	196.5	533	242.0	633	287.5	733	332.5	833	378.0
34	15.5	134	61.0	234	106.0	334	151.5	434	197.0	534	242.5	634	288.0	734	333.0	834	378.5
35	16.0	135	61.5	235	106.5	335	152.0	435	197.0	535	243.0	635	288.0	735	333.5	835	379.0
36	16.5	136	62.0	236	107.0	336	152.5	436	198.0	536	243.0	636	288.5	736	334.0	836	379.5
37	17.0	137	62.5	237	107.5	337	153.0	437	198.5	537	243.5	637	289.0	737	334.5	837	380.0
38	17.5	138	63.0	238	108.0	338	153.5	438	199.0	538	244.0	638	289.5	738	335.0	838	380.5
39	18.0	139	63.5	239	108.5	339	154.0	439	199.5	539	244.5	639	290.0	739	335.5	839	381.0
40	18.5	140	64.0	240	109.0	340	154.5	440	200.0	540	245.0	640	290.5	740	336.0	840	381.0
41	19.0	141	64.0	241	109.5	341	155.0	441	200.5	541	245.5	641	291.0	741	336.5	841	381.5
42	19.5	142	64.5	242	110.0	342	155.0	442	200.5	542	246.0	642	291.5	742	337.0	842	382.0
43	20.0	143	65.0	243	110.5	343	155.5	443	201.0	543	246.5	643	292.0	743	337.0	843	382.5
44	20.0	144	65.5	244	111.0	344	156.0	444	201.5	544	247.0	644	292.5	744	337.5	844	383.0
45	20.5	145	66.0	245	111.5	345	156.5	445	202.0	545	247.0	645	292.5	745	338.0	845	383.5
46	21.0	146	66.5	246	111.5	346	157.0	446	202.5	546	247.5	646	293.0	746	338.5	846	384.0
47	21.5	147	67.0	247	112.0	347	157.5	447	203.0	547	248.0	647	293.5	747	339.0	847	384.5
48	22.0	148	67.5	248	112.5	348	158.0	448	203.5	548	248.5	648	294.0	748	339.5	848	385.0
49	22.5	149	68.0	249	113.0	349	158.5	449	204.0	549	249.0	649	294.5	749	340.0	849	385.5
50	23.0	150	68.5	250	113.5	350	159.0	450	204.5	550	249.5	650	295.0	750	340.5	850	386.0

lb	kg
901	409.5
902	409.5
903	410.0
904	410.5
905	411.0
906	411.0
907	411.5
908	412.0
909	412.5
910	413.0
911	413.5
912	414.0
913	414.5
914	415.0
915	415.5
916	416.0
917	416.5
918	416.5
919	417.0
920	417.5
921	418.0
922	418.5
923	419.0
924	419.5
925	420.0
926	420.5
927	421.0
928	421.0
929	421.5
930	422.0
931	422.5
932	423.0
933	423.5
934	424.0
935	424.5
936	425.0
937	425.5
938	425.5
939	426.0
940	426.5
941	427.0
942	427.5
943	428.0
944	428.5
945	429.0
946	429.5
947	429.5
948	430.0
949	430.5
950	431.0

A2-8

51	23.5	151	68.5	251	114.0	351	159.5	451	205.0	551	250.0	651	295.5	751	341.0	851	386.5	951	431.5
52	24.0	152	69.0	262	114.5	352	160.0	452	205.5	552	250.5	652	296.0	752	341.5	852	386.5	952	432.0
53	24.5	153	69.5	253	115.0	353	160.5	453	205.5	553	251.0	653	296.5	753	342.0	853	387.0	953	432.5
54	24.5	154	70.0	254	115.5	354	161.0	454	206.0	554	251.5	654	297.0	754	342.5	854	387.5	954	433.0
55	25.0	155	70.5	255	116.0	355	161.5	455	206.5	555	252.0	655	297.5	755	342.5	855	388.0	955	433.5
56	25.5	156	71.0	256	116.5	356	161.5	456	207.0	556	252.5	656	298.0	756	343.0	856	388.5	956	434.0
57	26.0	157	71.5	257	117.0	357	162.0	457	207.5	557	253.0	657	298.5	757	343.5	857	389.0	957	434.5
58	26.5	158	72.0	258	117.5	358	162.5	458	208.0	558	253.5	658	298.5	758	344.0	858	389.5	958	435.0
59	27.0	159	72.5	259	117.5	359	163.0	459	208.5	559	254.0	659	299.0	759	344.5	859	390.0	959	435.0
60	27.5	160	73.0	260	118.0	360	163.5	460	209.0	560	254.5	660	299.5	760	345.0	860	390.5	960	435.5
61	28.0	161	73.5	261	118.5	361	164.0	461	209.5	561	254.5	661	300.0	761	345.5	861	391.0	961	436.0
62	28.5	162	73.5	262	119.0	362	164.5	462	210.0	562	255.0	662	300.5	762	346.0	862	391.5	962	436.5
63	29.0	163	74.0	263	119.5	363	165.0	463	210.5	563	255.5	663	301.0	763	346.5	863	391.5	963	437.0
64	29.5	164	74.5	264	120.0	364	165.5	464	210.5	564	256.0	664	301.5	764	347.0	864	392.0	964	437.5
65	29.5	165	75.0	265	120.5	365	166.0	465	211.0	565	256.5	665	302.0	765	347.5	865	392.5	965	438.0
66	30.0	166	75.5	266	121.0	366	166.5	466	211.5	566	257.0	666	302.5	766	347.5	866	393.0	966	438.5
67	30.5	167	76.0	267	121.5	367	166.5	467	212.0	567	257.5	667	303.0	767	348.0	867	393.5	967	439.0
68	31.0	168	76.5	268	122.0	368	167.0	468	212.5	568	258.0	668	303.5	768	348.5	868	394.0	968	439.5
69	31.5	169	77.0	269	122.5	369	167.5	469	213.0	569	258.5	669	303.5	769	349.0	869	394.5	969	440.0
70	32.0	170	77.5	270	122.5	370	168.0	470	213.5	570	259.0	670	304.0	770	349.5	870	395.0	970	440.0
71	32.5	171	78.0	271	123.0	371	168.5	471	214.0	571	259.5	671	304.5	771	350.0	871	395.5	971	440.5
72	33.0	172	78.5	272	123.5	372	169.0	472	214.5	572	259.5	672	305.0	772	350.5	872	396.0	972	441.0
73	33.5	173	78.5	273	124.0	373	169.5	473	215.0	573	260.0	673	305.5	773	351.0	873	396.0	973	441.5
74	34.0	174	79.0	274	124.5	374	170.0	474	215.5	574	260.5	674	306.0	774	351.5	874	396.5	974	442.0
75	34.5	175	79.5	275	125.0	375	170.5	475	215.5	575	261.0	675	306.5	775	352.0	875	397.0	975	442.5
76	34.5	176	80.0	276	125.5	376	171.0	476	216.0	576	261.5	676	307.0	776	352.0	876	397.5	976	443.0
77	35.0	177	80.5	277	126.0	377	171.5	477	216.5	577	262.0	677	307.5	777	352.5	877	398.0	977	443.5
78	35.5	178	81.0	278	126.5	378	171.5	478	217.0	578	262.5	678	308.0	778	353.0	878	398.5	978	444.0
79	36.0	179	81.5	279	127.0	379	172.0	479	217.5	579	263.0	679	308.0	779	353.5	879	399.0	979	444.5
80	36.5	180	82.0	280	127.5	380	172.5	480	218.0	580	263.5	680	308.5	780	354.0	880	399.5	980	445.0
81	37.0	181	82.5	281	127.5	381	173.0	481	218.5	581	264.0	681	309.0	781	354.5	881	400.0	981	445.0
82	37.5	182	83.0	282	128.0	382	173.5	482	219.0	582	264.0	682	309.5	782	355.0	882	400.5	982	445.5
83	38.0	183	83.5	283	128.5	383	174.0	483	219.5	583	264.5	683	310.0	783	355.5	883	401.0	983	446.0
84	38.5	184	83.5	284	129.0	384	174.5	484	220.0	584	265.0	684	310.5	784	356.0	884	401.0	984	446.5
85	39.0	185	84.0	285	129.5	385	175.0	485	220.0	585	265.5	685	311.0	785	356.5	885	401.5	985	447.0
86	39.5	186	84.5	286	130.0	386	175.5	486	220.5	586	266.0	686	311.5	786	357.0	886	402.0	986	447.5
87	39.5	187	85.0	287	130.5	387	176.0	487	221.0	587	266.5	687	312.0	787	357.0	887	402.5	987	448.0
88	40.0	188	85.5	288	131.0	388	176.0	488	221.5	588	267.0	688	312.5	788	357.5	888	403.0	988	448.5
89	40.5	189	86.0	289	131.5	389	176.5	489	222.0	589	267.5	689	313.0	789	358.0	889	403.5	989	449.0
90	41.0	190	86.5	290	132.0	390	177.0	490	222.5	590	268.0	690	313.0	790	358.5	890	404.0	990	449.5
91	41.5	191	87.0	291	132.0	391	177.5	491	223.0	591	268.5	691	313.5	791	359.0	891	404.5	991	450.0
92	42.0	192	87.5	292	132.5	392	178.0	492	223.5	592	269.0	692	314.0	792	359.5	892	405.0	992	450.0
93	42.5	193	88.0	293	133.0	393	178.5	493	224.0	593	269.0	693	314.5	793	360.0	893	405.5	993	450.5
94	43.0	194	88.0	294	133.5	394	179.0	494	224.5	594	269.5	694	315.0	794	360.5	894	406.0	994	451.0
95	43.5	195	88.5	295	134.0	395	179.5	495	225.0	595	270.0	695	315.5	795	361.0	895	406.0	995	451.5
96	44.0	196	89.0	296	134.5	396	180.0	496	225.0	596	270.5	696	316.0	796	361.5	896	406.5	996	452.0
97	44.0	197	89.5	297	135.0	397	180.5	497	225.5	597	271.0	697	316.5	797	362.0	897	407.0	997	452.5
98	44.5	198	90.0	298	135.5	398	181.0	498	226.0	598	271.5	698	317.0	798	362.0	898	407.5	998	453.0
99	45.0	199	90.5	299	136.0	399	181.0	499	226.5	599	272.0	699	317.5	799	362.5	899	408.0	999	453.5
100	45.5	200	91.0	300	136.5	400	181.5	500	227.0	600	272.5	700	318.0	800	363.0	900	408.5	1000	454.0

IT-28

Temperature Conversions

°F	°C	°F	°C	°F	°C
40	40	320	160	1040	560
30		340		1060	
20	30	360	180	1080	580
−10		380		1100	
0	20	400	200	1120	600
10		420		1140	
20	−10	440	220	1160	620
30		460		1180	
40	0	480	240	1200	640
50	10	500	260	1220	660
60		520		1240	
70	20	540	280	1260	680
80		560		1280	
90	30	580	300	1300	700
100		600		1320	
110	40	620	320	1340	720
120	50	640	340	1360	740
130		660		1380	
140	60	680	360	1400	760
150		700		1420	
160	70	720	380	1440	780
170		740		1460	
180	80	760	400	1480	800
190	90	780	420	1500	820
200		800		1520	
210	100	820	440	1540	840
220		840		1560	
230	110	860	460	1580	860
240		880		1600	
250	120	900	480	1620	880
260		920		1640	
270	130	940	500	1660	900
280	140	960	520	1680	920
290		980		1700	
300	150	1000	540	1720	940
310		1020		1740	
320	160	1040	560	1760	960

APPENDIX
Section A3 — WRITING SI

A3.1 SCOPE — This section establishes the general rules writing SI units on drawings and in text form.

A3.2 APPLICABLE DOCUMENTS —
ISO 1000   SI Units and Recommendations for the Use of Their Multiples and of Certain Other Units.

A3.3 DEFINITIONS — NOT APPLICABLE

A3.4 SPELLING — Because SI is international in scope, the international spelling using "re" is recommended for the spelling of the words, "litre" and for "metre" (when used as a unit of length). Note: NBS SP 330 1977 (3rd Edition) has reverted back to "liter" and "meter" spelling.

A3.5 OBSOLETE TERMS, UNITS AND PREFIXES —

A3.5.1 DON'T use obsolete terms, units or prefixes listed in paragraphs A1.7 and A1.7.1

A3.5.2 DON'T use the words "billion" or "trillion". These words have different values in other countries. Say, "thousand million" or "million million". See A1.8.2

A3.6 WRITING UNIT NAMES AND UNIT SYMBOLS —

A3.6.1 DON'T capitalize a unit name, except Celsius, unless the unit name begins a sentence. (Ref. A1.9.1) writing unit symbols.

A3.6.2 DO capitalize the first letter of a unit symbol when the unit is derived from a person's name (Ref A1.9.1)

A3.6.3 DON'T capitalize the first letter of a unit symbol when the unit is not derived from a person's name (Ref. A1.9.2)

CORRECT	INCORRECT
50 amperes	50 Amperes
50 A	50 a
37 degrees Celsius	37 degrees celsius
37 °C	37 °c
110 volts	110 Volts
110 V	110 v

A3.7 WRITING PREFIX NAMES AND SYMBOLS —

A3.7.1 Except when beginning a sentence, ALL prefix names are written in lower case letters (Ref. A1.9.2).

A3.7.2 Six prefix symbols are written in capital letters: E, L, P, T, G, and M. Reference A1.9.1.4)

A3.7.3 All other prefix symbols are written in lower case letters. (Reference A1.9.2) DON'T let a prefix stand alone, as: 1 kilo.

A3.7.4 ALWAYS accompany the prefix with the unit it is intended to modify, as: 1 kilogram, 5 millimetres, 700 megahertz, etc.

A3.7.5 AVOID using multiple prefixes when forming compound SI units.

CORRECT	INCORRECT
73 picofarads	73 micro microfarads
73 pF	73 μ μ F

A3.7.6 DON'T use a prefix in a denominator. The exception is the base unit, kilogram.

CORRECT	INCORRECT
km/s	m/ms

A3.8 WRITING PLURALS —

A.3.8.1 The names of units are made plural in the usual way by adding a plural s. As: metres, farads, candelas, etc.

A3.8.2 The unit, Hertz, is both singular and plural—no "s" shall be added. All symbols are both singular and plural.

A3.8.3 DON'T add an s to a symbol.

CORRECT	INCORRECT
1 m, 500 m	500 ms (would mean milliseconds)

A3.9 PROPER SPACING —

A3.9.1 DO leave a space between the numerical value and the symbol.

CORRECT   INCORRECT
 9 mm   9mm
 220 V   220V

A3.9.2 DO leave a space between the numerical value and the degree symbol of "degree Celsius."

CORRECT   INCORRECT
 37 °C   37° C
                    37°C

A3.10 INDICATING DIVISION —

A3.10.1 DO use a slash to indicate division, but DON'T use more than one slash in a combination.

A3.10.2 An alternate method is to show the denominator as a negative power with the raised dot.

CORRECT   NOT PREFERRED   INCORRECT
 m/s $\frac{m}{s}$ m ÷ s
 m•s^{-1}

A3.11 MIXING WORDS, SYMBOLS AND UNITS —

A3.11.1 DON'T mix symbols and words. When using symbols, write:

CORRECT   INCORRECT
 m/s   m/second

A3.11.2 DON'T mix words with symbols. When writing words, write:

CORRECT   INCORRECT
 meters per second m/second

A3.11.3 DON'T mix units:

CORRECT   INCORRECT
 10.77 m   10 m 77 cm

A3.12 USE OF LINEAR DIMENSIONS ON DRAWINGS — On an engineering drawing, DO state linear dimensions in terms of millimetres.

A3.13 USE OF THE PERIOD —

A3.13.1 DO use a period to indicate a decimal.

A3.13.2 DON'T use a period after a symbol unless it concludes a sentence.

CORRECT	INCORRECT
1.57 m	1.57 m.
8.05 kg	8.05 k.g.

A3.14 USE OF RAISED DOT —

A3.14.1 DO use a raised dot to indicate multiplication.

CORRECT	INCORRECT
N·m	Nm (meaningless)
50 l·m	50 lm
(50 **liters** times **meters**)	(50 lumen-not the value intended)

A3.15 USE OF THE COMMA —

A3.15.1 DON'T use a comma to indicate a decimal.

A3.15.2 DON'T use a comma to separate numbers in groups of three.

A3.15.3 Use of a space:
When writing engineering reports, etc. it is advantageous to use a space to separate large numbers into groups of three. In the case of four digits, spacing is optional, or determined by digit spacing in a column. (Ref A1.8.2).

CORRECT	INCORRECT
1 234 567.89	1,234,567.89
12 345.67	12,345.67
1234.56	

A3.15.4 Spacing of numbers into groups of three is NOT recommended ON AN ENGINEERING DRAWING which will be duplicated. The empty space may be mistaken for a missing decimal point.

A3.16 WRITING DECIMALS —

A3.16.1 DO use a period to indicate a decimal.

CORRECT   INCORRECT
52.67   52,67

A3.16.2 DO use a decimal to show units less than one. DON'T use a fraction.

CORRECT   INCORRECT
52.9   52 9/10

A3.16.3 DON'T place a decimal point after a whole number.

CORRECT   INCORRECT
75   75.

A3.16.4 DO use a zero to the left of the decimal point when the value is less than one.

CORRECT   INCORRECT
0.59   .59
0.075   .075

A3.16.5 DON'T use a zero to the right of a decimal point UNLESS THE ZERO IS SIGNIFICANT.

CORRECT   INCORRECT
75   75.0 (Unless significant)

A3.16.6 DO use zeros for uniformity, to provide the same number of decimal places on both plus and minus tolerances and on limit dimensioning.

CORRECT   INCORRECT
57 $^{+0.25}_{-0.10}$ 57 $^{+0.25}_{-0.1}$

NOTES

APPENDIX
Section A4 — CONVERSION OF AN INCH DRAWING TO A METRIC DRAWING

A4.1 MODIFY EXISTING DRAWING FORM —

A4.1.1 The existing drawing form shall be modified to indicate dual-dimensioning.

A4.1.2 Line out existing note which states that all dimensions are in inches (see figure 1).

A4.1.3 Identify existing tolerances, in tolerance block, as inch tolerances (see figure 1).

A4.1.4 Add a metric tolerance block to the drawing. The metric tolerance block shall specify the metric tolerance of all items mentioned in the original tolerance block; tolerances on decimal dimensions, angular dimensions, hole diameters, etc. (Figure 1).

A4.1.5 Add the following note to the general drawing notes: (see figure 3).

"DIMENSIONS AND VALUES IN BRACKETS ARE METRIC"

A4.2 IDENTIFY AS A METRIC DRAWING — Many countries use first angle projection. Third angle projection is normal in the United States. Therefore, identify by showing the international projection symbol, and the words "THIRD ANGLE" near the title block (see figure 2)

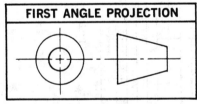

FIRST ANGLE PROJECTION

Normally used in Europe

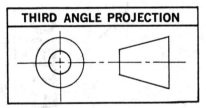

THIRD ANGLE PROJECTION

Normally used in U.S.A.

FIGURE 2

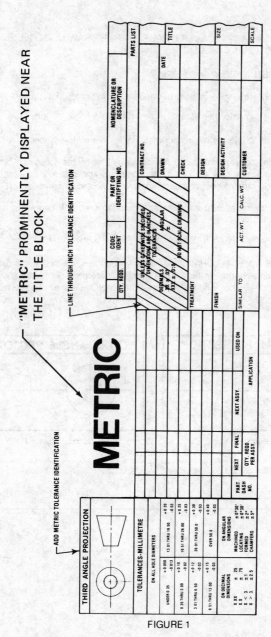

FIGURE 1

A4.3 EQUIVALENT DIMENSIONS — Metric dimensions and values shall be equated as accurately as practical, while remaining within the specified tolerance limits of the inch drawing.
NOTE:
Inspection may be performed using either customary or metric measurements.

A4.3.1 On the field of the drawing, metric dimensions may be shown in brackets [ ] below or to the right of the inch dimension, or a conversion table may be used on the field of the drawing. (See figures 3 and 3a).
NOTE:
Do not use parenthesis around metric dimensions. Parenthesis are reserved for reference dimensions.

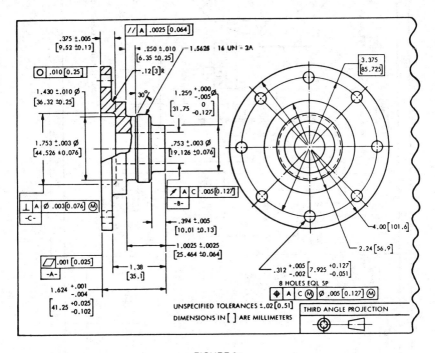

FIGURE 3a

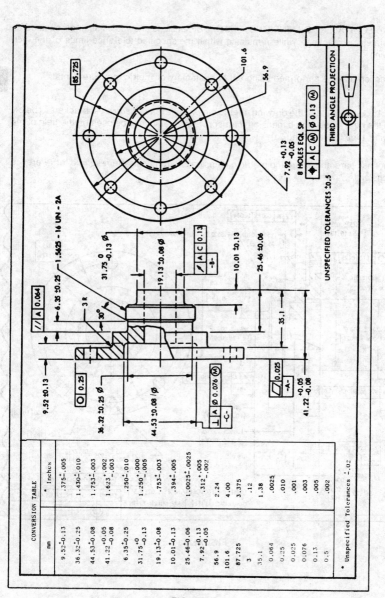

FIGURE 5

A4.3.2 Either the plus and minus, or the limit tolerance method may be used. However, the method used in the metric conversion shall be the same as the method used in the inch drawing. (see figure 4.)

A4.3.3 Only one of the methods of identification of units shown in Figure 4 is permitted throughout a single drawing.

A4.3.3.1 The dimension line may be used to separate the inch and the metric dimensions.

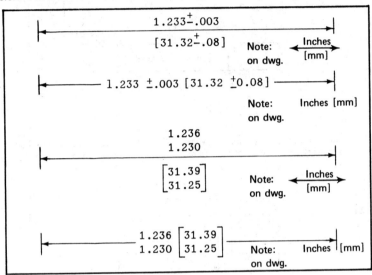

Figure 4

A4.3.4 Dual dimensions on drawings is discouraged except where the dimensions must interface with existing installation. If necessary, the use of the conversion table illustrated in Figure 5 is the preferred method. Once a method of inscribing dual dimensions is selected, it must be clearly described in a note on the drawing.

A4.3.5 Alternate Method — The reciprocal of what is illustrated in Figure 4 is permissable.

A4.4 UNITS OTHER THAN LINEAR DIMENSIONS —

A4.4.1 Angles presented in degrees, seconds and minutes, or in degrees and decimal parts of a degree, are common to inch and metric systems of measurement.

A4.4.2 Nominal sizes such as screw threads, pipe sizes, wood cross section, and other estimated standard sizes shall NOT be converted to metric dimensions.

A4.4.3 Military, industry, or company standard parts may be used without converting to metric units.

A4—5

A4.4.4 Interface dimensions for holes, cutouts, etc. are to be dual dimensioned except as described in the paragraph above.

A4.4.5 Units other than linear dimensions that appear on a drawing shall be converted to the appropriate SI unit. Example: Gallons per hour, to litres per hour.

A4.4.6 Drawings that show geometric symbols for form, position, etc. shall have the metric value added. See figures 3 and 5.

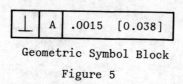

Geometric Symbol Block

Figure 5

A4.4.7 Welding symbols containing numerical values shall have metric values added. The drawing shall clearly indicate the customary and the SI metric units. (Figure 6)

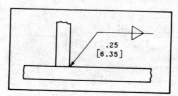

Metric Value Added to Welding Symbol

Figure 6

A4.4.8 Taper callout may be converted from inch per inch of taper to **millimeter** per millimetre of taper, or may be expressed as a ratio, as .05:1 (See figure 7).

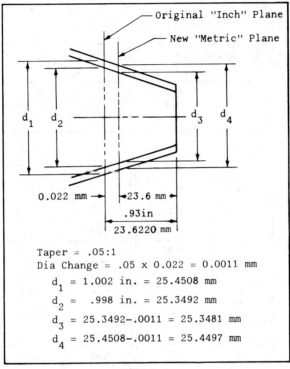

Taper = .05:1
Dia Change = .05 x 0.022 = 0.0011 mm
d_1 = 1.002 in. = 25.4508 mm
d_2 = .998 in. = 25.3492 mm
d_3 = 25.3492 - .0011 = 25.3481 mm
d_4 = 25.4508 - .0011 = 25.4497 mm

Figure 7

A cone of taper .05:1 has a diameter of 1.000 ±.002 inch in a reference plane located by the non-toleranced dimension .9300 inch. By virtue of the taper of the cone, the limits of the tolerance zone depend on the position of the reference plane. Consequently, if the dimension .9300 in. equals 23.6220 mm is rounded off to 23.600 mm (a reduction of 0.022 mm), each of the two original limits, when converted exactly into millimeters, must be corrected by 0.022 x .05 = 0.0011 mm before being rounded off. (See Figure 7)

A4.4.9 EQUIVALENT METRIC VALUES FOR SURFACE ROUGHNESS ARE GIVEN IN TABLE 1.

STANDARD VALUES FOR ROUGHNESS HEIGHT RATINGS

Roughness Value		Roughness Value	
Micro-inches	Micrometers [μm]	Micro-inches	Micrometers [μm]
* 1	0.025	50	1.25
* 2	0.050	* 63	1.6
* 4	0.100	80	2.0
5	0.125	90	2.2
6	0.15	100	2.5
* 8	0.20	* 125	3.2
10	0.25	* 250	6.3
13	0.32	320	8.0
* 16	0.40	400	10.0
20	0.50	* 500	12.5
25	0.63	800	20.0
* 32	0.80	*1000	25.0
40	1.00	*2000	50.0

* Preferred ratings.

TABLE 1

A4.4.9.1 APPLY METRIC VALUES FOR SURFACE ROUGHNESS TO DRAWING AS SHOWN IN FIGURE 8.

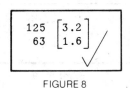

FIGURE 8

A4.4.10 EQUIVALENT METRIC STANDARD VALUES FOR WAVINESS HEIGHT RATING ARE GIVEN IN TABLE 2.

Inches	Millimeters	Inches	Millimeters
*.00002	0.0005	*.0010	0.025
.00003	0.0008	*.002	0.05
*.00005	0.0012	.003	0.08
.00008	0.0020	*.005	0.12
*.00010	0.0025	.008	0.20
*.0002	0.005	*.010	0.25
.0003	0.008	.015	0.38
*.0005	0.012	.020	0.50
.0008	0.020	*.030	0.80

TABLE 2

*PREFERRED VALUES

A4.4.10.1 APPLY METRIC VALUES FOR WAVINESS HEIGHT RATING AS SHOWN IN FIGURE 9.

Waviness Height Rating

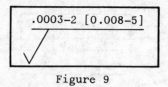

Figure 9

A4.4.11 Metric values for roughness width cutoff are given in table 3.

Standard Values for Roughness Width Cutoff

	Inches	Millimeters
	.010	0.25
*	.030	0.80
	.100	2.5

TABLE 3
*PREFERRED RATINGS

A4.4.11.1 APPLY metric values for roughness width cutoff as shown in Figure 10.

Roughness Width Cutoff

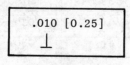

FIGURE 10

APPENDIX
Section A5 — MISC METRIC INFORMATION

A5.1   METRIC PAPER SIZES — ISO Standard size paper is cut to a ratio of $\sqrt{2}$ to 1 the large size, "AO", has an area equal to one square meter. Smaller sizes are made by dividing a sheet into halves, as shown in figure 1. Sheet measurements are listed in table 1.

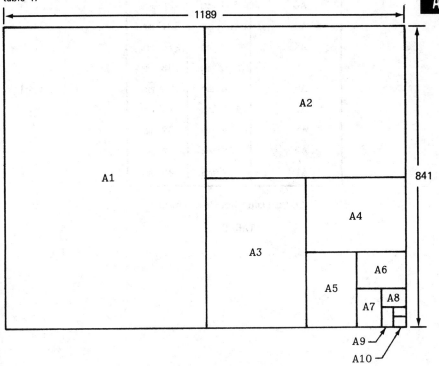

DESIGNATION   AO (841 x 1189 mm)

Division of Metric Paper

Figure 1

A5-1

DESIGNATION	LENGTH	WIDTH
A0	1189 mm	841 mm
A1	841 mm	594 mm
A2	594 mm	420 mm
A3	420 mm	297 mm
A4	297 mm	210 mm
A5	210 mm	148 mm
A6	148 mm	105 mm
A7	105 mm	74 mm
A8	74 mm	52 mm
A9	52 mm	37 mm
A10	37 mm	26 mm

ISO Standard Sheet Measurements

TABLE 1

APPENDIX 2
METRIC GLOSSARY

acceleration — symbol: m/s^2
  The SI unit for acceleration is **meter** per second squared.

ampere — (AM-peer) pl. amperes symbol: A
  The ampere is the base unit for electrical current.

area — symbol: m^2
  The SI unit for area is the square **meter**.

atto — (AT-tow) symbol: a
  The metric prefix for one millionth millionth millionth of the unit with which it is used.

candela — (CAN-del-a) pl. candelas symbol: cd
  The SI base unit for luminous intensity.

Celsius — (SELL-see-us) pl. Celsius symbol: C
  The temperature scale formerly called, "centigrade." The Celsius scale has 100 equal divisions between the freezing point and boiling point of pure water, at sea level.

centi — (SENN-tea) symbol: c
  The metric prefix for one hundredth of the unit with which it is used.

centigrade — One hundredth of a gradient. An obsolete term for Celsius temperature.

coulomb — (KOO-lahm) pl. coulombs symbol: C
  The coulomb is the derived unit of electrical charge.

cubic centimetre symbol: cm^3
  A volume equal to that of a cube, one centimetre on each edge. Obsolete symbol: cc.

deci — (DESS-see) symbol: d
  The metric prefix for one tenth of the unit with which it is used.

degree Celsius — pl. degrees Celsius symbol: °C
  One degree Celsius is one temperature gradient on the Celsius scale.

degree (geometric) symbol: °
  One degree of angle is equal to 1/90 part of a right angle.

deka — (DECK-uh) symbol: da
  The metric prefix for ten units. "deca" is an obsolete spelling.

exa — (X-uh) Symbol: E
  The metric prefix for one million million million times the unit with which it is used.

farad — (FARE-add) pl. farads   symbol: F
The farad is the derived unit for electric capacitance.

femto — (FEM-tow)   symbol: f
The metric prefix for one thousandth millionth millionth of the unit with which it is used.

giga — (JIG-uh)   symbol: G
Giga is the metric prefix for one thousand million times the unit with which it is used.

hecto — (HECK-tow)   symbol: h
Hecto is the metric symbol for one hundred of the units with which it is used.

henry — (HEN-re) pl. henries   symbol: H
The henry is the derived unit for electrical inductance.

hertz — (HEHRTZ) pl. hertz   symbol: Hz
A derived unit for frequency. Generally associated with radio frequency—formerly called, "cycles per second."

joule — (JOOL) pl. joules   symbol: J
The joule is the derived unit for quantity of energy.

kelvin — (KELL-vin) pl. kelvins   symbol: K
The SI base unit for thermodynamic (absolute) temperature.

kilo — (KILL-oh)   symbol: k
Kilo is the metric prefix for one thousand of the units with which it is used.

liter— (LEE-tur) pl. **liters**   symbol: **L ("ell")**
Although the litre is not an SI unit, it may be used as a measure of liquid volume equal to 1000 cubic centimetres, or one cubic decimetre.

lumen — (LU-men) pl. lumens   symbol: lm
The lumen is the SI derived unit for luminous flux.

luminance — (LU-men-ance)
The SI derived unit for luminance is one candela per square metre. The symbol for candela per square **meter** is cd/m^2.

lux — (LUCKS) pl. lux   symbol: lx
The derived unit for illuminance.

mass — (MAS) pl. masses
The SI unit of mass is the kilogram. The symbol for the kilogram is: kg.

mega — (MEGG-uh)   symbol: M
Mega is the metric prefix for one million times the unit with which it is used.

meter — (MEET-er) pl. meters
   An instrument for measuring, as a gas meter, a volt meter, a taxi meter, etc.

meter — (MEET-er) pl. metres   symbol: m
   The SI base unit of length.

micro — (MIKE-row)   symbol: μ
   The metric prefix for one millionth of the unit with which it is used.

milli — (MILL-ee)   symbol: m
   Milli is the metric prefix for one thousandth of the unit with which it is used.

mole — (MOLE) pl. moles   symbol: mol
   Mole is the SI base unit for amount of substance.

nano — (NAN-oh)   symbol: n
   Nano is the metric prefix for one thousandth millionth of the unit with which it is used.

newton — (NEW-ton) pl. newtons   symbol: N
   The newton is the SI derived unit of force. The newton is the amount of force that accelerates one kilogram of mass, one metre per second, per second.

ohm — (OH-mmm) pl. ohms   symbol: Ω
   The ohm is the derived unit of electric resistance.

peta — (PEA-ta)   symbol: P
   The metric prefix for one thousand million million times the unit with which it is used.

pascal — (PASS-kul) pl. pascals   symbol: Pa
   The pascal is the SI derived unit for pressure, or stress.

pico — (PEA-ko)   symbol: p
   Pico is the metric prefix for one millionth millionth of the unit with which it is used.

radian — (RAY-de-ann) pl. radians symbol: rad
   The SI supplementary unit for plane angle. One radian equals 57 degrees, 17 minutes and 46 seconds of arc.

second — (SEK-und) pl. seconds   symbol: "
   Angle: 1/60 part of a minute of angle.

second — (SEK-und) pl. seconds   symbol: s
   Time: 1/60 part of a minute of time.

siemens — (ZEE-munz) pl. siemens   symbol: S
   The SI derived unit for electrical conductance.

B1

steradian — (STEER-ray-de-ann) pl. -s   symbol: sr
    The steradian is the supplementary unit for the solid angle at the apex of a conical section of a sphere, when the spherical area of the base of the conical section is equal to the square of the radius of the sphere.

tera — (TEAR-uh)   symbol: T
    Tera is the SI prefix for one million million times the unit with which it is used.

tesla — (TESS-la) pl. teslas   symbol: T
    Tesla is the SI derived unit for magnetic flux density.

ton — (TUN) pl. tons   symbol: t
    One metric ton is equal to 1000 kilograms.

tonne — (TUN) pl. tonnes   symbol: t
    An alternate spelling for "ton."

volt — (VOHLT) pl. volts   symbol: V
    The SI derived unit for electromotive force, or electrical potential difference.

volume — (VOL-ume)
    The SI unit for volume is the cubic **meter**. The symbol for cubic **meter** is: m^3.

watt — (WAHT) pl. watts   symbol: W
    The watt is the SI derived unit of power.

weber — (WEB-er) pl. webers   symbol: Wb
    The weber is the SI derived unit for magnetic flux.

NOTES

NOTES

NOTES

NOTES

NOTES

NOTES